DESIGN METHODOLOGIES FOR VLSI CIRCUITS

NATO ADVANCED STUDY INSTITUTES SERIES

Proceedings of the Advanced Study Institute Programme, which aims at the dissemination of advanced knowledge and the formation of contacts among scientists from different countries.

The series is published by an international board of publishers in conjunction with NATO Scientific Affairs Division

A	Life Sciences	Plenum Publishing Corporation
B	Physics	London and New York
C	Mathematical and Physical Sciences	D. Reidel Publishing Company Dordrecht and Boston
D	Behavioural and Social Sciences	Sijthoff & Noordhoff International Publishers B.V.
E	Applied Sciences	Alphen aan den Rijn, The Netherlands and Rockville, Md., U.S.A.

Series E: Applied Sciences – No. 47

DESIGN METHODOLOGIES FOR VLSI CIRCUITS

edited by

Paul G. Jespers
Université Catholique de Louvain
Laboratoire de Microélectronique
Louvain-la-Neuve
Belgium

Carlo H. Sequin
University of California
Department of Electronical Engineering
and Computer Sciences
Berkeley, California
U.S.A.

Fernand van de Wiele
Université Catholique de Louvain
Laboratoire de Microélectronique
Louvain-la-Neuve
Belgium

Sijthoff & Noordhoff 1982
Alphen aan den Rijn, The Netherlands
Rockville, Maryland, U.S.A.

Proceedings of the NATO Advanced Study Institute on
"Design Methodologies for VLSI Circuits"
Louvain-la-Neuve, Belgium
July 8-18, 1980

ISBN 90 286 2781 2

Printed in The Netherlands.

TABLE OF CONTENTS

PREFACE

An Advanced Study Institute on Design Methodologies for VLSI Circuits was held in Louvain-la-Neuve, Belgium, on July 8-18, 1980 under the auspices of the Scientific Affairs Division of NATO. The institute was organized by a scientific committee consisting of Professor P. Jespers of the Université Catholique de Louvain, Louvain-la-Neuve, Professor C. Sequin of the University of California, Berkeley and Professor F. Van de Wiele of the Université Catholique de Louvain, Louvain-la-Neuve. This book contains most of the contributions of the lectures. The well known book of C. Mead and L. Conway : "Introduction to VLSI Systems" (Addison Wesley, 1980) was also used during the institute.

This book illustrates the fact that the rapidly increasing complexity of modern IC's requires an effort to interdisciplinary nature. A review of potential technologies for VLSI circuits is presented first with a survey of MOS transistor models. Structured design methodologies are presented further as a mean to design efficiently rather complex integrated circuits. Along these lines, the problem of interchange formats is dealt with in two chapters. Means to reach larger complexities such as CAD, tools for reliability improvements, and design for testability are considered also. Problems encountered in designing microprocessors and microcomputers are described in three chapters. Multiprocessor system are considered furtheron. The book ends with some views on VLSI analog circuitry.

The subject matter of this book is suitable for a wide range of interests from the advanced student to the system designers. Portions of this book may be used as textbooks since the chapters are instructional and self-contained.

The editors want to express their appreciation to Monseigneur Ed. Massaux, Rector of the Université Catholique de Louvain, for the facilities and accomodations which made the institute a success, and to Dr. Di Lullo of the Scientific Affairs Division of NATO for his encouragement and support. The personnal of the Université Catholique de Louvain are to be commanded for their help during the institute, especially Mrs. Mercenier, Mrs. and Mr. Trullemans, Mr. Windal, Mr. Herent and Mr. Iserentant as well as Professor Van Vyvere and Professor Mercier of the Institut Supérieur d'Archéologie et d'Histoire de l'Art for their contribution to the cultural aspects of the institute.

TRENDS IN VERY LARGE SCALE INTEGRATION AND THEIR IMPACT ON THE DESIGNER

Carlo H. Séquin

Computer Science Division
Electrical Engineering and Computer Sciences
University of California, Berkeley, CA 94720

ABSTRACT. *A classification of integrated circuits by complexity and a definition of VLSI is attempted. The driving forces behind the exponential evolution of integrated circuits and their trends for the near future are discussed. Effects of VLSI on the IC designer are analyzed and the need for a new IC design methodology is established.*

1. INCREASING CHIP FUNCTIONALITY

In the last two decades the total functional complexity of single-chip integrated circuits has risen from a few transistors to about 200,000 devices (transistors and capacitors) in a 64k bit random access memory. While originally complexity has doubled about every 12 months, more recently this evolutionary rate has slowed down. Nevertheless increases in chip complexity will continue at an awesome pace for the foreseeable future.

Several factors have contributed to this exponential increase of chip functionality. They can be grouped into (a) technological advances, which have lead to larger manufacturable chips and to an increased feature density, and (b) circuit and layout cleverness, which has reduced the number of individual circuit elements necessary to perform a specific function [Moore 75]. The factors in group (a) are expected to sustain their influence for the near future, while (b) can only make further contributions to the growth of IC chip complexity, if the way in which integrated circuits are constructed is changed in a significant way.

2. COMPLEXITY CLASSIFICATION

Many voices claim that entering the domain of VLSI will significantly change design and fabrication methodologies for integrated circuits. However it is not obvious where the threshold for VLSI really lies. According to the TTL data book the boundaries between SSI, MSI, and LSI are at levels of 12 and 100 gates, respectively, or roughly at 100 and 1000 devices per chip. Extrapolating this simple measure, we would have to place the threshold to VLSI around 10,000 devices, a number passed long ago with one of the early microprocessors.

A better classification of the various levels of integration may be derived from the functionality of the chips. While SSI chips contain individual logic gates, MSI parts offer complete register level functions such as register slices, decoders, multiplexers. LSI then provided bit-slices of complete data paths, including ALU and register file, and also made simple processors on a chip possible.

Somewhere in the range between 100,000 and one million devices, it will be possible to put a complete computer on a chip, including a processor and a substantial amount of "main" memory. This threshold may best fit the current usage of the term VLSI.

However, this level of complexity not only permits the fabrication of these high-level general-purpose building blocks, i.e. single-chip computers, but also the construction of complete customized systems of substantial complexity on a single chip. The latter will be a severe test for our design tools and for our methodology to deal with very large complexity. Table 1 reviews the classification of the various levels of integration during the last 20 years.

CLASSIFICATION OF INTEGRATION LEVELS				
Name	Device Count			Chip Functions
SSI	0	-	100	gates
MSI	100	-	1000	register parts
LSI	1000	-	10,000	bit slices
		-	100,000	processors
VLSI	100,000	-	?00,000	computers
		-	?00,000	systems

Table 1. Classification of integrated circuits by complexity and functionality.

3. TECHNOLOGICAL ADVANCES

In the two decades since the introduction of the planar technology, the number of devices that can be realized with reasonable yield on a single silicon chip has increased by five orders of magnitude. More than half of this exponential evolution is due to technological advances. Refined photolithographic procedures have permitted the minimum feasible feature dimension to shrink to a few microns, and correspondingly have led to an increase in the number of circuits per unit area of silicon. In parallel, a continuously improving processing technology of the silicon wafer has reduced the number of defects per unit area that are introduced in each step. Thus it has become possible to accept more complicated processing sequences and larger chip areas without reducing the overall manufacturing yields below commercially viable values.

3.1. Reduction in Feature Size

Linear feature dimensions such as line widths and spacings of integrated circuits have been reduced from 1 mil to 4 microns in the last twenty years. It is quite probable that this trend will continue and lead to 1 micron features in the 1990's. Powerful lithographic tools are being developed in many laboratories. Visible or UV light, which limits the resolution of optical systems, is replaced with X-rays or electrons. X-ray lithography has demonstrated the feasibility of submicron features on laboratory test patterns [Feder 75]. Its attraction is based on the fact that the exposure does not need to take place in vacuum and that the use of the strongly penetrating X-rays results in a certain insensitivity to dust particles on the mask. Particular problems to be solved have to do with the fabrication of masks of good stability, and with their alignment to the pattern on the wafer during exposure.

Alternatively, there are a number of competing electron exposure systems. The demagnifying transmission-electron projection system is the analog of an optical projection system, and promises the feasibility of 0.3 micron lines with an accuracy of 0.1 micron over a field of 3 x 3 mm. However, it is a complicated system with several lenses [Heritage 75]. Again, the minimum useful feature dimensions in these exposure systems will be limited by the reregistration margins of subsequent patterns and by image distortions or runout errors resulting from insufficient flatness of the wafer.

The scanning electron-beam system offers more flexibility. Corrective local distortion can be introduced so that tight registration can be maintained over large areas. This system also offers shorter overall production cycles in the development of new devices, since the wafer can be processed directly from a software description of the desired pattern. This system has already gained significance for photomask fabrication [Herriott 75]. Its main disadvantage as an exposure system for devices in production quantity is the long exposure time, which currently is on the order of half an hour for a four-inch wafer with a 0.5 micron raster.

All the mentioned systems permit, in principle, the fabrication of submicron features. Studies of the effect of geometrical scaling on the performance of devices indicate [Dennard 74], and are corroborated by experimental results, that most devices can profit of such a size reduction, since the fundamental physical limitations are still a long way off. The real question will be whether wafer processing can live up to these reduced feature sizes. Defects of proportionally smaller sizes, which normally occur with steeply increased density, will result in circuit failure. Furthermore, sensitivity, adherence and thickness uniformity of the electron resist need further development. Etching techniques, such as ion milling, that avoid undercutting the resist pattern have to be refined.

3.2. Increase in Chip Size

Complementing the increased feature density, chip sizes themselves have also grown by a factor of two about every four years. The maximally tolerable active area of a commercially viable chip is directly linked to the achievable yield. Substantial yield increases have been attributable to dry processing steps such as plasma etching and ion implantation. Better step coverage has been obtained by flowing the isolating glass layer before metal is deposited. Other approaches to achieve better yield resort to redundant processing, i.e. double exposure at certain mask levels or the combined subsequent use of positive and negative resists to define a set of features.

For very large and complicated integrated circuits the concept of redundancy can be applied to the circuit itself. Memory chips in the above-100 k-bit range may only be feasible with viable yield if subcircuits which contain faulty storage cells can be disconnected selectively.

4. CIRCUIT AND LAYOUT CLEVERNESS

The above mentioned two factors, chip size and feature density, can only account for a fraction of the actual growth in on-chip circuit complexity. A little more than one third of the observed advances must to be attributed to the ingenuity of the designers. By making better use of properly tailored characteristics of semiconductor devices, the number of devices that are required to perform a certain function has been reduced. In addition, these devices have been packed together in a more efficient manner, which has further diminished the number of individual geometric features per desired function.

4.1. Circuit Simplification

The trend leading to simpler circuits with fewer devices is well demonstrated by the evolution of the unit cell of random access semiconductor memories. Starting with a static flip-flop using six transistors, the basic memory cell was reduced in several steps to a single transistor and a capacitor [Panousis 70]. A major landmark along this path was the switch from a static cell to a dynamic three-transistor cell. Such a step will usually lead to a change in the overall systems approach, e.g. a dynamic cell requires the implementation of a refresh cycle. From the viewpoint of the overall system, the additional complexity of the refresh circuitry is still worthwhile if it increases the overall function density. Similarly, the use of a single-transistor cell has shifted more of the circuit complexity to the periphery of the device; more sophisticated, balanced sense amplifiers are required to detect the small amounts of charge that are dumped passively onto the large bit line capacitances. However, the complexity of this sense amplifier is shared by many memory cells, and this approach thus results in an overall increase in storage density.

A corresponding reduction in circuit complexity is also encountered in serial memories or shift registers. The static master-slave flip-flop with 16 MOSFET's has been reduced to dynamic stages with six transistors which in turn has lead to integrated bucket brigade structures [Sangster 69] with only two transistors per stage (Fig.1). However, it appears that the full potential of this trend has been exhausted and that further simplification of the two circuits described is not possible.

4.2. Layout Ingenuity

In parallel with the above development, structural improvements have lead to a more compact implementation of a particular circuit. Techniques for self-aligned source and drain areas have eliminated the need to include extra area in the layout to accommodate registration tolerances. Contact windows, which use up area without providing any useful function from a signal-processing point of view, are being eliminated as much as possible. And similarly, the 'useless' isolation area between adjacent devices is being minimized, by the development of new techniques such as silicon anodization [Cook 74] and by the design of circuits that allow several transistors to be placed into the same isolation well.

An important factor in the evolution of integrated circuits was the merging of the intrinsic electrode terminals of connected transistors. Not only is the density of active devices increased, but at the same time, by eliminating extra interconnects and the necessary contact interfaces, yield and reliability are augmented (Fig.2).

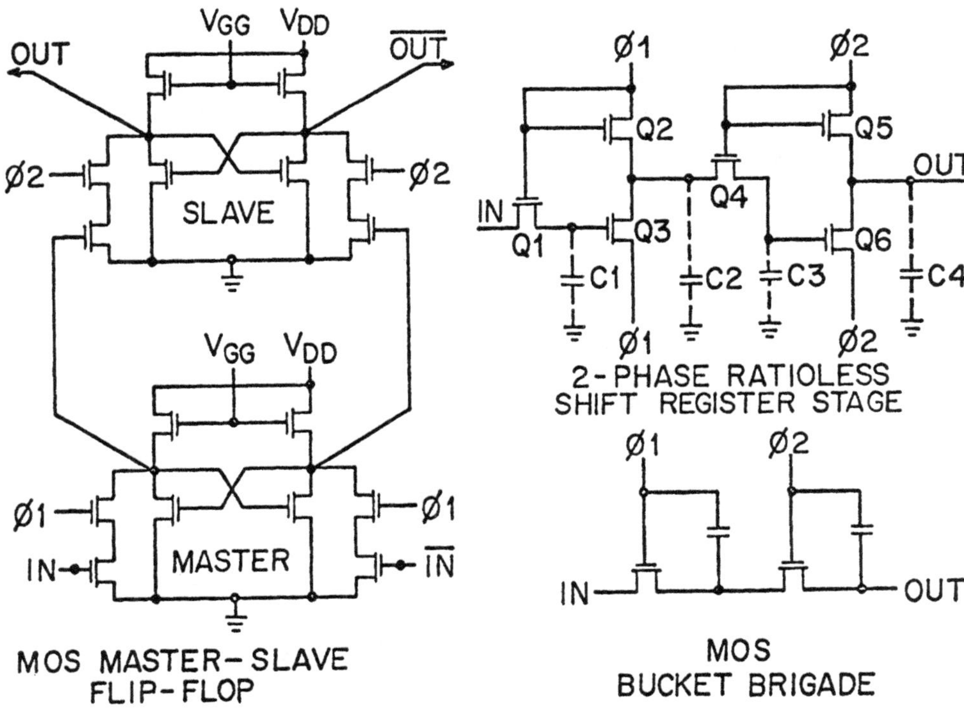

Figure 1. Reduction of circuit complexity demonstrated with the example of shift registers.

For MOS circuits the most important contribution to an efficient merging technique has come from the introduction of two separate levels of MOS electrodes (Fig.3). This can eliminate the need for an extra source/ drain area between two adjacent MOS gates. In the extreme this approach leads to charge coupled devices [Boyle 71] in which large active areas are covered with MOS electrodes using no intermediate source/drain diffusions at all. At this point, individual devices have lost their identity and have merged into fully integrated functional blocks optimized to perform a particular task. With respect to MOS RAMs this tendency has culminated in a cell which uses only a single electrode to realize the storage capacitor and the readout gate. In practice, the two areas derive their different functional behavior from a localized ion implantation under part of the electrode [Tasch 75].

5. FUTURE EVOLUTIONARY TRENDS

In view of these developments, one can try to project the growth of integrated circuit complexity into the future. The developments described in Section 3 indicate that increases in feature density will continue to contribute to overall IC complexity during the next ten years at least. Future increases in chip sizes appear to be less certain,

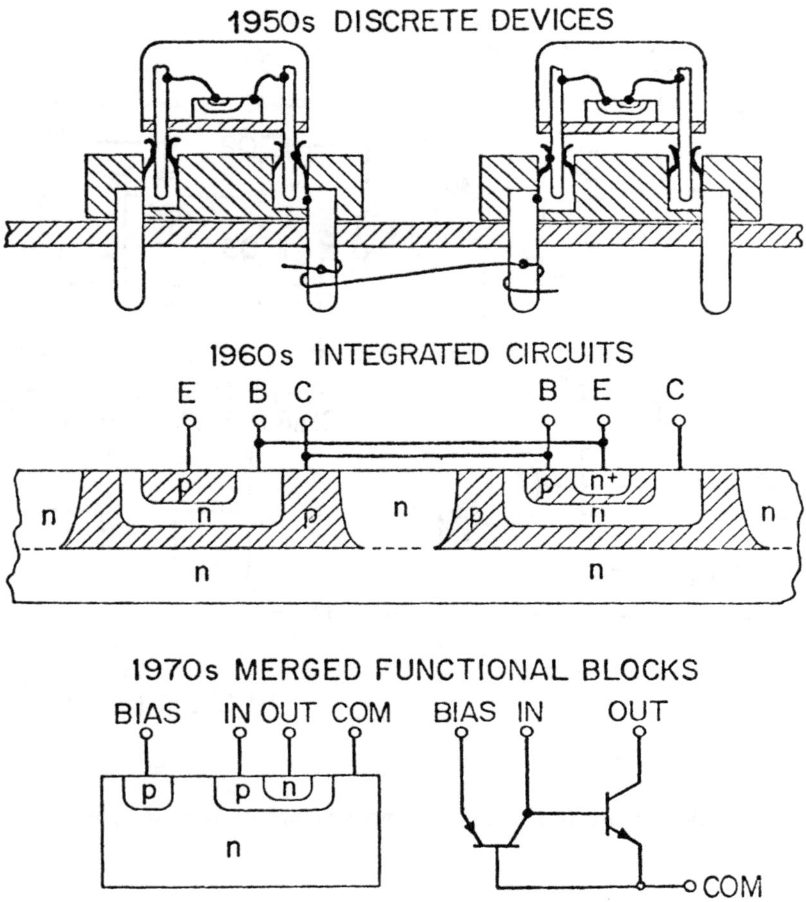

Figure 2. Reduction of interconnection interfaces in integrated circuits.

because of problems with power dissipation and packaging. Also, the gain due to chip area may be partially offset by the incorporation of redundant circuitry.

It is more difficult to predict what creativity in circuit design will contribute. It appears that we have come to a logical end of what appears now a reasonably straightforward path of circuit or layout simplification, and it is difficult to imagine how a memory cell can be reduced to less than the currently required few squares of minimum feature size. A conceivable way out of the apparent dead-end may rely on a more efficient usage of the third dimension. By stacking more features vertically, on-chip functional density could be further increased. The main contribution of circuit cleverness may lie at the systems level, providing ways to tolerate a finite number of defects. Built-in circuit redundancy or error correcting codes may permit one to push the size of an IC beyond the complexity that could be implemented with a

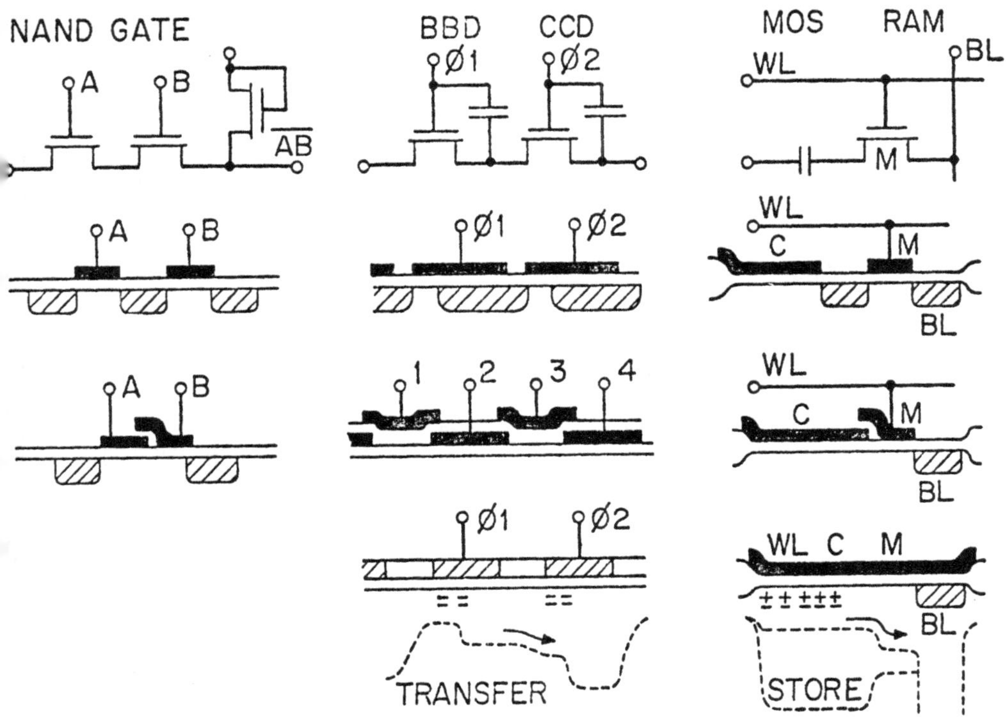

Figure 3. Merging of electrodes and features in MOS circuits, illustrated with a NAND gate, a dynamic shift register cell (BBD and CCD), and a single-transistor RAM cell.

commercially viable yield using an ordinary approach.

Figure 4 shows the historical trend of IC complexity roughly partitioned into is basic three contributing factors. Originally the three factors added up to a doubling of complexity every year. In the mid 1970's circuit cleverness ceased to be a significant factor after charge coupled devices had been developed and the single-transistor RAM cell had matured to the point where its layout took not much more area than a CCD cell. A more limited increase in effective chip size may further reduce the evolutionary rate.

In any case, integrated functional devices with a complexity corresponding to a megabit memory or 100,000 logic gates can be expected towards the end of this decade. This will have a significant impact on computer architecture and on the design methodology for such chips.

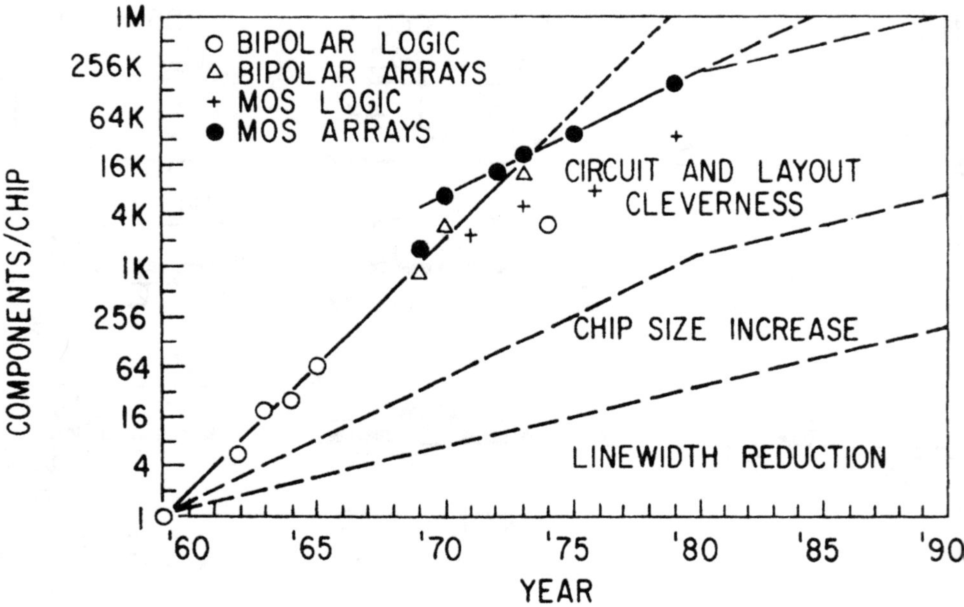

Figure 4. Trends in functional complexity of integrated circuits, and in the basic contributing factors.

6. IMPACT ON THE DESIGNER

During the time when SSI and MSI circuits dominated, the task of an IC designer was quite clear, and relatively narrow. Given a particular technology and a set of layout rules, he had to implement one of the elementary boolean logic functions or a particular flip-flop, register or counter. Already during the area of bit slice parts this task became significantly more involved. Not only was the layout larger and more diverse, the definition of the function itself required a lot more knowledge and good judgement. An understanding of the microarchitectures that the customer was likely to implement was a prerequisite for the design of a useful chip.

Now, in the realm of VLSI, the task has become almost superhuman. Even for general purpose components, i.e. single-chip computers, questions about communication protocol, memory management, and operating systems start to play a significant role. For complete systems on a chip, knowledge of the whole application area covered by the chip have to be understood by the designer.

In addition, while the upper boundary of design responsibility has moved higher in the definition of the overall system, the lower limit has actually decreased, thus extending the span at both ends. While at the level of the MSI circuit the logic gate could often be used as a lowest level primitive, some VLSI designs will clearly require that the designer consider circuit implementation at the device level or even below, taking as primitives the rectangles and wires from which circuit elements are composed. Such a thinking at the geometrical level is often required to get the best layout for a crucial functional block in the VLSI system. Thus the designer of a VLSI system may have to span a range of knowledge and considerations unprecedented in the history of integrated circuit design.

In addition the complexity of a VLSI design is of such magnitude that the task of managing all the associated information can only be accomplished in a satisfactory manner, if the designer is supported by a good set of software tools. Yet another definition of VLSI may thus refer to the level of integration where humans alone can no longer handle the design of the chip.

REFERENCES

[Moore 75]
G.E. Moore: "Progress in Digital Integrated Electronics", *Tech. Digest, Intl. Electron Devices Meeting,* Washington D.C., Dec. 1975, pp 11-13.

[Feder 75]
R. Feder, E. Spiller and J. Topalian: "Replication of 1.0um Geometries with X-ray Lithography", *Jour. Vac. Sci. Technol. Vol.12,* 1975, pp 1332-1335.

[Heritage 75]
M.B. Heritage: "Electron-Projection Microfabrication System", *Jour. Vac. Sci. Technol. Vol.12* 1975, pp 1135-1140.

[Herriott 75]
D.R. Herriott, R.J. Collier, D.S. Alles and J.W. Stafford: "EBES, A Practical Electron Lithographic System", *IEEE Trans. Electron Devices ED-22,* 1975, pp 385-392.

[Dennard 74]
R.H. Dennard, F.H. Gaensslen, H.N. Yu, V.L. Rideout, E. Bassous, and A.R. LeBlanc: "Design of Ion-Implanted MOSFET's with Very Small Physical Dimensions" *IEEE Jour. Solid-State Circuits SC-9,* 1974, pp 256-268.

[Panousis 70]
P.T. Panousis: "A TRIM Bipolar Charge Storage Memory", *Intl. Electron Devices Meeting,* Washington D.C., Dec. 1970, Abstract 2.4.

[Sangster 69]
F.L.J. Sangster and K. Teer: "Bucket Brigade Electronics - New Possibilities for Delay, Time-Axis Conversion and Scanning", *IEEE Jour. Solid-State Circuits SC-4,* 1969, pp 131-136.

[Cook 74]
B. Cook and R.E. Scherrer: "Anodic Aluminum Isolation", *Tech. Digest, Intl. Electron Devices Meeting,* Dec. 1974, pp 266-269.

[Boyle 71]
W.S. Boyle and G.E. Smith: "Charge-Coupled Devices - A New Approach to MIS Device Structures", *IEEE Spectrum 8,* 1971, pp 18-27.

[Tasch 75]
A.F. Tasch, H.S. Fu, and R.C. Frye: "The Charge-Coupled RAM Cell - A New MOS Dynamic RAM Cell Simpler than the One-Transistor Cell", *Intl. Electron Devices Meeting,* Washington D.C., Dec. 1975, Abstract 14.7.

TECHNOLOGICAL PROCESSES FOR VLSI

E. DEMOULIN

Université Catholique de Louvain, Laboratoire de Microélectronique, Place du Levant 3, 1348 Louvain-la-Neuve, Belgium.

I. INTRODUCTION

I-1 Trends

As pointed out by Gordon Moore in a widely quoted presentation at the 1975 International Electron Device Meeting |1|, IC complexity is roughly doubling every year (Figure 1). This progress proceeds from a number of reasons (Figure 2) among which

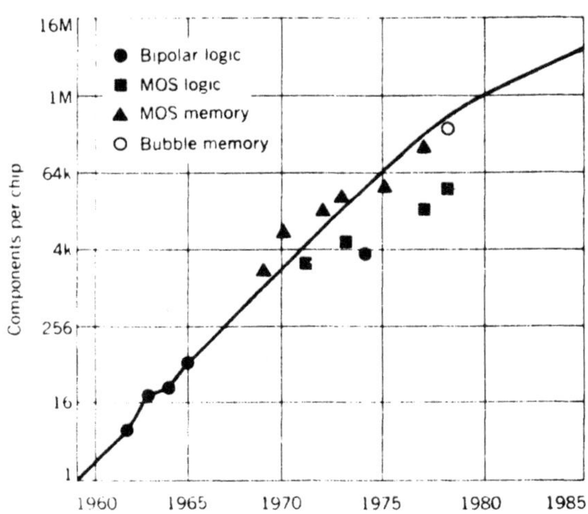

Fig.1.-Approximate IC complexity vs. year of introduction |1|

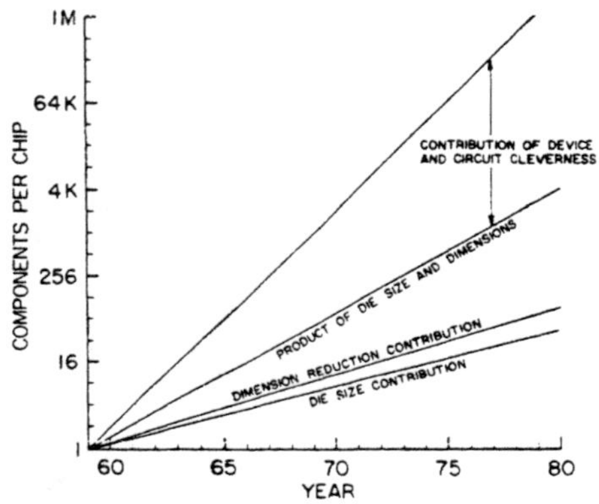

Fig.2.- Decomposition of the complexity curve into various components [1]

increases in overall chip size, decreases in minimum features sizes and improved circuit cleverness are mostly emphasized. Moore predicted that the latter contribution would be exhausted in the late 70s; consequently the growth of IC complexity would slow down to perhaps a doubling of complexity every two years. This prediction seems to be borne out during the last years. On the other hand the first two contributions to the growth of IC complexity look quite healthy at the advent of the 80s.

Shrinking device dimensions is considered more than ever as a unique tool for increasing device density. In addition the smaller feature devices also provide faster response and consume less power. The successful fabrication of these devices results from a variety of technological improvements, particularly in the area of lithographic techniques for patterning of submicron structures (section III). In relation with scaled-down devices fabrications, several difficulties have shown up at various levels (section II-1) : insulation integrity, material resistivity, mechanical behavior of the substrate, defect density, process control and reproducibility,... The solutions to these problems (sections IV through VI) are draining huge amounts of efforts in the R & D laboratories.

The other way to satisfy the ambitious demands of systems people for integrated systems on a chip involves increasing the chip size in order to reach levels of hundreds of thousands of devices per chip, and even more. Related problems (section II-2) involve high resolution exposure over wide fields, resistivity of interconnecting materials, defect density and overall fabrication yield.

Finally the economical aspects push the use of ever larger substrates. 4" and 5" silicon wafers are becoming standard substrates in IC production houses. 6" wafers are on the way. Severe requirements arise from the necessity of submicron alignment over a full wafer. To relieve the problem of mechanical distortion (section II-3) the technologists are discovering the virtues but also the vices of low temperature processing, setting some hopes on beam processing (section V).

I-2 Technologies for VLSI

Initiating a discussion on the technologies for Very Large Scale Integration (VLSI) can lead to a very exciting time, if you select properly the audience. As examples, gather "silicon-people" and "Gallium Arsenide-enthusiastics"... or put "bipolar-" and "MOS-people" around the same table... or within the "MOS" group invite "NMOS followers" and "CMOS partisans" for a friendly discussion...

This contribution is not devoted to the comparison of the various candidates for VLSI technology |2|, but aims at discussing selected problems of the envisaged VLSI technologies along with the solutions which are presently under development.

In this respect only MOS technologies will be considered since without any doubt NMOS and CMOS technologies will be part of the VLSI technological arsenal. To fix the ideas, let's review the basic steps in a standard NMOS technology (Figure 3).

The P-type silicon substrate (usually of <100> orientation) is oxidized and covered by a thin silicon nitride layer, which is opened in a first photolithography step (Figure 3.a) : the active area of the MOS transistors are protected by the nitride layer, while the insulating region between transistors, the so-called field-region, are no longer covered by nitride but, on the other hand, are implanted with boron in order to increase the threshold voltage of the parasitic "field transistor" and avoid parasitic conduction between otherwise isolated devices. After this an oxidation step gives a thick oxide (typically around 1 micron) in the field region while the active regions remain unaltered because the nitride is practically not oxidized and also is a good mask

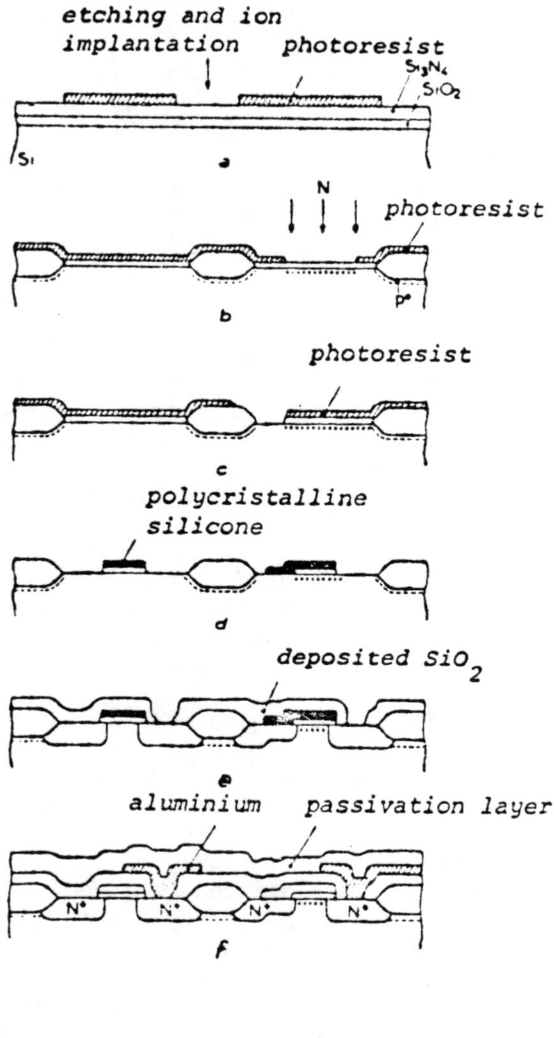

Fig.3.- Main fabrication steps in NMOS technology.

against the diffusion of oxidizing species. Oxide is thus grown locally in a so-called LOCOS (for LOCal Oxidation of Silicon |3|) process that isolates the active devices with a minimum spacing.

After removal of the silicon nitride and thin silicon dioxide layers, the gate oxide is grown in severely controlled conditions (cleanliness, gas composition, temperature). Typical thicknesses are in the range of 500 to 1000 Angströms. Two implantations are than performed to adjust the threshold voltages (these voltages depend on the supply voltage; the proposed values are typical of a 5V power supply) :

1. *an unmasked Boron implantation to set the enhancement threshold voltage at about 0.75 volts;*
2. *a masked Phosphorus or Arsenic implantation (Figure 3.b) to set the depletion threshold voltage in the range of -2 to -3 volts.*

Windows are than opened in the thin oxide (Figure 3.c) to allow a direct contact between the polycrystalline silicon (called "polysilicon") used for the gates and the N^+ sources or drains as we will see later (Figure 3.e). The next step consists in the deposition of a thin polycrystalline silicon layer (typically, 0.5 micron) which is heavily doped with Phosphorus to give a low resistivity (10 to 50 Ω/square, depending on the doping process) : the resistance of polysilicon interconnection lines must be kept as low as possible in order to achieve low propagation delays of the electrical signal and, thereby, high operating frequencies.

The polysilicon is then etched to define at the same time the gates of the transistors and the interconnection lines (Figure 3.d). The following Arsenic implantation realizes N^+ sources and drains, self-aligned with respect to the gate, thereby minimizing the interelectrode capacitances. The N^+ regions show a sheet resistance of 10 to 20 Ω/square.

An isolating oxide layer is then deposited and opened to provide contacts to N^+ diffused and polysilicon regions (Figure 3.e). An Aluminium layer is deposited and delineated to give the required interconnection pattern. A passivation oxide layer is finally deposited over the whole circuit wih openings over Aluminium pads where thin wires are soldered to insure connection to the package leads.

I-3 Widely accepted techniques

Whatever technology one selects for the fabrication of LSI circuits, several basic techniques have been accepted in the 70s as a common denominator to nearly all the available technologies. They are : projection lithography, chlorine oxidation, local oxidation with nitride protection for the field oxide growth, ion implantation for the introduction of dopant species into the silicon, low pressure chemical vapor deposition (LPCVD) of thin films (SiO_2, Si_3N_4, polysilicon), "dry" plasma etching instead of "wet" chemical etching for pattern replication onto the wafer. They offer overwhelming advantages with respect to previous techniques.

Most of these techniques will still be used in the VLSI technologies. Since they are reasonably well known, they will not be discussed in this paper unless improvements are to be brought in order to maintain them alive in the VLSI arsenal.

II. VLSI PROBLEMS

The trends in VLSI technology are raising a variety of problems which will be reviewed in this chapter before going through the techniques envisaged for their solutions. Substantial progresses in the areas described below are probably the key of success for making integrated systems on a chip, and are looked upon a challenges by technology-oriented people in the IC business.

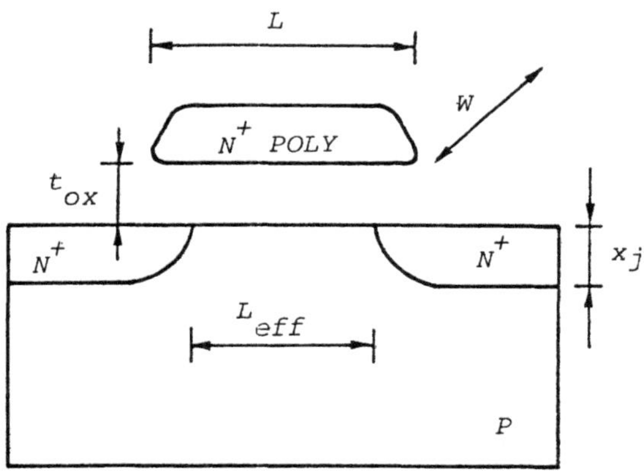

MAINTAINING CONSTANT THE ELECTRIC FIELD IMPLIES :

		SCALING FACTOR
1.	DIMENSIONS (t_{ox} , L , L_{eff} , W , x_j)	1/S
2.	SUBSTRATE DOPING (N_A)	S
3.	SUPPLY VOLTAGE	1/S
4.	SUPPLY CURRENT	1/S
5.	GATE CAPACITANCE (WL/t_{ox})	1/S
6.	THRESHOLD VOLTAGE (V_{TH})	1/S
7.	GATE DELAY	1/S
8.	POWER DISSIPATION	$1/S^2$
9.	POWER SPEED PRODUCT	$1/S^3$

Fig.4.- Scaling rules for an MOS transistor.

II-1 Scaled down devices and related problems

Scaling rules for MOS transistors have been originally proposed by Dennard et al. |4,5| : these rules are based upon maintaining constant electrical fields in the structure, independently of length or width. Figure 4 reproduces the relevant features of this set of rules. Several technological problems with respect to present state-of-the art, arise from the envisaged shrinking in dimensions.

II-1-1 Generating finer geometries

The generation of smaller features at the Silicon level first implies new developments in the photolithographic art. Today the processing engineer has a wide range of printing methods to choose from : contact, proximity, 1:1 projection,... At the present time, contact and proximity printing are still in heavy use, but they are gradually phased out because of their poor yield : the defect density of chips made with both these techniques is extremely high due to the mask damage and wafer contamination caused by contact between wafer and mask. This problem of defect density can be neatly circumvented by projection systems, which offer higher resolution than proximity printing together with large separation between mask and wafer. Well-established 1:1 reflective optical systems have resolution in the 2- to 3- µm range, alignment accuracies of 0.5 to 1 µm, and throughputs of about 60 wafers per hour : they are in use at almost every major IC manufacturer.

For all their popularity, the 1:1 projection machines do have a serious limitation : if a wafer is distorted in processing, it will throw layer-to-layer registration of successive masks on the wafer out of specification, reducing the IC yield. In addition the 3-µm resolution can no longer satisfy VLSI designers. Consequently new lithographic techniques are under development, both at the resist level (new sensitive materials) and the printing machine level (deep UV, wafer steppers, X-ray, E-beam). Section III-1 will deal with these new techniques.

Generating small size patterns in a resist layer represents only part of the complete lithographic process. Replication of those patterns in a thin layer (SiO_2, Si_3N_4, poly Si, Al...) involves an etching step of the layer material. Because of their isotropic behavior neither chemical etching nor plasma etching in barrel reactors are suitable for replication of micronsize features in a layer of thickness in the 0.2... 0.5 µm range (Figure 5). New etching techniques including plasma etching in planar reactor and reactive-ion-etching will be described in section III-2.

Fig.5.- Isotropic etching of 5 µm and 1 µm lines.

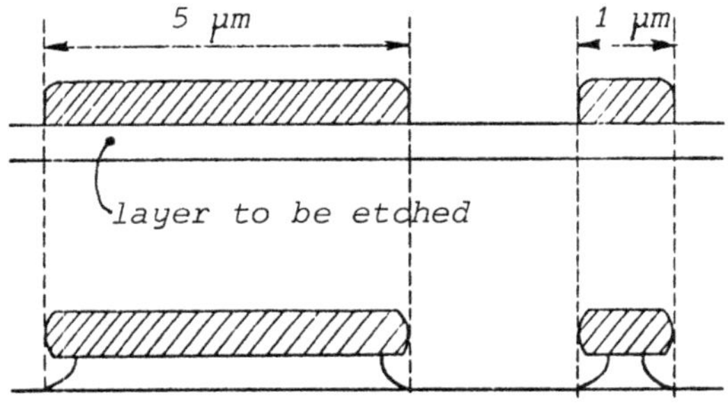

The issues in lithography not only concern the development of new machines capable of producing submicron lines with high throughput. A major difficulty in the fabrication of small size devices consists in the severe requirements in control and reproducibility of lateral processes : a lateral uncertainly of 0.25 um represents only 5% in a 5 µm technology but obviously 25% in a 1 µm technology. This means that in most cases present production techniques will have to be refined or drastically modified in order to cope with these requirements. Furthermore several two-dimensional processing effects become of importance and have to be understood and properly taken into account.

For instance, critical lateral oxidation occurs in a standard N-channel silicon-gate technology during two steps : the oxidation of the field region (Figure 6.a) and the oxidation of the polysilicon gate (Figure 6.b) after the source and drain implantation. The shape of the bird's beak and the length l of the encroachment in selective oxidation (Figure 6.a) raise problems width control both for the field region (length of a parasitic field transistor) and the active region (actual width of a narrow MOS transistor |6|).

A further example lies in the difficulties arising when one has to etch a layer whose grain size is of the same order of magnitude as the width of the stripe to be delineated. This might be the case for heavily doped polysilicon layer whose grain size can reach several thousands of Angströms, while submicron lines have to be defined for VLSI devices (Figure 7). Two ways can be taken to tackle this problem, either by managing the previous processes to keep the grain size as small as possible or by using an anisotropic etching system, insensitive to the grain boundaries.

Additional efforts are required in these areas among others, regarding not only the evolution of processing equipments but also an increased understanding of the basic material properties. As far as the last point is concerned, this should also induce ad-

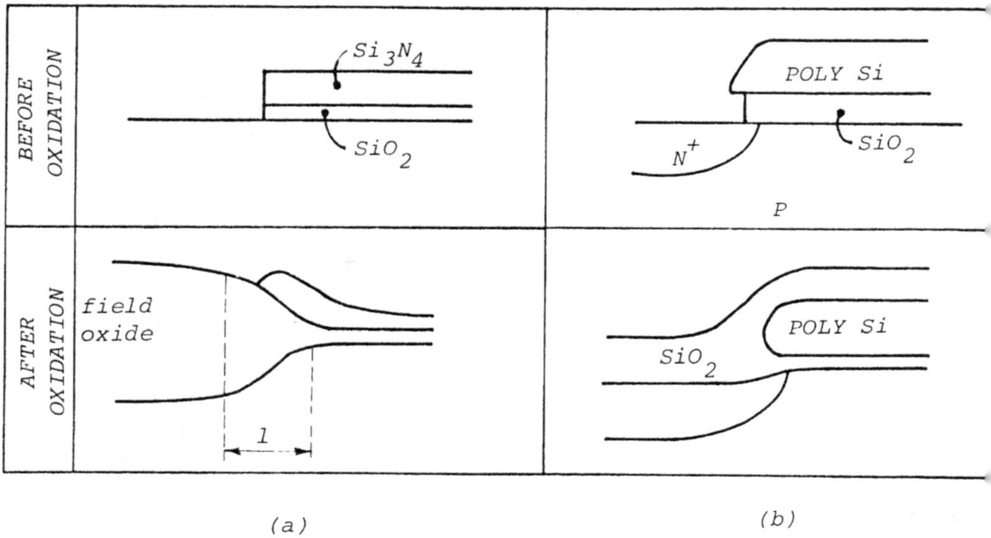

Fig.6.- Examples of two-dimensional oxidation : (a) LOCOS ; (b) source, drain and polysilicon gate reoxidation.

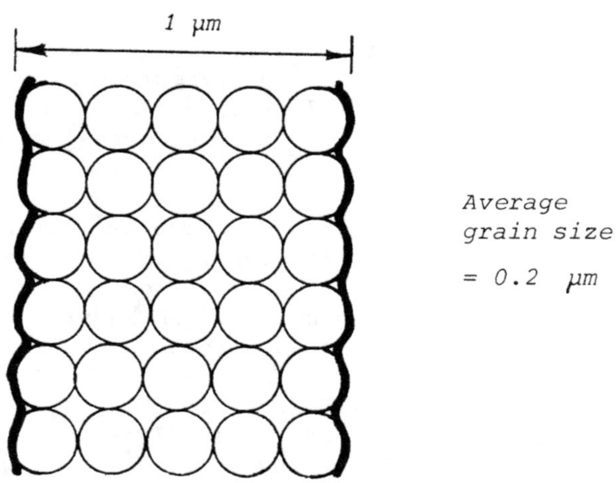

Fig.7.- Edge uncertainty in etching a large grain material.

ditional developments in two domains, process modeling and process control, which are relevant when the fine tuning of a new technology is in progress.

II-1-2 Using thinner layers

Scaling of MOS transistors means reduction of the thicknesses of all layers involved in device fabrication. This is required either to maintain the same electrical behavior or to keep the wafer surface relatively smooth and consequently relieve some difficulties in lithography.

Decreasing the gate oxide thickness has been a constant driving force in the development of new technologies, only restricted by problems of reliability. Thicknesses of 400 A... 300 A or less are under consideration for submicron devices, provided their integrity can be assured : better control of the surface (improved cleaning procedure), low defect density substrate and improved oxidation process (by the addition of chorine) are the envisaged solutions to the problem. At the same time the controlability issue has come into view : these thicknesses are not that much different from the so-called "initial thickness" used to characterize an oxidation process, i.e. some kind of an apparent initial thickness necessary to fit an experimental oxide growth cuve to the theoretical one |7|. Great benefits for the control of oxide thickness in the range 100-300 A should follow from a better understanding of the initial oxide growth. Furthermore a better process control can be obtained at lower temperatures; this will be described as a general trend for VLSI processing in section V. Finally it is worth mentionning that the oxide integrity problem has led to the prospective utilization of new insulators for the gates; silicon oxynitride (section IV-2) looks quite attractive to this respect.

The field oxide shows the same trend as the gate oxide : typically a field to gate oxide ratio of 10 to 1 gives proper relations between parasitics and device capacitances. A thinner field oxide rules out -to some extent- some worries of the process engineer : on one hand etching is performed more easily while on the other hand the oxidation conditions might be less severe (lower temperature, shorter time) leading to a lower density of induced defects. Moreover the reduction of the bird's beak encroachment (Figure 8) will increase the density of integration.

The polysilicon layer classically uses as the gate material |8| should also be thinned. Unfortunately since its doping level cannot be increased substantially, this will result in larger sheet resistances. Coupled with the need for longer interconnection lines in larger chips, this fact is tackling the development

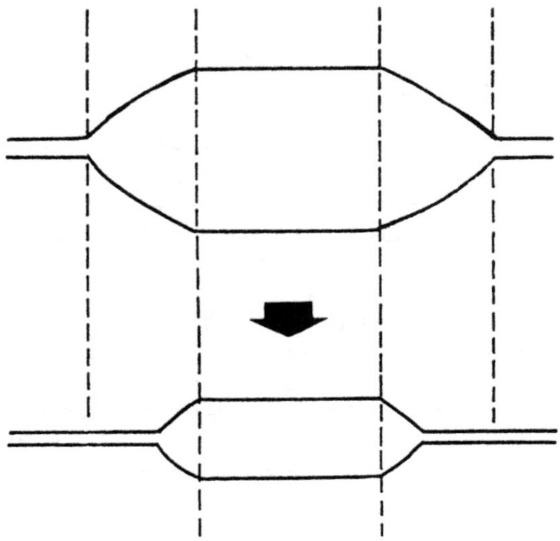

Fig.8.- Reduction of bird's beak encroachment when the oxide thickness is decreased.

of new gate materials with the same process capabilities but with a sheet resistivity lower than 1 Ω/□ : refractory metal silicides appear to be good candidates to this respect (Section IV-3).

Finally a last difficulty appears as far as reliability is concerned when the size of the metal (usually Aluminium) interconnections is decreased : the electromigration sets an upper bound to the current density.

II-1-3 Diffusing shallower junctions

Strong short channel effects, such as the reduction of the threshold voltage for decreasing channel length, are alleviated by the use of shallower sources and drains (see |6| for a review of these effects). Present junction depths of 3 to 1.5 μm have to be reduced to values of 0.5... 0.3... μm. This imposes practically unambiguous processing conditions.

Preferential dopant will be Arsenic for source and drain diffusions in an N channel technology; scaling of CMOS technology will suffer from the fact that no P-type slow diffusant species is available. In any case, ion implantation, which allows shallow profiles with high concentration will be the unrivalled predeposition technique. Further heat treatments mainly for oxidation or annealing have to be performed at the lowest temperature and for the shortest time possible. This has prompted the technologists to look for accelerated oxidation techniques, for instance pressure oxidation or plasma-assisted oxidation. Substitution annealing techniques like laser or E-beam annealing seem to have fine prospects before them. The low-temperature processes will be discussed in Section V.

However it should be pointed out that the shallow junction realization is accompanied by a severe drawback concerning the contact between the metal layer and the (100) silicon substrate. It is well known |9| that, during the final annealing step in the temperature range of 400°C to 475°C, a reaction occurs at the interface between Aluminium and Silicon; Aluminium spikes are formed in the Silicon (Figure 9) up to 1 μm deep, increasing the leakage current of the junctions, if not causing a short circuit.

Fig.9.- Formation of spikes below Al-Si contacts.

Several solutions have been proposed, such as |10| :

- *adding 1.5 to 2% Silicon in the Aluminium to prevent Silicon dissolution in the Aluminium film;*
- *diffusing a deeper junction just beneath the contact hole;*
- *lowering the post-metallization annealing temperature.*

Satisfactory results have been reported with these proposals. Nevertheless, further improvements in the contact technology should be appreciated by VLSI technology designers.

II-1-4 Reducing the defect density

Scaled devices with their very small area and reduced current level will push the requirements upon defect density to a much lower level than it does for present LSI devices. The quality of the starting material has been and still must be enhanced substantially, by reducing both the concentration of impurities and the density of crystallographic defects. The prevention of impurities contamination during cleaning procedures, processing steps and handling of the wafer goes through :

- *the use of pure and ultrafiltrated products (chemicals, gases...).*
- *the improvement of process equipments as well as handling material : the trend to automatic systems solves this problem with the bonus of increased productivity;*
- *the control of plant atmosphere by using better clean room designs.*

In addition process engineers have developed or try to develop techniques or tricks which either prevent the formation of process-induced defects (oxidation-induced stacking faults, dislocations,...) or remove unavoidable defects or contaminants (gettering of metallic impurities, annealing of implantation damage,...).

Extra benefit results from these techniques as an increased fabrication yield, which should at least partially compensate for the decrease in yield following the chip growth.

II-2 Increasing the chip size

In addition to the increased difficulty of making good quality reticles and/or masks for enlarged chips with a growing number of devices, the natural trend to larger chips faces two main problems, which appear to be as many challenges to process-oriented engineers.

The first one is still waiting for a solution in research environment : the question is, that the length of interconnection lines increases as the chip size does, while a concomitant objective is to reduce the width of the line and the thickness of the conductive layer. Much higher sheet resistivity results from these three factors. Even if the node capacitance follows the scaling principle, the line response time will somewhat increases. Especially for long interconnection lines, the propagation delay might become excessive. Furthermore a larger current density will flow in the fully scaled conductive line than in the unscaled one (if both width and thickness are scaled down) : the reliability problem of electromigration may be more severe. These two points force technologists in looking for new ways for the interconnection issue, particularly in order to replace the otherwise successful polysilicon : replacing the polysilicon with a metal silicide or, more drastically, with a refractory metal are presently evaluated techniques |11| (Section IV-3). However, these solutions will not help in places where aluminium is used.

The second -and much more restrictive- limitation to the chip size inflation resides in the dramatic reduction of the fabrication yield in the hypothesis of constant defect density (Figure 10). As it has already been pointed out previously,

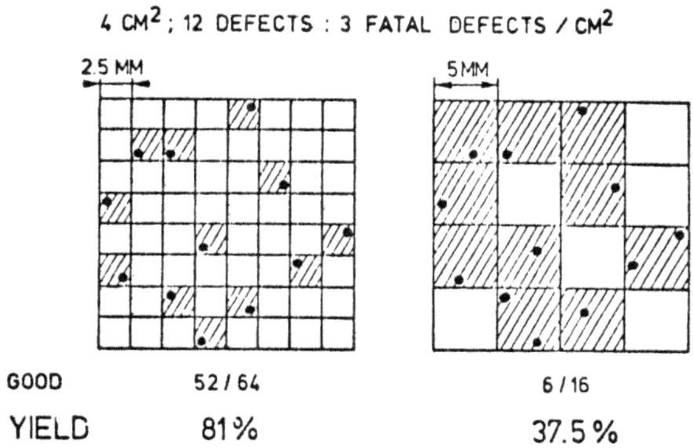

Fig. 10.- Effect of chip size on yield.

reduction in defect density is expected in order to make the VLSI circuits an economical success. Since this reduction is anyway a main driving force in the IC business, no one doubts about the vitality of the efforts aiming at bringing the solution to this problem, both at the processing and the environmental level (Sections V and VI).

II-3 Processing larger wafers

Even if it is not specific to VLSI, the adoption of ever larger wafers in IC production lines, coupled to the reduction in feature size of the devices, has generated many difficulties in the lithography process. Uniformity of both exposure and photosensitive material must be maintained over the entire wafer -in spite of increasing diameters (4", 5", 6",...)- in order to keep linewidth control within acceptable tolerances. This has justified part of the evolution in lithography equipment described in Section III-1, namely in illumination sources and optical parts.

As pointed out previously, mechanical distortion of the wafer prevents accurate layer-to-layer alignment over the wafer. At the same time the ultimate resolution in exposure cannot be maintained if the depth of focus no longer covers the flatness errors common in wafers after high temperature processing. These problems stress the importance of low temperature processes that will limit wafer in-plane distortion and bowing (Section V). They have also thrusted forward printing systems which can expose with optimum resolution small fields of the wafer; the wafer is realigned and refocused for every exposure until the entire area has been processed. Step-and-repeat projection systems have been widely accepted by most of the IC manufacturers for their first generation of VLSI circuits (Section III).

III. ADVANCED PATTERNING TECHNIQUES

III-1 Lithography

The complete characterization of a lithographic process must address the following aspects :

- *characteristics of the photosensitive material;*
- *preparation of the photoresist films over eventually non-planar surfaces;*
- *exposure of the photoresist on limited areas that must be positionned with respect to existing patterns;*
- *revelation of the exposed regions through resist development;*
- *stripping of the resist material after the patterned images have been transferred accurately to the underlying substrate in an etching step.*

Recent progresses as well as some prospects in these fields are now reviewed.

The different pattern transfer methods are illustrated in Figure 11. The starting material is the pattern data created on a lay-out generation system and generally stored on a magnetic tape. The different ways in which these data are transformed into a light pattern on the wafer are described. A few remarks will be mentionned on the issue of mask and on developments in photosensitive material.

III-1-1 Optical lithography

Optical lithography comprises the formation of images in photoresist on wafers by contact, proximity, or projection printing using visible or near ultraviolet light. These are old techniques which are in heavy use in all production lines and are still undergoing development to satisfy needs for higher resolution and tighter registration tolerances.

In contact printing, a glass mask bearing an emulsion or chromium -film pattern is first aligned with reference points on a resist-coated wafer, then pressed against the wafer; exposure is by light of wavelength near 0.4 µm. Very high resolution -$\lesssim$ 1 µm line width- is possible. However, because of spatial nonuniformity of the contact, resolution may vary considerably across the wafer. But, anyhow, the major limitation of contact printing comes from the defects produced by the tigth contact in mask and wafer. The accrued damage and particles of resist

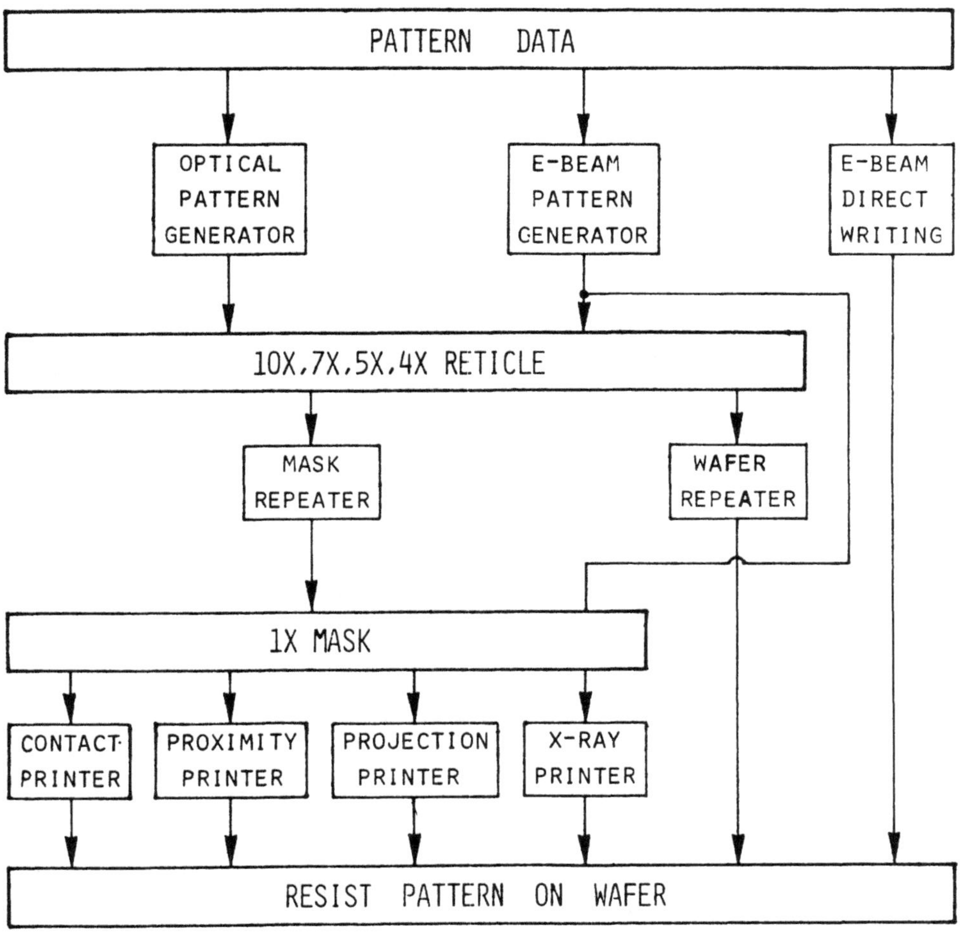

Fig.11.- Pattern transfer methods in IC fabrication.

adhering to the mask are printed on the following exposures causing a rapid buildup of defects, so that masks have to be discarded after a limited number of uses. Mask defect densities increasing to 37 defects/cm^2 after 15 exposures from 13 after 5 exposures have been reported |12|.

Nevertheless contact printers continue in wide use, for their good resolution, generally high throughput (about 60 wafers per hour) and low price of the basic equipment, even though the cost of producing working copy masks or cleaning them is non-trivial and tends to dominate the cost of operation.

The diffraction of light at the mask windows limits the resolution of the contact printing systems (Figure 12). Fresnel's diffraction theory predicts the following law for the minimum transferable period $2\,b_{min}$:

$$2\,b_{min} \simeq k\,\sqrt{\lambda s}$$

where λ *is the wavelength and* s *represents the gap between mask and wafer (for intimate contact an effective gap of about half the resist thickness should be considered) while* k *is an empirical factor approximately equal to 3 in good contact printing conditions |13|. This law justifies the move towards deep UV illumina-*

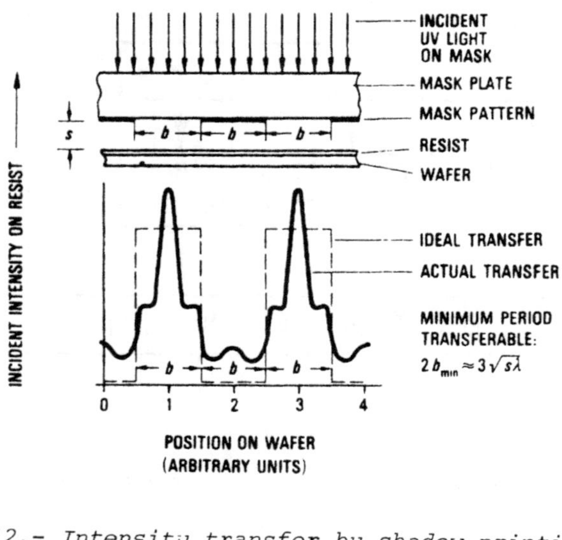

Fig.12.- Intensity transfer by shadow printing.

tion. For instance 0.25 µm features generated with a 0.20-026 µm radiation have been reported |14|.

In <u>proximity printing</u>, a gap of several microns divides mask from wafer; this ensures a longer mask life (the defect density can be decreased by a factor of 10 with respect to contact printing). However, this is obtained at the expense of resolution and linewidth control. 4 µm lines can be achieved on commercial printers with standard UV source. The distortion of the individual resist features depends on the actual mask-to-wafer spacing which is variable across the wafer; temperature induced and process induced distortions are especially troublesome.

Wafer throughput and overall cost of proximity printing are comparable to contact printing.

<u>Projection printing</u> offers higher resolution than proximity printing together with large separation between mask and wafer; it can be performed by using either refractive or reflective optics. A well established 1:1 reflective optical system has become a practically standard equipment through the IC industry |15|. It uses a unique optical system to form a 1:1 image of an arc-shaped area of the mask onto the corresponding part of the wafer (Figure 13). This strip of image pattern is scanned over the field to expose the entire area in one pass. The resulting image quality is high and uniform over the field.

The resolution of optical projection printers is limited by diffraction at the apertures, by non-ideal lenses or mirrors and by defocusing. It can reach 3 µm on a routine basis with present 1:1 aligners, while several improvements move the design goals for the next equipment generation down to a 2 µm level |15|. Achievements of $\pm$ 0.5 µm overlay have also been reported; the overlay errors are caused by thermal mismatch between the mask and the wafer and by process-induced wafer distortions. This again stresses the importance of low temperature processing. Defect density is generally very low as compared to contact or proximity printing. And, finally, the throughput of commercially available systems is quite high (30 to 60 wafers per hour can be aligned and exposed). Overall cost is about twice higher than for proximity systems.

Some improvements are also under scope : they include the production of new optics, the use of deep-UV illumination,... The question is whether 1:1 projection over the whole wafer in one pass can be improved enough to meet VLSI fabrication need, for example at a 1 µm level. No definite answer can be given for the moment.

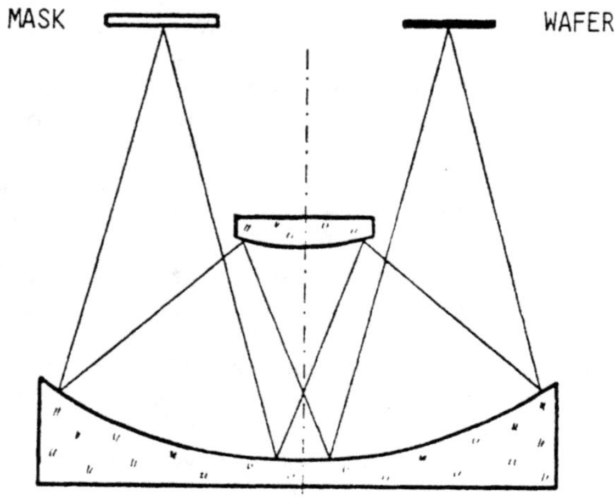

Fig.13.- ***Basic optical*** *system of a reflective 1:1 projection printer.*

Nevertheless, it seems that the problem of layer-to-layer registration of successive masks even on slightly distorted wafers has led to the wide adoption of a new class of projection aligners the so-called *<u>wafer steppers</u>* *(or wafer repeaters, or reduction projection machines, or direct-step-on-wafer machines). They all operate on the same principle (Figure 14). A high-quality image can be formed through a high-resolution lens, but only with a limited field; hence, only one chip or a few chips can be exposed in one step. Therefore the wafer must be moved, realigned and refocused for every exposure in order to cover its full area. An important benefit resulting from this step-and-repeat procedure is that it allows compensation for nonflatness and in-plane distortions of wafers : very high resolution with excellent overall overlay accuracy can be achieved even on large wafers. This method is attractive only if a very fast automatic alignment and focusing system is available, in order to keep short enough the time taken to expose the entire wafer.*

An important variable is the demagnification ratio which allows

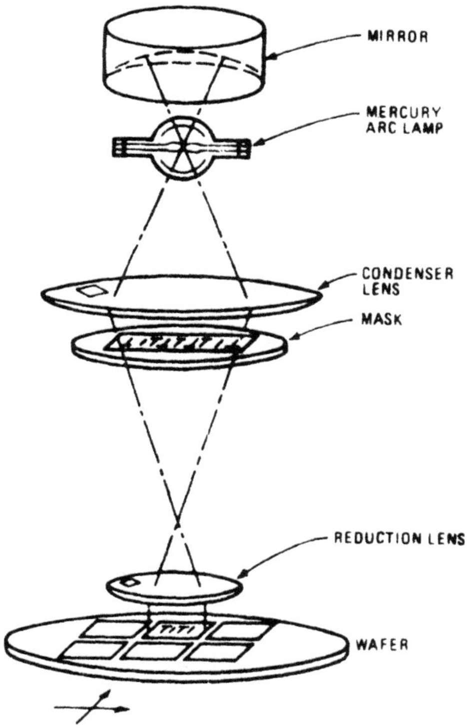

Fig. 14.- A step-and-repeat projection system.

trading performance of the printer against throughput, for a given price. A larger demagnification ratio increases the number of steps, and therefore the total time, but allows a larger aperture, and consequently better resolution and realignment capability. Typical demagnification factors are 10:1, 7:1, 5:1, 4:1, and 2:1. Commercially available systems with 10:1 ratio provide 1.25 μm resolution with 0.25 μm registration; throughput depends on the chip size but stays in the range of 10-50 wafers per hour |16 |.

The defect problem can be dramatic with a wafer stepper : for instance if only one chip is repeatedly exposed, a single defect on the reticle may be sufficient to cause zero yield. The problem is relieved to some extent with a high demagnification ratio, since small defects will not be imaged on the wafer. On the other hand demagnification ratio provides additional benefits, since linewidths errors and positional errors of the geometries on the reticle are reduced by the same factor, therefore making the requirements for the reticles less stringent.

In conclusion, the step-and-repeat projection technique looks capable of driving the lithography art in the 1-µm and even submicron region. Significant limitations to this method are the cost of existing printers (explained by the high level of sophistication) and the relatively slow throughput, as compared to other optical pattern transfer techniques. However it must be mentionned that both limitations are much more severe in the case of electron beam lithography.

A quick look through the literature on lithography a couple of years ago is quite instructive. The question was : "Will the optical systems do the job for 1-µm lines ?". The answer was "no" in most cases, while electron beam and X-ray lithography were receiving the most enthusiastic greetings. At the beginning of the 80s the answer is generally "yes", due to wafer steppers, which are getting a very large audience in the IC industry.

III-1-2 Accessories : resists, masks and wafers

Resists used in the manufacture of integrated circuits are generally organic polymers and are divided into two main classes, positive and negative, depending on their response to the exposing radiation. Positive resists become more soluble in the irradiated area relative to the unexposed area whereas negative resists become less soluble.

The resists requirements for VLSI applications are similar regardless of the exposure technology :

1. *Sensitivity : high sensitivity means large throughput, but it generally appears that enhanced sensitivity is attained only at the expense of other properties such as etch resistance.*

2. *Resolution : features of less than 1 µm should be accurately patterned in the comming years. Factors affecting the resolution, such as swelling during development, constitute a severe limitation. Negative resists are seriously affected by the swelling problem.*

3. *Adhesion : adhesion ceases to be a problem if wet etching is replaced by dry etching techniques (see Section III-2).*

4. *Etch resistance : the resist layer must withstand the etching environment, without degradation in resolution. This is perhaps the most difficult requirement to achieve, particularly for positive resists.*

5. *Defect density : the demand for low defect densities pushes the use of thicker resist film, which makes more difficult the achievement of 1-µm features, although this appears to be primarily a limitation of negative resists : the swelling limits the resolution to 2-3 times the initial film thickness.*

Up to now the trend, while decreasing the dimensions in lithography, has been to use conventional photoresists. However negative resists present limitations in resolution and a reduced sensitivity; it is unlikely that current negative photoresists can be used for 1-µm features with thickness sufficient for low defect densities. Furthermore no substantial developments in negative resists are expected in a near future particularly in the field of deep-UV exposure. Thus, more likely, the LSI and VLSI lithography will see the use of positive photoresists which are capable of 1.0 µm resolution or better. However, their adhesion and sensitivity are marginal and may need improvement. As far as deep UV (∿0.25 µm) applications are concerned, conventional photoresists are not very suited even if some success can be obtained |17|; nevertheless the lack of high power sources in the deep UV region places a requirement of high sensitivity on the resist. Among the newly developed resists, the PMMA (polymethylmethacrylate), with some additives which increase its sensitivity shows an excellent resolution but does not solve fully the question of sensitivy, and exhibit a rather low etch resistance.

In optical lithography with 3-4 µm ground rules, the <u>masks</u> used to be fabricated in a two-step technique (Figure 11). A 10X reticle is first generated by means of an optical pattern generator; this machine exposes pattern elements to form the chip image at 10X scale, making typically 50 to 100 flashes per minute. The reticle is used in a step-and-repeat camera which repeats the reduced image to fill the desired mask area. The master mask produced in this way may be used in a nondestructive projection printer or copied, and the copies used in a contact printer. Typical overlay error of a master mask with a second mask of a set will be $\pm$ 0.8 µm for a 10 cm diameter area; for the copy the error will be $\leq$ 1 µm. Linewidth uniformity on the master will be ∿ $\pm$ 0.15 µm and linewidth tolerance ∿ $\pm$ 0.3 µm.

With respect to the previous scheme the main problem for VLSI will reside in the time required for generating a reticle. For example, a 64K MOS RAM has in average $\sim 10^5$ pattern elements per level; pattern generation times may range from 10 to 30 hours depending on the number of flashes and the system being used. After that, the reticle must be mounted and aligned manually on the step-and-repeat camera; the master mask is thus obtained after a second photographic step (exposure and development).

New techniques are presently used to generate directly the master mask (Figure 11) : they take advantage of the high resolution attainable by electron lithography because of the small wavelength of the 10-25 keV electrons used (the resolution of electron systems will be described in a following section). Commercially available E-beam systems can generate master masks with overlay error $\leq \pm$ 0.5 µm and linewidth tolerance $\pm$ 0.25 µm for a 10 cm array. These achievements supersede those obtained with optical equipments. But much more impressive is the patterning time spent in generating such masks : 30 to 60 min. leading to fabrication times two orders of magnitude smaller on an E-beam pattern generator with respect to an optical one. These two facts explain why E-beam mask generators are becoming more and more widely used in mask shops. The short turn-around time is of particular interest, in addition, when multi-chip masks are wanted, since the fabrication time remains constant, independently on the member of chips, whereas an optical system requires as many reticles as different chips. E-beam generated masks will likely be the only choice for VLSI optical printers, not only for 1:1 projection printers but also for wafer steppers.

Several properties related to the <u>wafer</u> itself can affect the overall performance of an advanced lithographic technique; they are now briefly discussed :

- *Chemical state of the surface : it is responsible for the adhesion of the resist film. In some critical cases an adhesion promoter might be recommended.*

- *Flatness : a several microns bowing appears, mainly during high temperature steps, due to internal stresses in the films on both sides of the wafer. This bowing leads to out-of-focus imaging in projection printing and to variable gap between wafer and mask in proximity printing; both effects cause linesize variations in photoresist patterns.*

- *In-plane distortions : internal stresses in thin films deposited on the surface(s) of the wafer also generate in-plane dis-*

tortions, which cause overlay errors if the amount of distortion differs from step to step in the fabrication process. These distortions arise from mechanical deformation of the wafers during heating or cooling phases; they are also associated to large densities of dislocations in the bulk, which is highly undesirable in LSI circuits. Part of the solution consists in ramping the temperature during both heating and cooling phases. This is increasingly more important when larger wafers are used in the process.

- *Steps on the surface : when spinned on the wafer with steps on its surface, the resist displays a non-uniform thickness near the step (Figure 15). This gives problems of linewidth variations, since the exposure and development on one hand, the etch resistance on the other hand are different near the step with respect to the flat zone. The difficulty is reduced with thinner layers, but enhanced by the trend towards thinner resist films wished for higher resolution. Spikes on some films (polysilicon, aluminium) may cause the same problem.*

- *Particules and dusts : obviously submicron devices and interconnections cannot afford pinholes or spots generated by particules or dusts during the lithography process.*

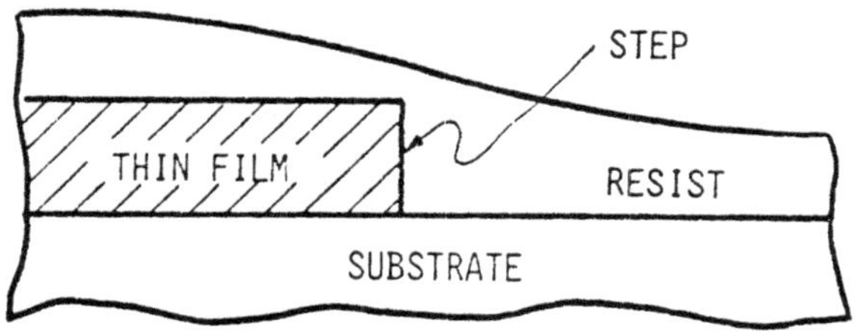

Fig.15.- Thickness variation of photoresist in the vicinity of a step.

III-1-3 Electron lithography

Even if it is generally accepted now that optical lithography will still dominate the IC production scene until 1985, mainly by taking advantage of the progresses in wafer steppers, it appears that the era of electron lithography will begin when IC design rules will move to 1 µm and below. This technique offers higher resolution than optical lithography because of the small wavelength of the 10-25 keV electrons used. The diffraction does not limit the resolution, which results from electron scattering and from limitations of the resist.

Two techniques are used to expose an electron-sensitive resist, either scanning or projection. The latter one remains for the moment just a promise whereas the former one has gained much more importance and is now available in commercial equipments. The first production application of electron lithography has been in photomask making, as discussed previously. Direct printing on the wafer is presently used in R&D laboratories, but has not yet been adopted in production for economic reasons.

The development of high quality electron-sensitive resists is a crucial problem to be solved before electron lithography can move to the production lines. Both positive and negative resists are studied today. Two major factors limit the resist resolution : swelling of the resist in the developer (more important for negative resists) and electron scattering. With swelling narrow lines present undulations, and adjacent lines can touch. Backscattering of electrons from the substrate is responsible for the "reduced" resolution; it depends on resist thickness, beam energy and substrate composition. The scattering also induces a proximity effect : an exposed pattern element adjacent to another one receives exposure not only from the incident beam but also the scattered electrons from the adjacent elements. The resolution of commercially available electron resists reaches the submicron level in all cases : 0.5 µm with PBS (polybutene - 1 - sulfone) and 0.1 µm with PMMS (polymethylmethacrylate). Unfortunately PBS is quite difficult to handle in wafer processing, while PMMA has a low sensitivity (this means long exposure time in present systems). In addition the resistance to etching conditions is rather poor. Substantial progresses are still expected in this area before wide adoption in production lines.

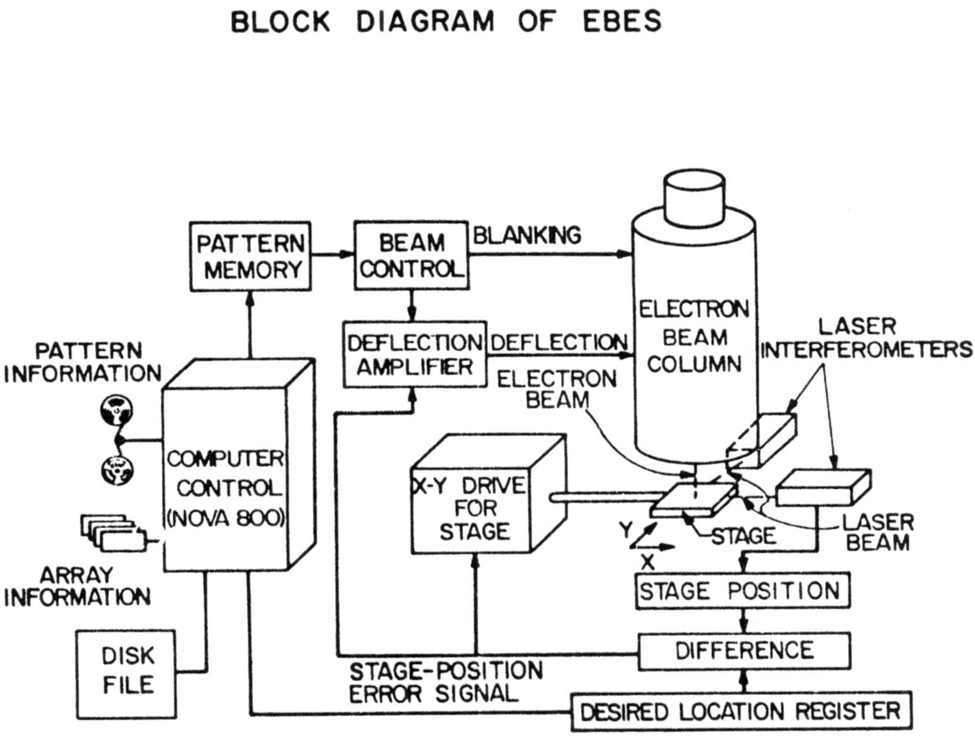

Fig.16.- Scanning electron beam system.

Several methods for pattern generation are used in existing electron beam writing machines. In a raster scan pattern generator, the beam is deflected repetitively over the exposure field, as in a television raster. The beam is turned on at various points in the scan to expose the desired pattern (Figure 16). The EBES machine |18| developed by Bell Laboratories became one of the first commercially available systems. Since the whole field is scanned this system is not really efficient.

Higher efficiency is obtained in a <u>*vector scan*</u> *system, where the beam is directed in random access manner to the regions of the chip to be exposed. The pattern is decomposed into basic elements, and each one is filled in by the small writing beam. Printing time, depending on the chip complexity, is still quite high in order to consider it for mass production. New exposure methods are developed in an attempt to decrease the printing time (independently from efforts related to improvements in electron sources and in resist sensitivity) : the pattern can be exposed by a square beam or -better- by a variable shaped rectangular beam (unfortunately this is realized at the expense of additional complexity of an already complicated system).*

Alignment marks are usually located in the four edges of a deflection field; they are either heavy metal marks or deep trenches into the substrate. Mark position is sensed by the variation in back-scattered and secondary electrons on scanning over the mark edges. Automatic chip-by-chip alignment can be achieved with this technique, a feature of interest since it can account for distortions caused by wafer warpage. Typical overlay accuracy lies in the range ± 0.1 µm to 0.5 µm in vector scan machines.

The inherently superior capabilities of E-beam system have been proven in some applications, such as mask generation and discretionary wiring. The last application is running for the personalization of logic master slices |19| : of the fourteen masking steps associated with a bipolar chip, the eleven fixed pattern levels are realized with an optical system while the three personality defining ones use E-beam lithography. In this case the fast turnaround joined to the small quantity of identical wafers provides an important benefit. E-beam direct exposure has also been demonstrated a valid tool for making submicron integrated circuits in a laboratory environment |20|.

However more widespread use of this technique has been prevented mainly by economic factors. The initial cost of a scanning electron beam machine is extremely high, while throughput remains in most cases limited to a few 4" wafers per hour. Undoubtedly the improved scanners of the future will cost more than today's slower units. In any case this means that exposure cost per wafer will be at least an order of magnitude higher than with optical systems. The only certain future of E-beam writing, at least with present perspectives, is in such applications where flexibility and fast turnaround prevail instead of throughput : mask making, personalization of prediffused wafers, prototype production...

III-1-4 X-Ray lithography

The difficulty of developing an E-beam system suitable for mass production has generated reniewed interest in X-ray lithography |21, 22| : it is simply a form of proximity printing in which an X-ray source replaces the UV source. This system offers submicron resolution with good linewidth control, while throughput should be comparable to that of present optical systems, provided that power source, resists, masks and manipulation systems are appropriate.

X-rays are produced by the interaction of incident electrons with a target material. With wavelength of 5 to 50 Å, diffraction effects can be ignored; this technique can, consequently, utilizes the full potentiality of E-beam generated masks. The masks constitute a major weakness of the X-ray lithography. They consist of a patterned metal absorber on a transmissive membrane substrate. The ratio of metal thickness to substrate thickness is not so small; this explains the inherent fragility of the X-ray masks; with questions remaining about their dimensional stability, the ease of handling and the minimum defect densities attainable.

Developments efforts in many laboratories involve more powerful source (synchrotron radiation and storage rings), more sensitive resists (PMMA is presently used in most cases), improved masks and automatic exposure system. Commercial X-ray systems are in preparation, but the overall cost of that technique is not yet well evaluated.

III-1-5 Ion beam lithography

Since ions can be controllably deflected, ion beam technique might be a substitute for fine line lithography in several senses :

- *Ion beam writing and ion beam projection are possible in the same way as electron lithography. Ions can be used to expose organic resists layer, but can also be used to modify locally the etch rate of a given layer, giving rise to a possibly new class of masking materials.*

- *Ion beams offer the possibility of either direct selective doping of wafers, or material sputtering, depending on the type and energy of incident ions.*

Ions present the advantage over electrons of smaller proximity effects, since they produce lower energy secondary electrons when incident on a target material; 0.1 μm features should be possible and even smaller ones. However, special problems regarding the ion sources and the ion optics have to be solved before the ion beam technique can be envisaged for IC fabrication. Particularly the writing speed with ion beam seems too low for replacing present ion implantation technique through selective mask.

III-2 Etching

The micron and submicron lines patterned in a resist layer are of little value unless their images can be transferred accurately to the underlying film or substrate. Traditionally the transfer process consisted of wet chemical etching, a completely isotropic technique which is no longer acceptable when narrow and deep patterns have to be etched (Figure 5); the poor resolution and undercutting of wet etchning have pushed the development of new techniques, which should offer the following qualities in order to be adapted for VLSI technology : high anisotropy (the layer is etched faster vertically than it is laterally), sufficient etch selectivity (the material etch rate must be higher than the etch of the mask or of the next underlying layer), good uniformity (in order to avoid uncontrolled underetching), no redeposition of etched material and maximum etch rate (to increase throughput). Dry etching techniques are good candidates for replacing the standard wet chemical etchants.

Dry etching techniques are divided in three main classes, depending on whether the etching process relies on physical effect, chemical mechanism or combination of both. Physical or non reactive techniques are used in sputter etching and ion milling |23| where the erosion of a surface is only due to ion bombardment (generally Ar^+). In spite of several advantages, many drawbacks preclude the use of physical etching techniques, at least with present lithography and layers in the IC technology; they are : very poor selectivity, generally low etch reate, redeposition of sputtered material, radiation damage,...

A chemical process takes place in plasma etching where the substrate is exposed to reactive chemical species generated in a plasma. These species (mostly neutral ones) interact with the material to be etched and form volatile products eliminated through the vacuum system. For instance, the following mechanism is used for silicon etching :

$$CF_4 \xrightarrow{\text{plasma}} CF_3 + F$$

$$Si + 4F \longrightarrow SiF_4(g)$$

Initial plasma etching experiments were done in cylindrical or barrel reactors (Figure 17). Although they have high throughput because of the vertical arrangement of the wafers, they present a serious drawback as far as fine pattern transfer is concerned : the etching is very non-uniform across the wafer and along the wafers carrier (because of gradients in the density of reactive species and in the temperature); in addition this purely chemical process is clearly isotropic. Plasma etching in barrel reactors will be confined to non-critical applications.

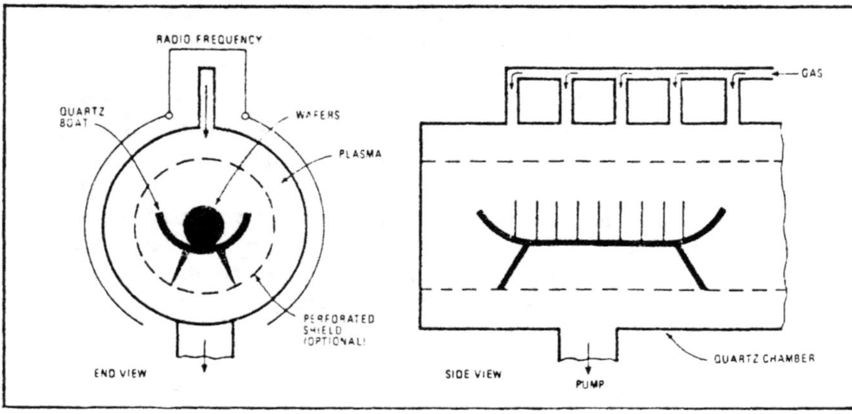

Fig. 17.- Barrel reactor for plasma etching.

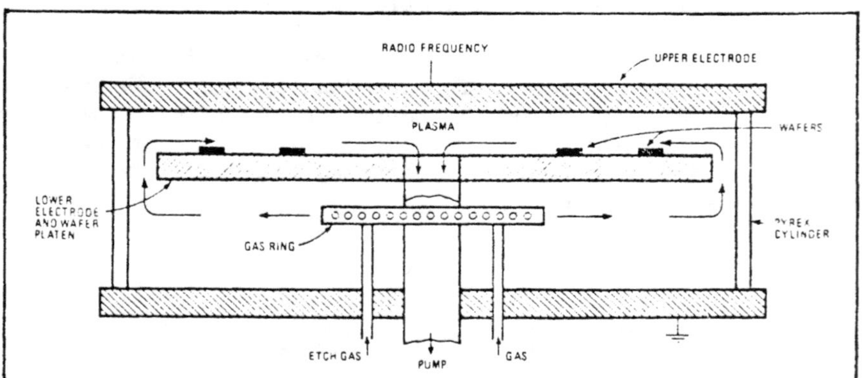

Fig. 18.- Parallel plate reactor for plasma etching.

Some improvement results from the adoption of a planar configuration in parallel plate reactors |23| (Figure 18). Although its throughput is smaller, this system allows for proper control of the substrate temperature and gives less undercutting. As a consequence commercial equipements are available; based on that arrangement some have been used successfully for etching 2 to 1 µm lines, even if the etch process is quite critical (it depends severely on the exact gas composition) and may induce some radiation damage.

The combination of both plasma etching and sputter etching gives rise to the so-called reaction sputter etching or reactive ion etching |25|. In this mode the substrates are placed on the powered electrode in a parallel plate system similar to the one shown in Figure 18. Etching is done with the same gases as used for plasma etching but in this case the charged species are important. Due to the lower pressure and the higher energies of the ions, the substrates are etched anisotropically while the selectivity is increased. The reactive ion etching technique is now extensively studied in the research laboratories of the major IC companies, and is generally considered as the most promising for the replication of the submicron details of a VLSI circuit.

It should be noticed, however, that the basic mechanism of reactive etching is not fully understood, even if it has been proven |26| that physical effects due to ion bombardment and chemical effects in the direct interaction of the reactive species with the substrate cannot be treated separately, but must be studied as a whole. From new developments in the basic understanding and improvements in the system design, it is expected that reactive ion etching will become a widely used equipment in VLSI lithography.

IV. NEW MATERIALS

IV-1 Substrates

Even if from time to time some prophets announce the advent of such or such material for VLSI, silicon remains the undisputed substrate for mass production of VLSI circuits. The only question concerns the future of thin silicon film over an insulating substrate against single-cristal substrate. Outstanding results have been obtained with SOS (Silicon-On-Sapphire) substrates, where the inherently lower parasitic capacitances allow much faster operation. Complementary MOS on Sapphire offers one of the best speed power of any technology (around 0.1 pJ). Unfortunately the high defect density makes SOS technology questionable for VLSI, even if remarkable success has been obtained, for instance a 7000-gate microprocessor (20000 transistors on 8.85 mm x 6.66 mm chip; typical speed : 2.3 times higher than for the bulk-silicon version) fabricated with acceptable yield |27|. A more severe drawback of the SOS technology lies in the process-development costs and above all higher substrate-startup costs. This last point, added to the superior capabilities of SOS technology, explains the activity around new ways of growing high quality crystalline silicon films on insulating substrates. Initial but promising results have been reported on graphoepitaxy |28| (a technique that uses artificial surface relief structures to induce crystallographic orientation in thin films) and on laser annealing of polycrystalline silicon on SiO_2 or Si_3N_4 underlying films |29,30|.

Nevertheless, it is obvious that VLSI at a production level will concern essentially single-crystal silicon substrates. Up to now the production yields were generally limited by "process induced defects". At the eve of the VLSI era, emphasis is placed on high yields on large area chips for evident economical reasons. Since at the same time MOS wafer processing technology has dramatically reduced the incidence of process induced defects, it is expected that material inhomogeneities (such as dopant nonuniformities and nonuniform distributions of impurity/point defect complexes) might become yield limiting factors. Mechanical requirements (flatness...) are also important ones, as this has been stressed in the section devoted to photolithography.

IV-2 Insulators

Submicron IGFET displays improved behavior when its gate insulator thickness is reduced |6| : insulator charges have less influence on threshold voltage and various short channel effects are minimized. The main problem, however, appears to be in growing reliable thin insulators : thin thermally grown silicon dioxide layers are not acceptable below the 200 Å barrier. First encouraging results have been reported on nitride |31,32| or oxynitride films. Thermally-grown nitride films look promising because of their highly dense structure ; even a thin nitride film acts as an effective barrier against dopant impurities diffusing from the gate. Low surface state densities ($\sim 10^{10}$ eV^{-1} cm^{-1} at midgap) have been achieved. High performance devices have been realized with a 70 Å nitride layer as gate insulator | 33| : the fabrication technology was a standard NMOS one, except for the substitution of the gate thermal oxidation by a gate thermal nitridation.

A remaining problem concerns the difficulty of the nitridation process : several hours (> 5 hours) at high temperature (> 1100°C) are required to grow a 50-60 Å nitride layer. This is not compatible with the general trends in MOS processing. A solution might come from a combination of oxidation and nitridation, resulting in thin oxynitride layers, in shorter times, suitable for MOS processing.

IV-3 Conductors

Propagation delays in interconnection lines are becoming a major problem in VLSI circuits : the resistance of the lines associated to their distributed capacitances will be limiting factor in VLSI circuits, where the individual device performances have been optimized by scaling and by improved processes. Polysilicon lines with their limited sheet resistivity ($\sim 20\,\Omega/\square$ square) for a 0.4 - 0.5 µm thickness) are obviously a hot topic. A second difficulty arises when fine patterns have to be defined : to decrease the sheet resistivity, the polysilicon grain size must be increased which in turn makes a fine etching quite questionable.

For both reasons new gate and interconnection materials are studied, that should offer the following qualities :

- *low sheet resistivity*
- *submicron patternability*
- *compatibility with NMOS process*
- *high reliability.*

Among the contenders are present :

- *refractory metals as molybdenum, tungsten, tantalum |11|*
- *silicides of refractory metal |34|*
- *sandwich layers of polysilicon and a metal (refractory metal, platinum ...).*

Most efforts are directed towards the silicide approach, since it seems feasible to replace the polysilicon layer by a silicide one without major alteration of the whole process. Sheet resitivities of about 1 Ω/square have been achieved with $MoSi_2$ and WSi_2 |34|; pure refractory metals allow for even lower values : 0.25 Ω/square for Mo and 0.4 Ω/square for W |11| with 0.35 µm thick layers. Patternability has been successfull with dry etching technique, down to a 1 µm level : the point is that the grain size in those layers is much smaller than in heavily doped polysilicon films. Finally silicides are more easily passivated than refractory metals, giving more probability of good stability.

Last but not least, for interconnection lines driving "high" currents (supply buses...) aluminium has always been the unique choice. No solution is envisaged presently to scale the aluminium interconnection lines.

V. LOW TEMPERATURE PROCESSES

V-1 Goals

The scaling rules for proper device operation and the feasability of fine line lithography impose to decrease the temperature of most process steps :

- *shallow junctions limit the short channel effects in a MOS transistor | 6 |.*
- *all thicknesses have to be decreased : gate oxide to decrease the small size effect in the transistor; field oxide, gate layer, aluminium interconnection,... to ease the lithography problem.*
- *for lithography considerations, too, the geometry of the wafer should remain constant throughout the whole fabrication process : bowing endangers the focus capability in an optical printer, while in-plane distortions increase the overlay error.*
- *the defect density should be maintained as low as possible, to keep the leakage currents at their minimum level.*

V-2 Arising problems

Except for the last point where some trade-off has to be found, the trend is clear : the temperature must be decreased with respect to standard values. However this cannot be realized without any major adaptation or even without changing some basic concepts.

- *Growing a 0.5 µm oxide in wet ambient takes 13 hours at 800°C and about 60 hours at 700°C... This means unacceptable oxidation time below 800°C.*
- *If a high temperature process introduces several types of defects in the substrate (oxidation induced stacking faults, impurities...) on the other hand it provides also benefits in defects removal : gettering of impurities by high temperature predeposition, annealing of lattice damage created by ion implantation, passivation of the Si-SiO_2 interface,...*

V-3 Envisaged solutions

Several techniques are presently experimented, in order to cope with the listed goals and problems in low temperature processing. Obviously ion implantation will be used for dopant introduction.

An acceleration of the oxidation process has been achieved by using high-pressure oxidation |35| : growing a 1.0 µm oxide takes 5 hours in wet ambient (1 atm) at 1000°C, while it also takes 5 hours in wet ambient at 20 atm but for a temperature reduced to 700°C. RAM circuits have been fabricated with a high pressure oxidation of the field region |36| : observed improvements in refresh time have been explained by the suppression of oxidation-induced stacking faults.

Plasma processing at relatively low temperature (300°C) is getting more and more attention. Plasma deposition of silicon dioxide and silicon nitride becomes an industrial solution. And plasma might make new entrances in various fields like plasma-enhanced oxidation, plasma development of exposed photoresist...

But the highest excitement lies in beam processing : laser, electron-beam or ion-beam processing. Laser processing |37|, the most advanced one, offers the following advantages :

- *Localized heating, with high temperature for only short time periods; therefore it can be considered as a "cold process" from the point of view of the whole wafer;*
- *improved material characteristics and consequently higher device performances (by repairing the lattice damage and activating the dopants after ion-implantation, without redistributing the dopant atoms and without mechanical stress in the wafer);*
- *ability to produce new material properties.*

Unfortunately since the energy absorptions depends on the surface characteristics, edge effects have prevented up to now the use of laser processing on wafer where the top layer is nonuniform |38|.

VI. PROCESSING CONDITIONS

As a conclusion it should also be mentionned that VLSI chips with increased dimensions stress the need for improved processing conditions and ambient, in order to maintain the fabrication yield at a reasonable level. The class 100 criterion commonly used today will probably be replaced by class 10 or even less. More severe checkings will be made on the products in contact with the wafers : chemicals, gases...

And finally the complete fabrication process will take place throughout much tighter controls at every step : even if it is possible to fabricate submicron MOS transistors with present equipments and techniques, it is obvious that the dispersion in basic characteristics precludes the fabrication of complex circuits on a commercial basis. Better controlled operations have to be set up before VLSI circuits can go to automated production lines.

REFERENCES

|1| G. MOORE, "Progress in Digital Integrated Electronics" IEDM Conf. Digest, pp. 11-13 (1975).

|2| For a detailed comparison the reader is referred to the contributions by C. Sequin and R. Brodersen in the NATO Advanced Study Institute, "Design Methodologies for VLSI Circuits", U.C.L., Louvain-la-Neuve, July 8-18, 1980.

|3| J.A. APPELS et al., Philips Res. Rep., 25, 118 (1970).

|4| R.H. DENNARD et al. "Design of Micron MOS Switching Devices" IEDM Washington DC, 1972.

|5| R.H. DENNARD et al., IEEE J. Solid-State Circuits, SC-9, 256 (1974).

|6| E. DEMOULIN, "Physics of Small Size MOS Devices", NATO Advanced Summer Institute, Louvain-la-Neuve, 1980.

|7| B.E. DEAL and A.S. GROVE, J. Appl. Phys. 36, 3770 (1965).

|8| L.L. VADASZ et al., IEEE Spectrum, 6, 28 (1969).

|9| R.J. ANSTEAD and S.R. FLOYD, IEEE Trans. Electron Devices, ED-16, 381 (1969).

|10| J.Mc CARTHY, Microel. Reliab. 9, 187 (1979).

|11| P.L. SHAH, IEEE Trans. Electron Devices, ED-26, 631 (1979).

|12| W.N. JONES, Proc. Kodak Microelectronics Seminar Interface (1975).

|13| D. WIDMANN, "Pattern Generation for Integrated Circuit Fabrication", NATO Advanced Study Institute, Louvain-la-Neuve, 1977.

|14| B.J. LIN, J. VAC Sci. Techn., 12, 1317 (1975).

|15| J.W. BOSSUNG and E.S. MURASKI, Solid-State Technology, 109, August(1979).

|16| G.L. RESOR and A.C. TOBEY, Solid-State Technology, 101 (1979).

|17| B.J. LIN, 8th Int. Conf. on Electron and Ion Beam Sci. and Techn., ECS Meeting, May 21-26, Seattle (1978).

|18| D.R. HERRIOTT, IEEE Trans. Electron Dev., ED-22, 385 (1975).

|19| H.S. YOURKE and E.V. WEBER, Techn. Digest IEDM, Washington DC, 431 (1976).

|20| H.N. YU et al., J. Vac. Sci. Technol., 12, 1297 (1975).

|21| D.L. SPEARS and H.I. SMITH, Electronics Letters, 8, 102 1972.

|22| R.K. WATTS, Solid-State Technology, p. 68, May(1979).

|23| C.M. MELLIAR-SMITH, J. Vac. Sci. Technol. 13, 1008 (1976).

|24| A.R. REINBERG, "Etching for Pattern Definition", H.G. Hughes and M.J. Rand Edit., Electrochemical Soc. Proceedings 91 (1976).

|25| N. HOSOKAWA et al., Jap. J. Appl. Phys. (suppl.), 2, 435 (1974).

|26| J.W. COBURN and H.F. WINTERS, J. Vac. Sci. Technol., 16 391 (1979).

|27| M. ISOBE et al., IEEE Trans. Electron Devices, ED-26, 588 (1979).

|28| M.W. GEIS et al., IEDM, abstr. 9.1, p. 210, Washington DC, (1979).

|29| K.F. LEE et al., Appl. Phys. Lett., 35, 173 (1979).

|30| T.I. KAMINS et al., IEEE Trans. Electron Devices, ED-27, 290 (1980).

|31| T. ITO et al., J. Electrochem. Soc., 125 448 (1978).

|32| S.P. MURARKA et al., J. Electrochem. Soc., 126, 996 (1979).

|33| T. ITO et al., ISSCC, Abstr. n° 7.3. p. 74, San Francisco (1980).

|34| B.L. CROWDER and S. ZIRINSKY, IEEE Trans. Electron Devices, ED-26, 369 (1979).

|35| R.J. ZETO et al., J. Electrochem. Soc., 122, 1409 (1975).

|36| *N. TSUBOUCHI et al., IEEE Trans. Electron Devices, <u>ED-26</u> 618 (1979).*

|37| *E. RIMINI and S.U. CAMPISANO, "Laser Annealing of Semiconductors", ESSDERC, Munich, September (1979).*

|38| *C. HILL, "Laser Annealing of Device Structures", Proc. Symp. on Laser and Electron Beam Processing of Electronic Materials, edited by C.L. Anderson et al., Electrochem. Soc., vol. 80-1, pp. 26-43 (1980).*

MODELS FOR MOS TRANSISTORS

F. Van de WIELE

Université Catholique de Louvain, Laboratoire de Microélectronique, Place du Levant 3, 1348 Louvain-la-Neuve, Belgium

1. INTRODUCTION : THE MOS TRANSISTOR

An appropriate positive voltage V_G applied on a metal gate, separated from a p-type semiconductor substrate by a thin oxide layer (see Fig. 1), induces a negative surface charge of free electrons, which is called inversion charge. The M(etal) -

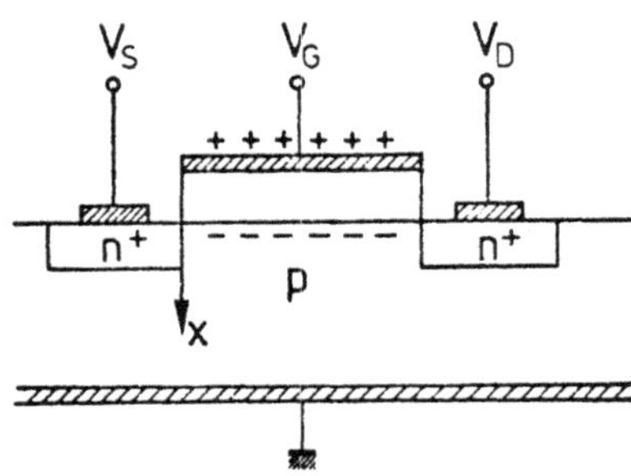

Fig. 1 - Structure of a MOS transistor

O(xide) - S(emiconductor) Transistor structure is obtained by adding a Source n^+-type contact and a Drain n^+-type contact to the MOS capacitor. Since the inversion charge at the oxide-semiconductor interface provides a conducting path between Drain and Source (also called conducting channel), a potential difference $V_{DS} = V_D - V_S$ between the drain and source contacts will result in an electron current I_D. Hence the MOS transistor behaves as a variable resistor from source to drain, which value is modulated by the gate potential. The relationship between the current I_D, the applied drain-to-source voltage V_{DS} and the gate-to-source voltage V_{GS} is sketched in Figure 2.

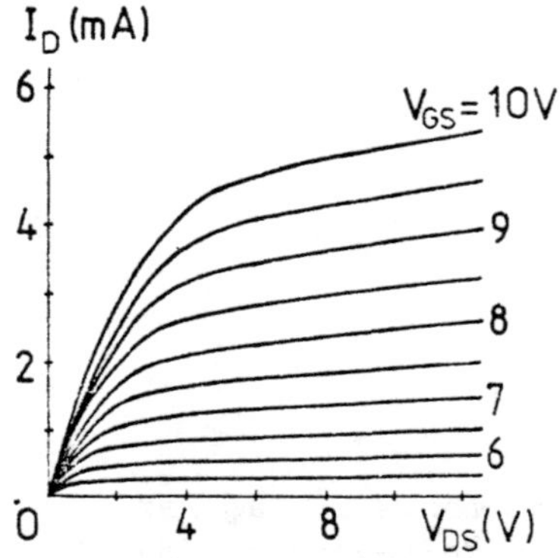

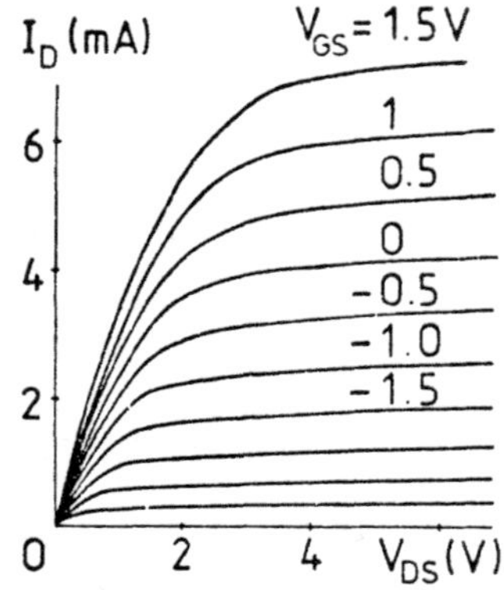

Fig. 2 - Current characteristics of an n-channel enhancement mode MOS transistor

Fig. 3 - Current characteristics of an n-channel depletion mode MOS transistor

For increasing values of V_{DS} the local inversion charge along the channel decreases and the equivalent resistance value becomes larger; this yields the non-linear relationship between I_D and V_{DS}.

The conducting channel between source and drain exists, if V_G exceeds a particular value, called threshold voltage V_{TH}. If V_{TH} is positive (for a p-type substrate) the device is called enhancement mode MOS transistor (Fig. 2). In the opposite case, the threshold voltage V_{TH} is negative, and a negative gate voltage V_G is required to switch off the drain current (see Fig. 3); the device is then called depletion mode MOS transistor. The exact value of the threshold voltage V_{TH} depends on the manufacturing process.

For large values of V_{DS} the drain current saturates (see Fig. 2 and 3). The difference between V_G and V_D is not large enough to maintain an electron inversion charge at the drain side of the channel. Hence a depletion layer is formed between the end of the conducting channel and the drain contact; the current becomes almost independent of V_{DS}.

For small values of V_{DS} one can calculate the drain current in the following way [1]. The gate voltage V_G in excess of the threshold voltage V_{TH} induces a total negative inversion charge given by :

$$Q = -WLC_{ox}(V_G - V_{TH}) \quad (1)$$

where C_{ox} *is the oxide capacitance per unit area*

$$C_{ox} = \varepsilon_{ox}/t_{ox} \quad (2)$$

ε_{ox} *being the oxide permittivity,* t_{ox} *the oxide thickness, L the channel length between source and drain and W the channel width.*

Calling τ *the transit time or average time required for an electron to move from source to drain, the drain current is given by :*

$$I_D = -Q/\tau \quad (3)$$

For small V_{DS} *values the electric field along the channel is almost constant and equal to* V_{DS}/L*; hence the electron velocity is :*

$$v_n = \mu_n V_{DS}/L \quad (4)$$

μ_n *being the electron mobility. The transit time* τ *is then*

$$\tau = L/v_n = L^2/\mu_n V_{DS} \quad (5)$$

As a consequence one obtains the following first order linear region approximation, valid only for small V_{DS} *values,*

$$I_D = \frac{W}{L}\mu_n C_{ox}(V_G - V_{TH})V_{DS} \quad (6)$$

The drain conductance for small V_{DS} *values is :*

$$g_D = \frac{W}{L}\mu_n C_{ox}(V_G - V_{TH}) \quad (7)$$

In the preceding description we considered a p-type substrate and a conducting channel of free electrons. The device is called an n-channel MOS transistor. It is obvious that with an n-type substrate one obtains a p-channel MOS transistor for which the channel carriers are free holes.

2. DIFFERENT MOS TRANSISTOR MODELS

According to semiconductor physics the hole and electron densities within the surface region of an n-channel MOS transistor are given by (assuming a quasi-Fermi level formulation) :

$$p = p_{po} \exp(-q\emptyset/kT) \quad (8)$$

$$n = (n_i^2/p_{po}) \exp\{q[\emptyset - V(y)]/kT\} \qquad (9)$$

n_i being the intrinsic carrier concentration ($1.38 \times 10^{10}/cm^3$ for Si at $T = 300°K$), $\emptyset$ the semiconductor potential and $V(y)$ the applied potential along the channel ($V_S \leq V(y) \leq V_D$). The hole density in the bulk of the semiconductor is given by :

$$p_{po} = \frac{1}{2}(N_A + \sqrt{N_A^2 + 4 n_i^2}) \simeq N_A \qquad (10)$$

N_A being the doping density of acceptors.

The semiconductor potential $\emptyset$, in the direction x normal to the oxide-semiconductor interface, is related to the local space charge density through Poisson's equation :

$$\partial^2\emptyset/\partial x^2 = -(q/\varepsilon_s)(p - n - N_A) \qquad (11)$$

q being the electron charge and ε_s the semiconductor permittivity.

Integration yields the space-charge density at each point y along the channel :

$$Q_{sc}(y) = -\lambda C_{ox}\Big\{\emptyset_s\left[1 - \exp(-2q\emptyset_F/kT)\right] \qquad (12)$$

$$+ (kT/q)\{\exp(-q\emptyset_s/kT) - 1$$

$$+ \exp\left[q(\emptyset_s - 2\emptyset_F - V)/kT - \exp\left[-q(2\emptyset_F + V)/kT\right]\right\}\Big\}^{1/2}$$

where the so-called body factor is given by :

$$\lambda = (2\varepsilon_s q p_{po})^{1/2}/C_{ox} \qquad (13)$$

and

$$\emptyset_F = (kT/q)\ln(p_{po}/n_i) \qquad (14)$$

The surface potential $\emptyset_s$ is related to the applied gate voltage V_G through Gauss' law

$$Q_{sc} = -C_{ox}(V_G - V_{FB} - \emptyset_s) \qquad (15)$$

The term V_{FB}, called flat-band potential, accounts for all fixed-charge and work functions effects due respectively to the oxide layer and the nature of the metal gate with respect to the semiconductor substrate.

The exact value of the surface potential $\emptyset_s$ must be obtained by solving the implicit equation resulting from (12) and (15).

Figure 4 illustrates the relationship between the surface potential $\emptyset_s$ and the potential V(y) along the channel, for different gate potentials V_G.

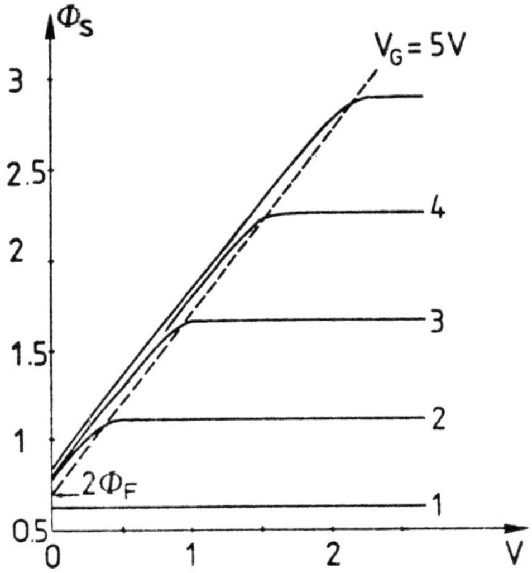

Fig. 4 - Surface potential as a function of the local potential V(y) along the channel ($N_A = 10^{16}$ cm^{-3}, t_{ox} = 1100 Å, $V_{FB} = -1.1$ V, $V_B = 0$).

Consider now the drain current calculation problem. The electron current density in the y direction along the channel is the result of a drift component, due to the electric field, and of a diffusion component, due to the gradient of free electrons, :

$$J_n(y) = -q\mu_n \left[-n\ (\partial\emptyset/\partial y) + (kT/q)\ (\partial n/\partial y)\right] \tag{16}$$

According to (9)

$$J_n(y) = qn\mu_n\ (dV/dy) \tag{17}$$

By integration over the active cross section area of the channel one obtains the drain current expression :

$$I_D = -\ W\mu_n\ Q_N(y)\ (dV/dy) \tag{18}$$

$Q_N(y)$ being the free charge density at a point y along the channel. A final integration along the channel of length L yields :

$$I_D = -\ \frac{W}{L}\ \mu^{\star} \int_{V_S}^{V_D} Q_N\ dV \tag{19}$$

μ^* being the effective or mean mobility along the channel. The fact that the current depends upon the ratio of W/L is of major importance in the design of MOS circuits.

The problem to be solved now is to obtain an explicit expression of $Q_N(V)$. According to (12), the charge density $Q_{sc}(y)$ at a point y under the gate electrode is, under the normal operation conditions of a MOS transistor, composed of the carrier charge density Q_N of free electrons and a depletion charge density Q_D due to the fixed acceptor ions :

$$Q_D = -\lambda C_{ox} \left(\emptyset_s - \frac{kT}{q}\right)^{1/2} \tag{20}$$

Hence, a good approximation is :

$$Q_N = Q_{sc} + \lambda C_{ox} \left(\emptyset_s - \frac{kT}{q}\right)^{1/2} \tag{21}$$

or, according to (15) :

$$Q_N = - C_{ox} \left[V_G - V_T(y)\right] \tag{22}$$

where :

$$V_T(y) = V_{FB} + \emptyset_s + \lambda \left(\emptyset_s - \frac{kT}{q}\right)^{1/2} \tag{23}$$

$V_T(y)$ can be considered as the local threshold voltage at the point y along the channel. The drain current expression becomes :

$$I_D = \frac{W}{L} \mu^* C_{ox} \int_{V_S}^{V_D} \left[V_G - V_T(y)\right] dV \tag{24}$$

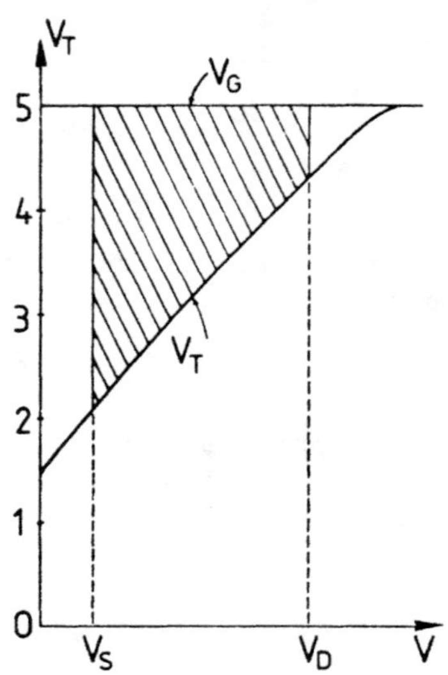

Fig. 5 - Local threshold voltage $V_T(y)$ as a function of the potential V(y) along channel for $V_G = 5$ V
($N_A = 10^{16}$ cm^{-3}, $t_{ox} = 1100$ Å, $V_{FB} = -1.1$ V, $V_B \cong 0$).

Figure 5 illustrates the relationship between V_T and the local potential $V(y)$ along the channel.

The threshold voltage V_{TH} of the device is defined as the particular gate voltage V_G for which the free carrier density at the source side is equal to zero, i.e.

$$Q_N(o) = - C_{ox} \left[V_{TH} - V_T(o)\right] = 0 \qquad (25)$$

or

$$V_{TH} = V_{FB} + \phi_s(o) + \lambda \left[\phi_s(o) - \frac{kT}{q}\right]^{1/2} \qquad (26)$$

$\phi_s(o)$ being the surface potential for $V(o) = V_S$. Although the current for an applied gate voltage $V_G < V_{TH}$ is not strictly zero (subthreshold current), for many applications V_{TH} may be considered as a separation between the active and passive mode.

According to the expression of the drain current I_D and Figure 5, saturation of the drain current occurs when the inversion charge Q_N becomes equal to zero at the drain side, i.e. when :

$$V_G - V_T(L) = 0 \qquad (27)$$

The main difficulty to calculate I_D relies on the fact that $V_T(y)$ is a function of ϕ_s, which is implicitly related to $V(y)$. Different approaches are possible. According to Figure 4 a first order approximation for the relationship ϕ_s - V is :

$$\phi_s(y) = 2\phi_F + V(y) \qquad (28)$$

This equation implies the well known condition that the electron density at the surface n(o) is equal to the bulk majority carrier density p_{po}. In this case one obtains :

$$V_T(y) = V_{TO} + V(y) + \lambda \left[\sqrt{2\phi_F - \frac{kT}{q} + V(y)} - \sqrt{2\phi_F - \frac{kT}{q}}\right] \qquad (29)$$

where :

$$V_{TO} = V_{FB} + 2\phi_F + \lambda\sqrt{2\phi_F - \frac{kT}{q}} \qquad (30)$$

1) In order to simplify the current relation (24), one can linearize the expression of $V_T(y)$ in the following way : (see Fig. 6a)

$$V_T^{(1)}(y) = V_{TO} + V(y) \qquad (31)$$

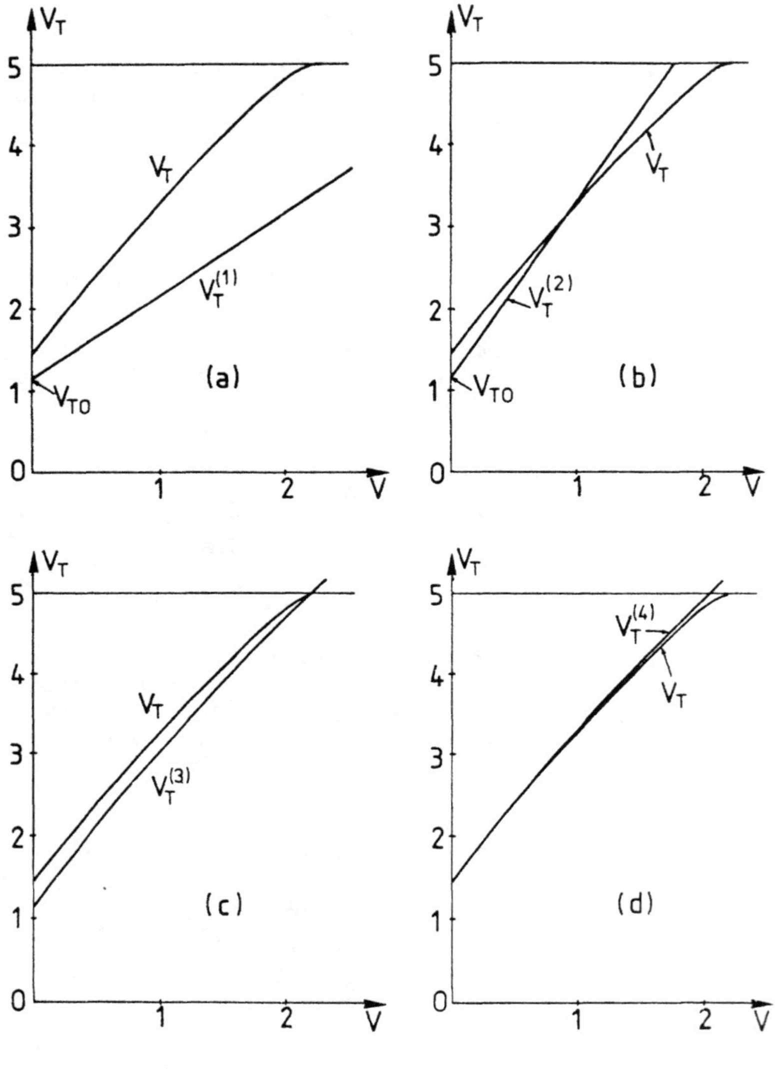

Fig. 6 – Local threshold voltage V_T as a function of the potential $V(y)$ along the channel. $V_T^{(i)}$: different approximations. ($N_A = 10^{16}$ cm^{-3}, $t_{ox} = 1100$ Å, $V_{FB} = -1.1.$ V, $V_G = 5$ V, $V_B = 0$).

In this case, one gets :

$$I_D^{(1)} = \frac{W}{L} \mu^{\star} C_{ox} \left[(V_G - V_{TO})(V_D - V_S) - \frac{1}{2} (V_D^2 - V_S^2) \right] \quad (32)$$

for $V_G \geq V_{TH}^{(1)}$ and $V_S \leq V_D \leq V_{D,sat}^{(1)}$, with

$$V_{TH}^{(1)} = V_{TO} + V_S \quad (33)$$

The current saturates for $V_D \geq V_{D,sat}^{(1)}$, with :

$$V_{D,sat}^{(1)} = V_G - V_{TO} \quad (34)$$

and

$$I_{D,sat}^{(1)} = \frac{1}{2} \frac{W}{L} \mu^{\star} C_{ox} (V_G - V_{TH}^{(1)})^2 \quad (35)$$

The linear region approximation of (32) yields the expression (6) derived in the introduction.

2) A slightly different linear expression for $V_T(y)$ is obtained by expanding (29) around V = 0 (see Fig. 6b) :

$$V_T^{(2)}(y) = V_{TO} + kV(y) \quad (36)$$

where

$$k = 1 + \frac{\lambda}{2} (2\emptyset_F - \frac{kT}{q})^{1/2} \quad (37)$$

In this case, one obtains for $V_G \geq V_{TH}^{(2)}$ and $V_S \leq V_D \leq V_{D,sat}^{(2)}$

$$I_D^{(2)} = \frac{W}{L} \mu^{\star} C_{ox} \left[(V_G - V_{TO})(V_D - V_S) - \frac{k}{2} (V_D^2 - V_S^2) \right] \quad (38)$$

$$V_{TH}^{(2)} = V_{TO} + k\, V_S \quad (39)$$

$$V_{D,sat}^{(2)} = \frac{1}{k} (V_G - V_{TO}) \quad (40)$$

The saturation current is given by :

$$I_{D,sat}^{(2)} = \frac{1}{2k} \frac{W}{L} \mu^{\star} C_{ox} (V_G - V_{TH}^{(2)})^2 \quad (41)$$

For both expressions (35) and (41), the saturated drain current varies with the square of $(V_G - V_{TH})$, a behaviour often found in practise.

3) By keeping the square root terms in the expression (29) of $V_T(y)$, one gets the well known 3/2 power dependence (see Fig. 6c)

$$I_D^{(3)} = \frac{W}{L}\mu^\star C_{ox}\,\{(V_G - V_{FB} - 2\phi_F)(V_D - V_S) - \frac{1}{2}(V_D^2 - V_S^2)$$

$$- \frac{2}{3}\lambda\left[(2\phi_F - \frac{kT}{q} + V_D)^{3/2} - (2\phi_F - \frac{kT}{q} + V_S)^{3/2}\right]\} \quad (42)$$

for $V_G \geq V_{TH}^{(3)}$ and $V_S \leq V_D \leq V_{D,sat}^{(3)}$, where :

$$V_{TH}^{(3)} = V_{TO} + V_S + \lambda\left(\sqrt{2\phi_F - \frac{kT}{q} + V_S} - \sqrt{2\phi_F - \frac{kT}{q}}\right) \quad (43)$$

$$V_{D,sat}^{(3)} = V_G - V_{FB} - 2\phi_F + \frac{\lambda^2}{2} - \lambda\,(V_G - V_{FB} - \frac{kT}{q} + \frac{\lambda^2}{4})^{1/2} \quad (44)$$

One may shown [2] that the current expression (42) is in error by at least 10% over the useful gate potential range.

4) According to Figure 4 a better result is obtained if the surface potential is approximated by :

$$\phi_s = \phi_{so} + V(y) \quad (45)$$

with ϕ_{so} calculated from the implicit equation (12), (15) for $V = 0$; the potential ϕ_{so} can be different in a significant manner from the usual value $2\phi_F$. One obtains for the non-saturation regime (see Fig. 6d) :

$$I_D^{(4)} = \frac{W}{L}\mu^\star C_{ox}\,\{(V_G - V_{FB} - \phi_{so})(V_D - V_S) - \frac{1}{2}(V_D^2 - V_S^2)$$

$$- \frac{2}{3}\lambda\left[(\phi_{so} - \frac{kT}{q} + V_D)^{3/2} - (\phi_{so} - \frac{kT}{q} + V_S)^{3/2}\right]\} \quad (46)$$

$$V_{TH}^{(4)} = V_{FB} + \phi_{so} + V_S + \lambda \,.\, \sqrt{\phi_{so} - \frac{kT}{q} + V_S} \quad (47)$$

$$V_{D,sat}^{(4)} = V_G - V_{FB} - \phi_{so} + \frac{\lambda^2}{2} - \lambda\sqrt{V_G - V_{FB} - \frac{kT}{q} + \frac{\lambda^2}{4}} \quad (48)$$

One can shown [2] that the current expression (46) is accurate within 5% for large values of V_G.

5) The main disadvantage of the preceding models, which are widely used in the literature, is that low voltage operation, which emphasizes the near-threshold regime of gate biases, is not accurately described. Consider again the general current expression (24). Since $V_T(y)$ is a function of the surface potential ϕ_s, one may write :

$$I_D = \frac{W}{L} \mu^\star C_{ox} \int_{\phi_s(o)}^{\phi_s(L)} (V_G - V_T) \frac{dV}{d\phi_s} d\phi_s \qquad (49)$$

The slope $d\phi_s/dV$ can be calculated from (12), (15). In the range of operation of the MOS transistor (depletion regime, weak inversion regime, strong inversion regime), one obtains the following expression for the current [2] :

$$I_D = \frac{W}{L} \mu^\star C_{ox} \int_{\phi_s(o)}^{\phi_s(L)} \left(V_G - V_T + \frac{kT}{q} + \frac{\lambda^2 + V_G - V_T}{2\lambda \sqrt{\phi_s - \frac{kT}{q} + V_G - V_T}}\right) d\phi_s \qquad (50)$$

Integration is possible and yields a very accurate model, as discussed by G. Baccarani et al. [3] and by F. Van de Wiele [2]; the resulting current relationship is however quite complicated. One may simplify the preceding relation in the following way :

$$I_D = \frac{W}{L} \mu^\star C_{ox} \int_{\phi_s(o)}^{\phi_s(L)} \left(V_G - V_T + \frac{kT}{q} + \frac{\lambda}{2\sqrt{\phi_s - \frac{kT}{q}}}\right) d\phi_s \qquad (51)$$

After integration one obtains the so-called charge-sheet model, as proposed by J.R. Brews [4],

$$I_D^{(5)} = \frac{W}{L} \mu^\star C_{ox} \{(V_G - V_{FB} + \frac{kT}{q}) \left[\phi_s(L) - \phi_s(o)\right] - \frac{1}{2}\left[\phi_s^2(L) - \phi_s^2(o)\right] - \frac{2}{3}\lambda(A^{3/2} - B^{3/2}) + \lambda \frac{kT}{q}(A^{1/2} - B^{1/2})\} \qquad (52)$$

where

$$A = \phi_s(L) - \frac{kT}{q} \quad , \quad B = \phi_s(o) - \frac{kT}{q} \qquad (53)$$

The values of $\phi_s(L)$ and $\phi_s(o)$ must be obtained from the implicit equation (12), (15) respectively for $V = V_D$ and $V = V_S$.

This general drain-current expression is valid in the saturation regime, the non-saturation regime and the subthreshold regime. One can shown [2] that the charge-sheet model (52) is in error by less than 1.5% over the entire voltage range, including the subthreshold region.

3. CARRIER MOBILITY CORRECTION

In the preceding models we considered an effective mobility μ^* for the entire channel. It is well known that mobility variations occur along the channel, depending on the local electric field. According to Merckel et al. [5] the carrier mobility dependence on the normal field and the lateral field in the channel can be approximated by :

$$\frac{1}{\mu_n} = \frac{1}{\mu_o} \{1 + \theta_s \left[V_G - V_T(y)\right] + \theta_c L \, (d\phi_s/dy)\} \qquad (54)$$

θ_s (θ_c) being the normal (lateral) field coefficient and μ_o the inversion layer carrier mobility. Integration along the channel yields the effective mobility expression :

$$\frac{1}{\mu^*} = \frac{1}{\mu_o} \{1 + \theta_s \left[V_G - V_{FB} - \frac{1}{L}\int_o^L (\phi_s - \lambda\sqrt{\phi_s - \frac{kT}{q}}) \, dy\right]$$

$$+ \theta_c \left[\phi_s(L) - \phi_s(o)\right]\} \qquad (55)$$

By considering, as suggested by M. White [6], a linear approximation for the surface potential, one obtains :

$$\frac{\mu_o}{\mu^*} = 1 + \theta_s \{V_G - V_{FB} - \frac{\phi_s(L) + \phi_s(o)}{2}$$

$$+ \frac{2}{3}\lambda \frac{\left[\phi_s(L) - \frac{kT}{q}\right]^{3/2} - \left[\phi_s(o) - \frac{kT}{q}\right]^{3/2}}{\phi_s(L) - \phi_s(o)}\}$$

$$+ \theta_c \left[\phi_s(L) - \phi_s(o)\right] \qquad (56)$$

The surface potential $\phi_s(L)$ and $\phi_s(o)$ are calculated from the implicit equation (12), (15) for the case of the charge-sheet model (52). For the simple models (32), (38), (42), one has :

$$\phi_s(L) = 2\phi_F + V_D \quad , \quad \phi_s(o) = 2\phi_F + V_S \qquad (57)$$

while for the model (46)

$$\phi_s(L) = \phi_{so} + V_D \quad , \quad \phi_s(o) = \phi_{so} + V_S \qquad (58)$$

A simplified expression, which is frequently used, is obtained by expanding (56) for small differences $V_D - V_S$:

$$\frac{\mu_o}{\mu^*} = 1 + \theta_s \left(V_G - V_S - V_{TO}\right) + \theta_c \left(V_D - V_S\right) \tag{59}$$

The mobility expression (56) together with the 3/2 power model (46) has been used by M. White et al [7] for the description of the experimental current characteristics of surface-channel enhancement-mode long channel MOS transistors; an accuracy of 1 à 3% was obtained. The set of modeling parameters may be determined accurately by examination of the drain conductance for small drain voltages.

Mobility corrections can also be taken into account by introducing modified V_{TO}^* and k^* parameters, as suggested by P. Jespers [8]. Consider the general current relation :

$$I_D = \frac{W}{L} C_{ox} \int_{V_S}^{V_D} \mu_n \left[V_G - V_T(y)\right] dV \tag{60}$$

and assume, according to (54), that :

$$\mu_n = \frac{\mu_o}{1 + \theta_s \left[V_G - V_T(y)\right]} \tag{61}$$

Define a new local threshold voltage $V_T^*(y)$ by means of the following relation :

$$V_G - V_T^*(y) = \frac{V_G - V_T(y)}{1 + \theta_s \left[V_G - V_T(y)\right]} \tag{62}$$

In this case, one obtains after substitution into (60) :

$$I_D = \frac{W}{L} C_{ox} \mu_o \int_{V_S}^{V_D} \left[V_G - V_T^*(y)\right] dV \tag{63}$$

Figure 7 illustrates the relationship between $V_T^*(y)$ and the local potential V(y) for a typical case. Clearly a good approximation is :

$$V_T^*(y) = V_{TO}^* + k^* V(y) \tag{64}$$

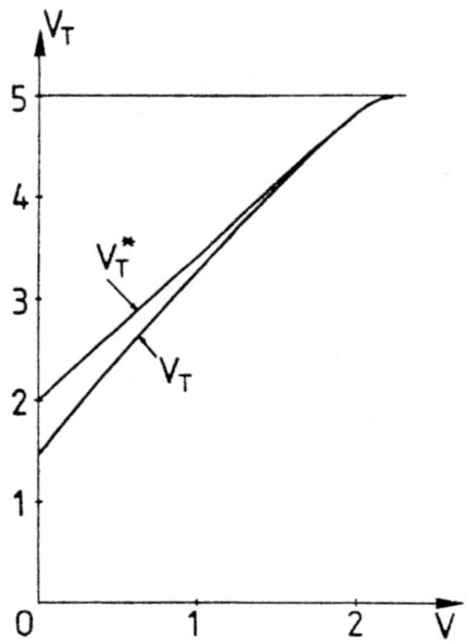

Fig. 7 - Mobility correction along the channel. V_T and V_T^* as a function of the potential $V(y)$.
$\theta_s = 0.05$
($N_A = 10^{16}$ cm^{-3}, $t_{ox} = 1100$ Å, $V_{FB} = -1.1$ V, $V_B = 0$, $V_G = 5$ V)

V_{TO}^* and k^* being appropriate device parameters, including the mobility corrections. This yields, after integration, the following current relation :

$$I_D^* = \frac{W}{L}\mu_o C_{ox}\left[(V_G - V_{TO}^*)(V_D - V_S) - \frac{k^*}{2}(V_D^2 - V_S^2)\right] \qquad (65)$$

for $V_G \geq V_{TH}^*$ and $V_S \leq V_D \leq V_{D,sat}^*$, where :

$$V_{TH}^* = V_{TO}^* + k^* V_S \qquad (66)$$

and

$$V_{D,sat}^* = \frac{1}{k^*}(V_G - V_{TO}^*) \qquad (67)$$

The saturation current is given by :

$$I_{D,sat}^* = \frac{1}{2k^*}\frac{W}{L}\mu_o C_{ox}(V_G - V_{TH}^*)^2 \qquad (68)$$

The parameters k^*, V_{TO}^* and μ_o are determined from the drain conductance characteristics for small drain voltages; an accuracy of a few percent can be obtained.

4. CHANNEL LENGTH MODULATION

According to the previous models the drain current remains constant for V_{DS} values larger than the saturation point $V_{D,sat}$. This result is correct only for long or moderate channel lengths ($L > 10\ \mu m$). For short channel devices the output conductance has a non-zero value. This is mainly due to the fact that, as the space-charge widens at the drain side, it tends to shorten the channel region. As the effective channel length becomes shorter for increasing V_{DS} values, the saturated drain current increases slightly.

Considering the simplest expression (35) for $I_{D,sat}$:

$$I_{D,sat}^{(1)} = \frac{1}{2}\,\frac{W}{L}\,\mu^{\star} C_{ox}\,(V_G - V_{TH}^{(1)})^2 \qquad (69)$$

the L value should become $L - \Delta L$, where ΔL accounts for the channel shortening effect,

$$I_{D,sat} = \frac{1}{2}\,\frac{W}{L - \Delta L}\,\mu^{\star} C_{ox}\,(V_G - V_{TH}^{(1)})^2 = \frac{I_{D,sat}^{(1)}}{1 - \frac{\Delta L}{L}} \qquad (70)$$

Calculations of the channel length modulation ΔL, as a function of the applied voltage V_{DS}, have been attempted by several authors [5, 9]. In most cases, the results are empirical; the current increase can only be calculated precisely by solving the two-dimensional Poisson equation in the channel region at the drain side. As a zero-order approximation of the effect one may write [10] :

$$\frac{\Delta L}{L} = \frac{V_D - V_{D,sat}}{V_E + V_D} \qquad (71)$$

V_E being the equivalent of an Early voltage (typical value : $V_E \simeq 75$ volt). Now for $V_D \ll V_E$, one has :

$$I_{D,sat} \simeq I_{D,sat}^{(1)}\left(1 + \frac{V_D - V_{D,sat}}{V_E + V_D}\right) \qquad (72)$$

The current characteristics in the saturation region have a slope equal to $I_{D,sat}^{(1)}/V_E$.

5. *BODY EFFECT*

The threshold voltage of a MOS transistor is changed when a bias voltage V_S *is applied between the source and the substrate or body of the transistor. This change in the characteristics is called body effect.*

According to (26) the threshold voltage V_{TH} *depends on* V_S *through the dependence of the surface potential* $\phi_s(o)$ *at the source side. By considering the usual approximation (28), or the relation (45), one may write :*

$$V_{TH}(V_S) = V_{TH}(V_S = 0) + \Delta V_{TH} \quad (73)$$

$$\Delta V_{TH} = V_S + \lambda\left(\sqrt{V_S + \phi_B} - \sqrt{\phi_B}\right) \quad (74)$$

where

$$\phi_B = \begin{cases} 2\phi_F - kT/q \text{ for } (28) & (75) \\ \phi_{so} - kT/q \text{ for } (45) & (76) \end{cases}$$

The body effect depends directly on the factor λ*, defined by (13) which is a function of the doping level of the substrate and of the thickness of the oxide layer. The body effect is of considerable importance in many circuit applications. The apparent threshold voltage is increased when the source is biased with respect to the substrate.*

REFERENCES

|1| *Carver Mead and Lynn Conway, Introduction to VLSI systems, Addison-Wesley Publishing Company, 1980.*

|2| *F. Van de Wiele, Solid-State Electron. 22, 991 (1979).*

|3| *G. Baccarini, M. Rudan and G. Spadini, IEE J. Solid-St. and Electron Dev. 2, 62 (1978).*

|4| *J.R. Brews, Solid-St. Electron. 21, 345 (1978).*

|5| *G. Merckel, J. Borel and J.Z. Cupcea, IEEE Trans. Electron. Dev. ED-19, 681 (1972).*

|6| *M. H. White, private communication.*

|7| *M.H. White, F. Van de Wiele, J.P. Lambot, IEEE Trans. Electron Dev. ED-27, 899 (1980).*

|8| *P. Jespers, private communication.*

|9| *F.M. Klaassen, Process and device modeling for integrated circuit design, edited by F. Van de Wiele, W. Engl, P. Jespers, Noordhoff, 1977.*

|10| *G. Merckel, Process and device modeling for integrated circuit design, edited by F. Van de Wiele, W. Engl, P. Jespers, Noordhoff, 1977.*

Basics of Structured Design Methodologies:

Data Path and Finite State Machines

Reiner W. Hartenstein

Kaiserslautern University, D-675 Kaiserslautern, F.R.G.

Preface

An important approach to cope with high design cost of complex integrated circuits is the synthesis of highly regular structures such as structures with a high degree of replication. The design of such structures in digital applications may also be tought to students with a limited background in technology. The chapter 3 of the text book "Introduction to VLSI Systems" by Carver MEAD and Lynn CONWAY gives a comprehensible intruduction to this style called "Structured Design", along with some basic examples. This paper gives a somehow different view on some examples used in the chapter 3 of this book. It especially stresses a more systematic register transfer level design specification. It maps examples, taken from chapter 3 of the book, concisely into the register level. It should be used together with this book, since not all details of it are repeated here in this paper.

Introduction

Currently the two main problems in design and application of highly complex integrated circuits are: the high design cost and the disability of many technology-oriented designers to meet the user's requirements. One way to cope with both of these two problems is the use of new design methods. Such design methodologies are a way to reduce design time and design cost in general, and, to make design more easy, so that IC design may also be tought to systems-type people with only a limited background in technology. Admittedly, widening the community of designers to include also systems specialists, needs more than just the application of new design methods. Teaching requires innovative text

I N F O R M A T I C S	P H Y S I C S
algorithms	materials
networks	semiconductors
structures	doping
logic	diffusion
sequences	devices
languages	device modelling
grammars	device simulation
parallelism	circuits
abstractions	differential equations
formalisms	matrix computations
verification	mask making
parallelism	pattern distortion
complexity	fabrication
computer aids	process parameters
file management	test circuits

Fig. 1. Two different worlds in VLSI development: illustrated by some of their keywords.

books /MC80/ and curricular modifications /SE80/, and chip implementation needs an improved designer/manufacturer interface (e. g. see /HR80/, /HE80/). Fig. 1 illustrates why systems level specialists urgently need access to the real world of silicon: the informatics world is that much different from physics, that technology-oriented designers really cannot thoroughly understand all relevant systems level issues to be covered in future VLSI circuit design.

A methodology to design complex digital integrated circuits, also called "Structured Design", proposes "top down design" using "good algorithms" to reduce design cost (the term "Structured Hardware Design" has been coined in 1973 /HA73/. A good definition /KF79/ says that "good algorithms" are such that, (1) most of the design consists of repetitive placement of a minimum of number of "cells" which are as simple as possible, (2) such cells are, whenever possible, designed in a way, that connections are implemented locally and automatically by mere placement of these cells and (3) a minimum of layout is designed individually. This means that most of the design verification is done by verification of a few types of simple cells. Most of such highly regular structures are extensible. This means, it is very easy, also to design differently dimensioned versions from such regular structures, such as, e. g., to design a 16 bits wide version from an 8 bits wide register or a similar block. Especially data path structures of digital systems, such as the operation part of digital processors, or, arrays of elementary storage cells such as register files and memories, which are inherently based on "good algorithms" in the sense of "Structured Design". This will be demonstrated by means of a few examples.

Another approach to reduce design time and design cost is the use of programmable structures, such as ROMs (Read-Only-Memories), PLAs (Programmable Logic Arrays). The use of such programmable structures is based on the use of predefined structures taken

methodological level		typical elements
Abbrev.	level name	
RT level	register transfer level	registers, (de)multiplexers functions, data paths, clock, storage arrays etc.
L level	logic level (gate level)	gates, flipflops
C level	circuit level	transistors, resistors, etc.
T level	topology level	e. g. stick diagram elements
G level	geometry level	material areas, their layers, placement, and geometry

Table 1.

from a library in a way, that the designer has to define only parts of a single mask layer. Such programmable structures form regularly structured micro architectures, which are predefined "regular structures" ready to use. Most of such programmable structures are extensible, what gives a considerable amount of flexibility in their use by the designer. One advantage of such programmable structures is, that the design time may be reduced considerably by designing the micro architecture and the high level architecture of a circuit concurrently (the term "micro architecture" has been coined 1975 /HZ75/). The "micro architecture" is the specification of the low level structure of a digital system, on which the achitecture of a system may be implemented. E. g. in a microprogrammable microcomputer, such as the 68000, the micro architecture is the structure, on top of which the instruction set is implemented by means of microprogramming. One important area of using programmable structures in design of digital systems is the implementation of finite state machines on PLAs.

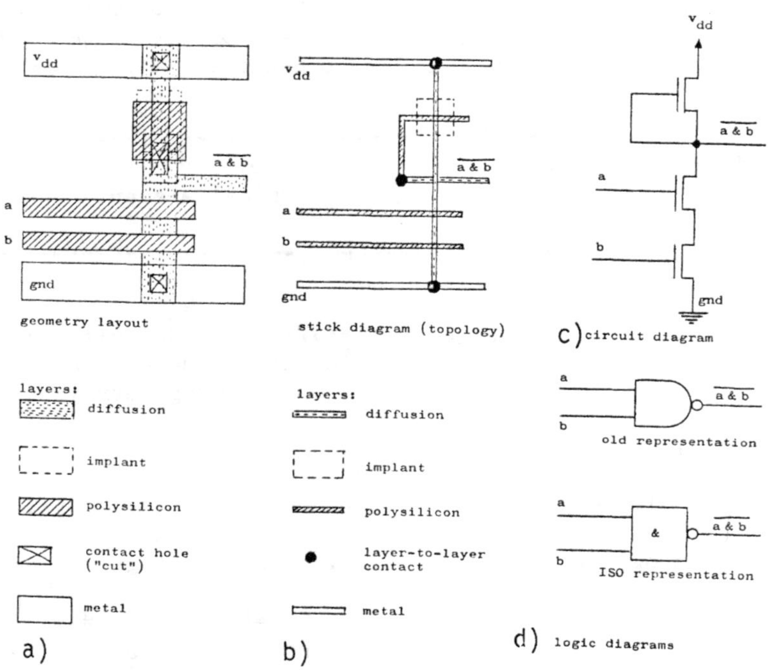

Fig. 2

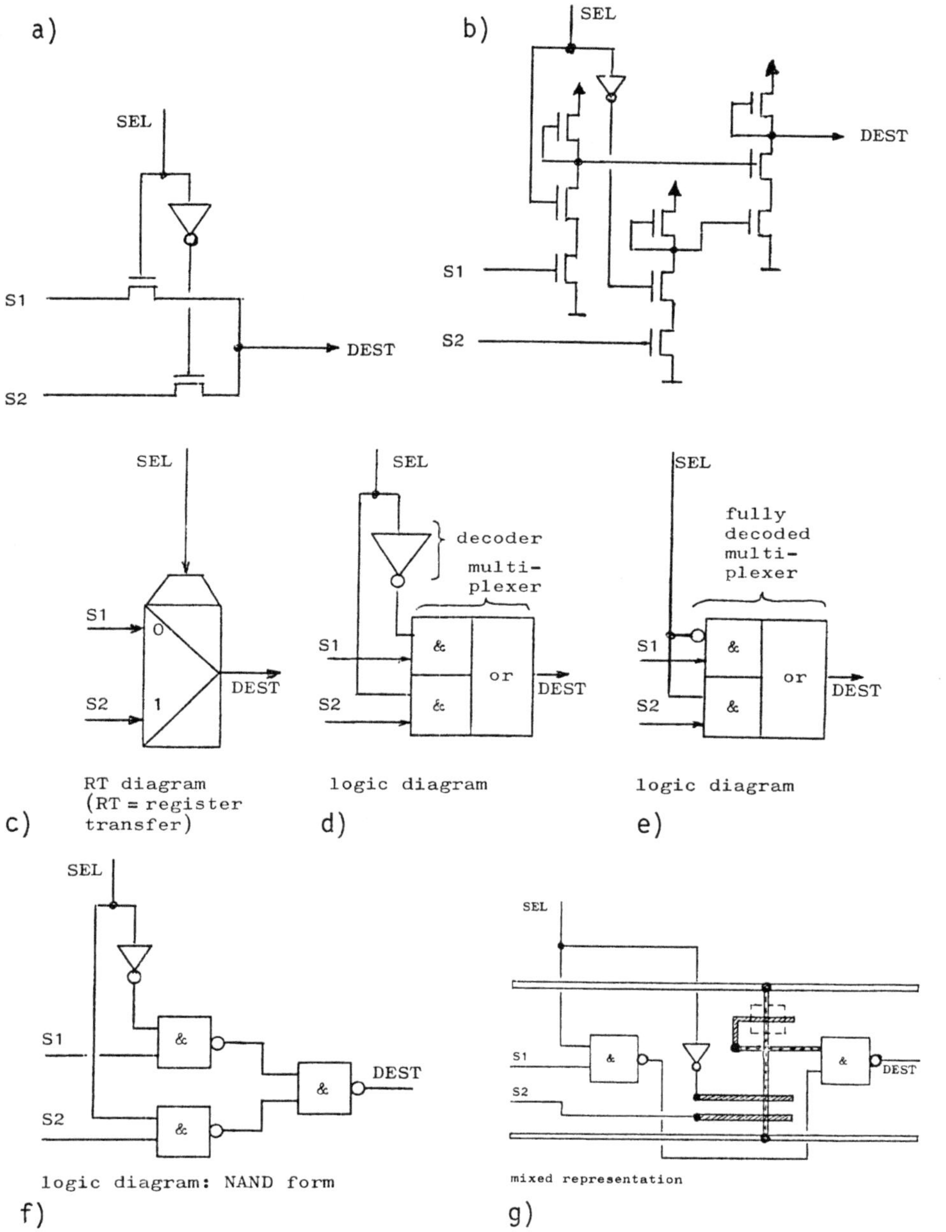
a)
SEL
S1
S2
DEST
b)
SEL
DEST
S1
S2
SEL
S1
S2
0
1
DEST
RT diagram
(RT = register transfer)
c)
SEL
decoder
multi-plexer
S1
S2
&
&
or
DEST
logic diagram
d)
SEL
fully decoded multi-plexer
S1
S2
&
&
or
DEST
logic diagram
e)
SEL
S1
S2
&
&
&
DEST
logic diagram: NAND form
f)
SEL
S1
S2
&
&
DEST
mixed representation
g)

Fig. 3

Levels and Notations

In digital integrated hardware system design and description we may distinguish several methodological levels. Each of these levels uses its typical methods of modelling and also its own notations for graphical and/or symbolic description. Going top-down the sequence of levels is: the installation level (IL or I level), block level (BL od B level), register transfer level (RTL or RT level), logic level (LL or L level), circuit level (CL or C level), topology level (TL or T level), and the geometry level (GL or G level). Each level has its typical primitives. From these primitives one may synthesize complexes, which are abstracted to be used as primitives in the next higher level. Or, looked into the opposite direction: Primitives of a particular level may be implemented by using the primitives of the next lower level.

The primitives of the B level are: data manipulators, storages, controllers, etc.(also see table 1 surveying the various levels and their primitives). Typical B level graphic representations are "block diagrams" using boxes. RT level primitives are: registers, array registers, functions, multiplexers, demultiplexers, data paths, etc. For example, a data manipulator, such as the RALU (register/ arithmetic/logic unit) of a computer, may be implemented by using primitives from the RT level. Typical RTL graphic representations use RT network diagrams, such as e.g. the ABL symbols used throughout this paper (ABL = A Block diagram Language, for the register transfer level is the graphic version of the symbolic language KARL /HA77/). Formal symbolic representations at RT level use "Computer Hardware Description Languages (CHDLs)", such as e.g. AHPL, CASSANDRE, DDL, ISP, KARL, or RTS. LL primitives are: gates and flipflops. CL primitives are: transistors, resistors, capacitors etc. The T level allows the representation of topology and distinguishes different materials` layers on the chip, whereas C level only allows connectivity representation. Stick diagrams (see fig. 2b) are T level representations. The geometry level finally shows the geometry and placement of various materials` areas on the chip, such as diffusion, polysilicon, ion implant, metal, contact holes, gate areas etc. (for an example see fig. 2a).

Fig. 2a shows the G level graphic representation of an NAND-gate using nMOS silicon gate depletion load technology (for more details see in the book /MC80/, just called "the book" in the remainder of this text). Fig. 2b shows the stick diagram representation of the same circuit (for stick diagrams also see in the book). It only shows topology, i. e. it does not show the size nor the absolute placement of parts such as transistors and cuts (contacts between materials` layers). Fig. 2c shows the circuit diagram and fig. 2d new and old standards of logic diagram representations of 2 different implementations. Fig. a shows a pass transistor implementation, whereas fig. b shows the NAND gate version of the multiplexer using 3 NAND gates for multiplexing. The inverter is the decoder (explained in detail later in this text: see section "the register transfer level"). Fig. 3c shows the RT level representation. Fig, 3d and e show the logic diagrams in 2 slightly different versions. Fig. 3 f shows the NAND/NAND gate implementation and its L level representation. Fig. 3g shows a mixed level representation of the version from fig. g, where one NAND gate is shown in stick diagram representation, and the remainder of the multiplexer uses L level graphic notation.

The Register Transfer Level

The RT level is a very useful level between L level and B level. One integrated circuit, being represented as one single box in the B level may consist of thousands of gates. So the B level description may be too imprecise and the L level description may be too complex to be comprehensible. So an intermediate level between B and L level is useful as a notation for intermediate design steps or for a more concise user-oriented description. In this section a few RT level primitive examples are explained. Multiplexers, demultiplexers, and decoders are important RT primitives.

A decoder interpretes its input to be a binary coded non-negative integer number and converts this number into its singulary-coded representation, which uses a 1-of-n representation. Let's repeat the evaluation of positional number representations. The value v of a binary coded non-negative number is

$$v = \sum_{i=0}^{h} d_i * 2^i$$

where i is the subscript of the bit position (i=0 for least significant bit), h is the highest subscript, and d_i is the value of the particular bit. The value v of a singulary coded number is

$$v = \sum_{i=0}^{h} d_i * i$$

with the restriction that all digits d_i but one must be of value zero. For example, a binary coded input 0101 (which is number 5) will be converted into the output 0000 0000 0010 0000, where subscripts i of bit positions right to left run from 0 to 3 in the input word, and from 0 thru 15 in the output. In the singulary

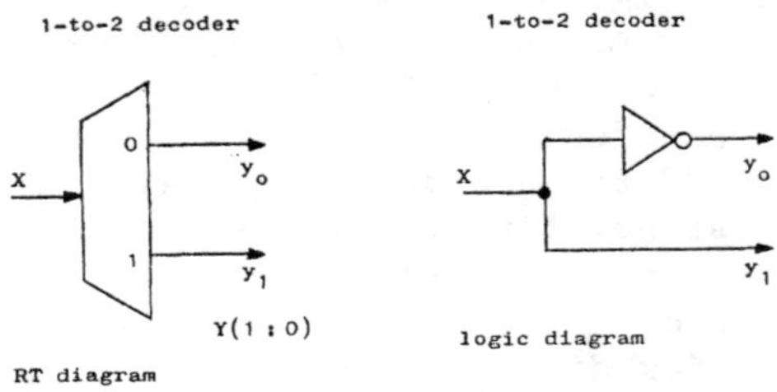

a) b)

decimal	binary coded i= 0	singulary coded i= 1 | 0
0	0	0 1
1	1	1 0

c)

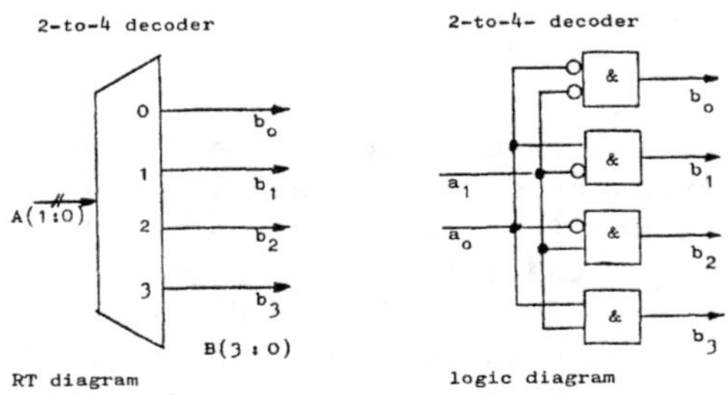

d) e)

decimal	binary coded i= 1 | 0	singulary coded i= 3 | 2 | 1 | 0
0	0 0	0 0 0 1
1	0 1	0 0 1 0
2	1 0	0 1 0 0
3	1 1	1 0 0 0

f)

Fig. 4

coded output the only 1 just flags the bit position the subscript of which equals the value of the number. In the above example the bits with i=0 thru 4 are 0, bit no. 5 is 1, and the rest is 0. This 1 in position i=5 indicates a value of v=5 for the number represented here.

Fig. 4 shows two simple examples of decoders. Fig. 4a shows the RT level graphic symbol of a 1-to-2 bit decoder which decodes a one bit binary number into 2 bits wide singulary representation. Fig. 4 c shows its conversion table, where the left colum serves as a comment by showing the decimal representation of the numbers in the other two colums of the table. Fig. 4b shows the L level of the 1-to-2 bits decoder, which means, that the implementation of that decoder is shown by its logic diagram. Fig. 4d shows the RT diagram of a 2-to-4 bit decoder, where the notation A(1 : 0) indicates a 2 bits wide input named A and its bits subscripted 1 thru 0. Also the 4 output bits b(3) thru b(0) may be bundled by representation of such a bit vector, which is B(3 : 0) in this example. This bundling of individual bits to form "bit vectors" is a typical means to make RT level representations as compact as possible. Especially in "structured design" environments such representations are an important means to reduce complexity of specifications and descriptions. Fig. 4 e shows the L level implementation of this decoder, and fig. f shows ist conversion table.

Other important RT level primitives are multiplexer and demultiplexer. Both are important means to route the ways of data flow within data path structures. Data sources and destinations may be selected by switching connections inside the path structure. Fig. 5 shows a simple example, where fig. a shows the RT diagram, fig. b illustrates the function. Controlled by a source address input SEL ("selector input") in this example one out of two sources S1 and S2 may be selected to be connected to the destination output DEST of the multiplexer. If SEL=0 S1 is selected (see fig.

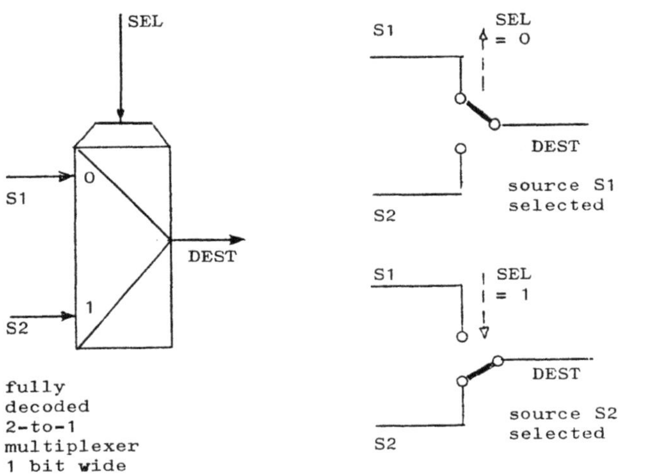

a) RT diagram

b) illustration of function

Fig. 5

b), otherwise S2 is selected. Fig. 3a and b show C level diagrams of 2 different implementations of the above multiplexer, where the inverter is the decoder. Fig. 6 shows an example of a demultiplexer, which selects one out of several destinations to be connected to a single source. In this example the data path width is 8 bits, subscripted from 7 thru 0, as indicated by the bit vector notations SE(7 : 0) (for the source), DA(7 : 0), and DB(7 : 0) (destinations no. a and no. b) in RT diagram in fig. 6 a. DS is the selector variable for addressing one of the two sources, where 1 bit is sufficient in this example, since only an address range from 0 thru 1 is required. Fig. 6 b is an illustration of the function of this example demultiplexer, also demonstrating the routing of the 8-wire bundle of the bit vector.

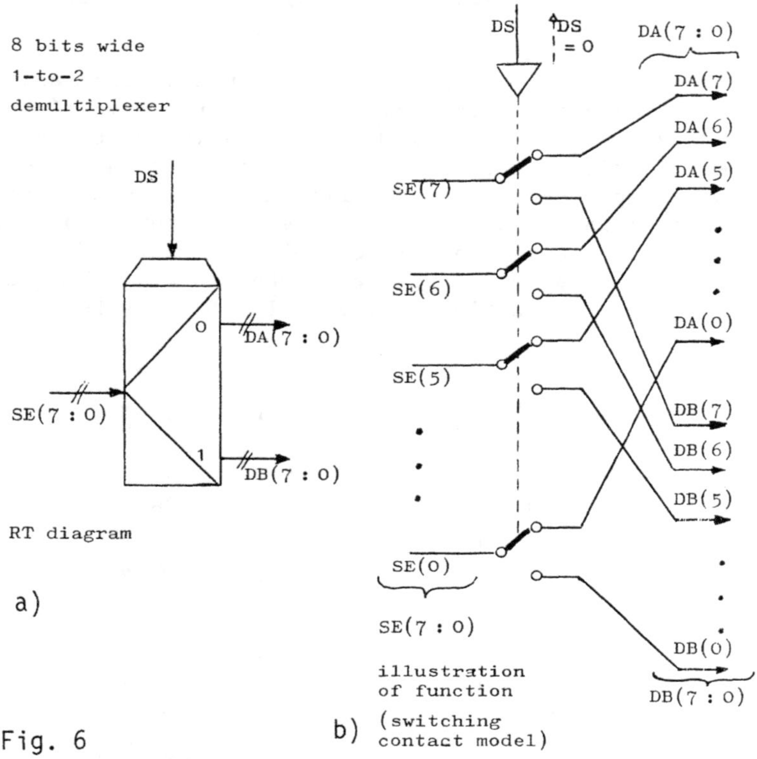

Fig. 6

Both example RT diagrams show a decoder symbol at the control input side of the entire diagram. This indicates, that the (de-) multiplexer is a completely decoded one which means, that it includes a decoder to decode the address input internally. Fig. 7a shows a RT diagram of an undecoded multiplexer. It does not have a decoder, so that a singulary coded address input is required, which is input C(3 : 0) in this example. Any non-singulary coded address input except 0000 would cause a malfunction of the multiplexer, since it would cause the selection of more than only one source at a time. Only address input 0000 would mean to select no source at all, so that a singulary input which consists of zeroes only means an "empty address word" to shut down the circuit. Fig. 7 b shows a AND/OR gate implementation of the 1 bits wide 4-to-1 multiplexer of fig. 7 a. Fig 7c shows a pass transistor implementation example, a C level description which also is a good explanation of the address input requirements mentioned above.

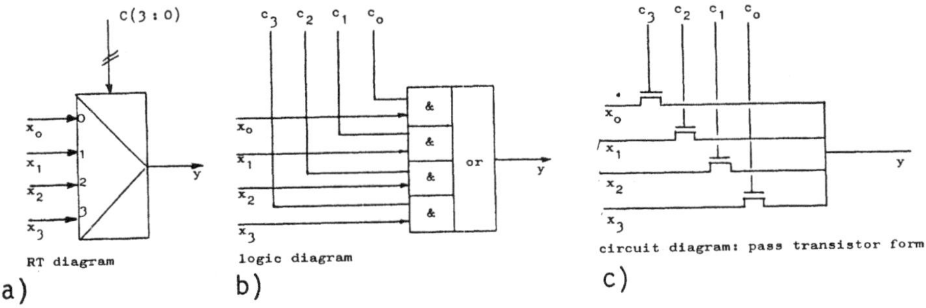

Fig. 7

Fig. 8 shows the fully decoded version of the multiplexer in fig. 7. Instead of a 4 bit singulary input only 2 input bits SS(1 : 0) are used, since the input now is binary coded (compare fig. 4). The logic diagram in fig. 8 b demonstrates the partitioning of the multiplexer into the decoder part and the multiplexer part. The SS(1 : 0) input is converted into the singulary coded representation C(3 : 0) which is used only internally (see fig. 8 b). Fig. 8 c shows the result of minimizing the version from fig. b, where only one switching network is doing both functions at the same time: the decoding and the multiplexing, so that only 5 gates are needed instead of 9 gates.

There is one interesting ambiguity in RT level primitives, which will be used later in this paper. As long as the data path is only one bit wide, a 1-to-n conditional decoder is funcitonally equivalent to a 1-to-n fully decoded demultiplexer. Let us first introduce the conditional decoder by means of a simple example. Fig 9 a shows the RT diagram of a conditional 1-to-2 decoder to convert the 1 bit binary input v into the 2 bit singulary output Y(1 : 0). The condition input w activates the decoder (if w = 0,

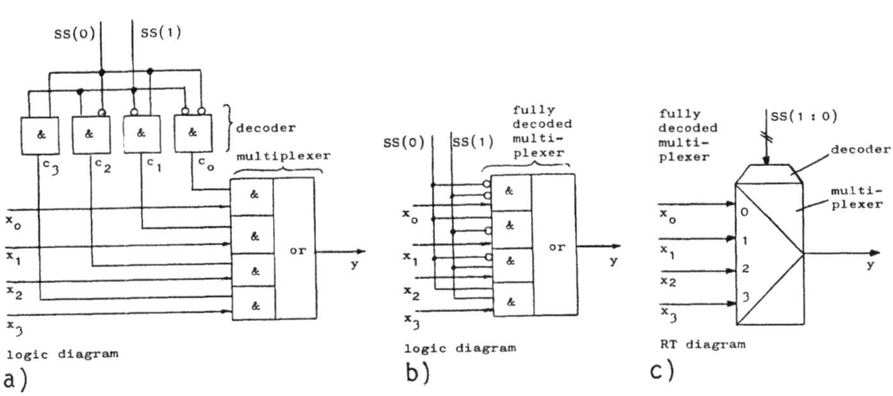

Fig. 8

i.e. condition w is true) or the decoder is shut down (if w = 1, i.e. condition w is false). When the decoder is disabled, it produces an empty word output (see above for "empty word"). Fig. 9 b shows the L level implementation logic diagram of this conditional decoder. Fig. 9 c shows the RT diagram and fig. d the logic diagram of a 1 bit wide 1-to-2 demultiplexer. We may distinguish the decoder part (which consists of an inverter, compare fig. 4b) and the demultiplexer part. Also note, that the logic diagrams of figs. 9 b and 9 d are equivalent, which proves the

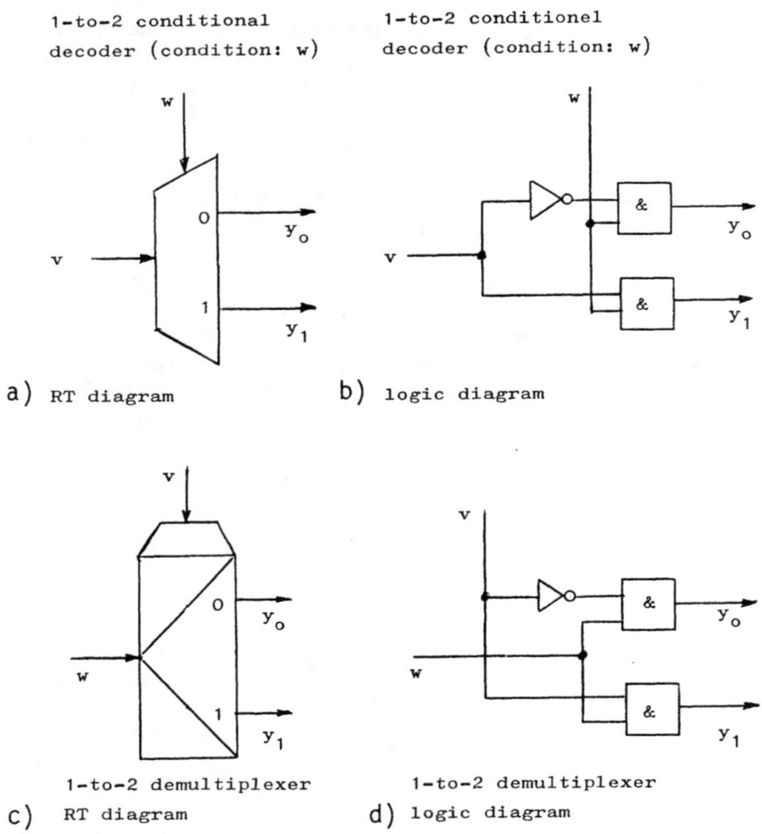

Fig. 9

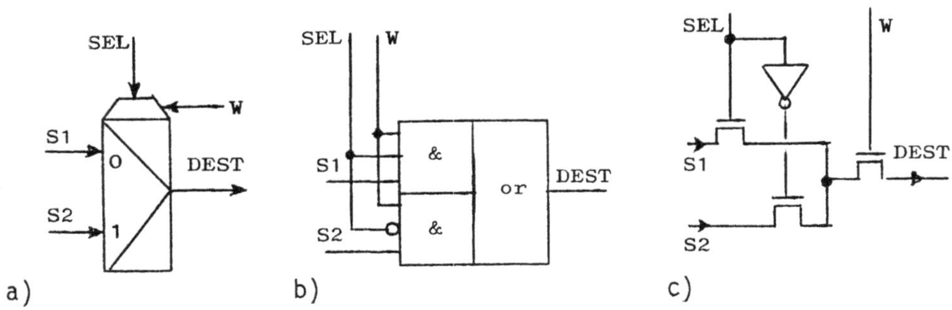

Fig. 10

functional equivalence of the RT primitives in figs. 9 a and c.

Also a multiplexer may use a conditional decoder to process its control input. Fig. 10 a shows the RT diagram of an example of such a conditional multiplexer: condition input W is the enable signal input to activate the multiplexer. One interesting application is clocked multiplexing. So in this example the use of a clock pulse at input W activates the data path only for a short period of time determined by the length of the clock pulse. So we may distinguish two types of control signals: data path selection by means of address input SEL, and timing by means of clock pulse at the enable input W. Fig. 10 b shows the logic diagram, and fig. c the mixed notation representation of an implementation of this conditional multiplexer, using a pass transistor tree, where the multiplexer output is anded with the clock pulse.

A Design Environment for Data Path and Control Structures

The examples make use of nMOS silicon gate depletion load technology, which is easy to understand also by many readers with a limited background in technology (in this case reading chapter 1 and 2 of the book /MC80/ before studying this paper may be advisable). Mainly non-overlapping two-phase clocks with phases $\varphi 1$ and $\varphi 2$ are used together with dynamic implementation of registers. Dynamic shift registers are preferrably implemented by chains of successive inverter stages. The two-phase clocking avoids a ripple-thru of data through such shift registers in an uncontrolled way (for explanation see sections 3.3 and 3.4 of the book).

Fig. 11 a shows the circuit diagram of a typical 2-phase clocked dynamic flipflop of such an environment. This circuit is similar to that of fig. 3.9 in the book, except the additional pass transistor Tr2 to be activated at clock phase 2. To hold the bit stored in this inverter chain Tr3 is activated by $\varphi 1$ and Tr2 is activated by clock $\varphi 2$. Transistor Tr3 feeds the output OUT back to the input of the first inverter to refresh (storage mechanism: see the book). To evoke the "write" funtion instead of the "hold" function during 2 the pass transistor Tr1 is activated instead of Tr3, such that the data bit at IN is fed through to the input of the first inverter. This means an "update" instead of a "refresh". Transistors Tr1 and Tr3 together form an undecoded multiplexer with a "load" input and a "hold" input. However, to keep to 2-phase clocking the static control variables L (for load) and H (for hold) have to be anded with clock $\varphi 1$ by some timing circuit outside the flipflop.

Fig. 11 b shows another implementation of this flipflop, where the control input is binary coded (L=1 for "load", L=0 for "hold"), and anding with $\varphi 1$ is done inside the conditional multiplexer. Fig. 11 c shows a mixed C/L level implementation of this flipflop. In case a large number of flipflops to form a register the circuit in fig. 11a saves chip area, since the centralized timing circuit is needed only once for a number of flipflops.

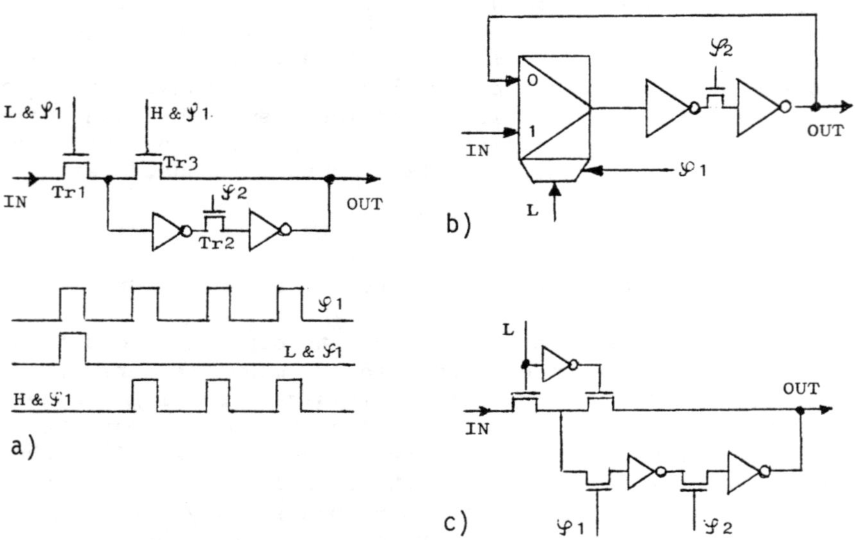

Fig. 11

An Example of a regular structure: a LIFO stack memory

The book shows the design of a LIFO stack memory, also called push down stack. The idea is to use several right/left shift registers as an array shift register. If, for instance, 8 shift registers are shifted simultaneously, 8 bit words are shifted at each clock cycle, where the 8 shift inputs form an 8 bit word shift input. When the array is rotated by an angle of 90°, right/left is turned into up/down, shift in is turned push down, and, shift out is turned pop up. The structure of the circuit is not affected by this rotation: only a renaming of functions occurred. For more details see chapter 3 of the book. Not to just repeat the text out of the book, some parts of the push down store will be modelled somehow differently, and not all details will be repeated here.

Fig. 12 a shows the C/L level mixed representation of the flip-flop cell used for the right/left shift array. Each cell includes 4 pass transistors, controlled by the signals SHR (shift right), SHL (shift left), TRL (transfer left), and, TRR (transfer right), where SHR and TRL have to be anded with $\varphi 1$, SHL and TRR with $\varphi 2$. TRR=TRL=1 during SHL=SHR=0 keeps the inverter chain in a 2-phase refresh sequence, timed by $\varphi 1$ and $\varphi 2$. To raise SHR instead of TRL during $\varphi 1$ evokes a shift right. To raise SHL instead of TRR during $\varphi 2$ evokes a shift left. As explained above the write/hold control is achieved by 2 multiplexers formed by the four pass transistors. Fig. 12 b shows a mixed RT/L level functional description of the cell.

Fig. 13 a shows the connections and placements within an array of such shift cells to form a push down store, where data flow and control flow are orthogonal to each other (illustrated by fig. 13 b). The stick diagram in fig. 13 c shows the left half of the the shift cell. The right half is identical, except that it is rotated by 180° (compare names of control inputs in fig. 13 c: turn the page upside down to read in parentheses the names of the control inputs of the right half cell; fig. 13 b shows the map of inputs for both halves together). Fig. 13 b demonstrates the orthogonality of data flow and control flow inside one flipflop cell of the array.

The control signals in this LIFO example are generated from the two clock signals $\varphi 1$ and $\varphi 2$, and from an operation code OP. The operation code OP is defined in the following way. During OP=0 the refresh function is maintained. OP=1 during $\varphi 1$ starts a pop operation (shift left), whereas OP=1 during $\varphi 2$ starts a push. Fig. 3.11 in the book shows a control signal generation circuit which generates SHL, TRR, TRL, and SHR. Fig. 14 here in this text is a more compact description of this circuit. Figures 15 a, b, and c indicate equivalence transformations which yield the even more compact representation in fig 16. The capacitors C1 and C2 are parasitic input capacitances of an inverter. These capacitors store OP & $\varphi 1$ resp. OP & $\varphi 2$ during the entire clock cycle.

Fig. 16 gives an operation example by means of pulse diagrams. As long as OP stays zero, capacitors C1 and C2 are not charged and the destination address of both demultiplexers, MUX1 and MUX2, remains 0. Thus $\varphi 2$ is the source of TRR and 1 the source of TRL so that refresh cycles are maintained in all cells of the array. Cycle no. 1 is such a refresh cycle with OP=0. During phase $\varphi 1$ of

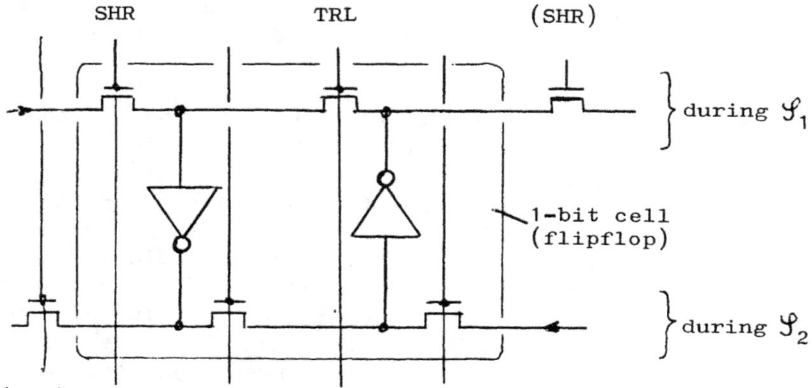

a)

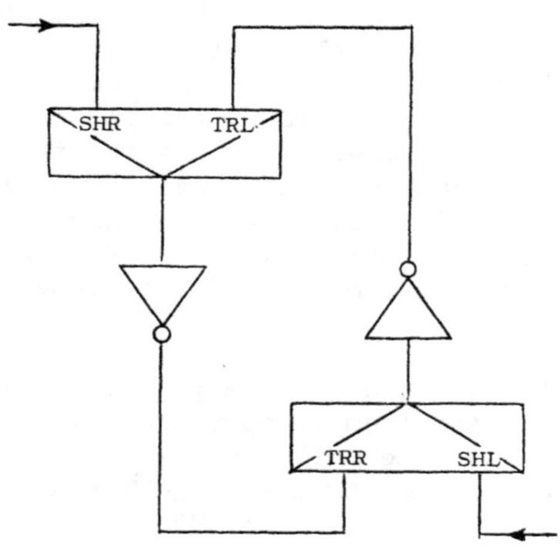

b)

DATA FLOW WITHIN PUSHDOWN STORE

cell-to-cell copy transfers:

SHR = shift right (push)

SHL = shift left (pop)

refresh transfers within cells:

TRL = xfer left

TRR = xfer right

c)

Fig. 12

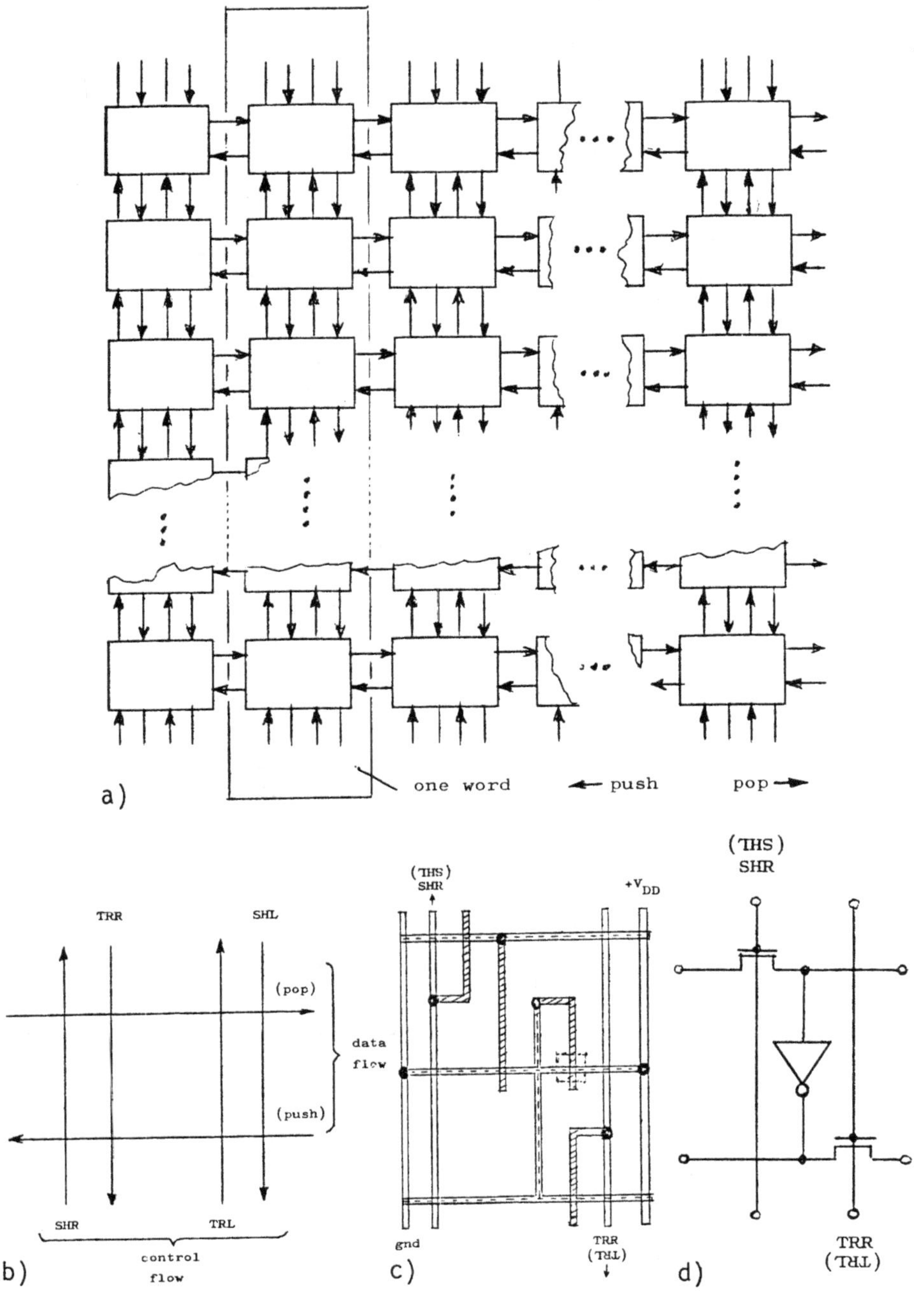
one word
push
pop
a)
TRR
SHL
(pop)
data
flow
(push)
SHR
TRL
control
flow
b)
(SHL)
SHR
$+V_{DD}$
gnd
TRR
(TRL)
c)
(SHL)
SHR
TRR
(TRL)
d)

Fig. 13

cycle no. 2 OP is raised and stored in C1. That's why at phase $\varphi 2$ of this cycle the demultiplexer MUX1 is switched over to connect the SHL output instead of TRR to source $\varphi 2$. So a push operation is evoked. At $\varphi 2$ of example cycle no. 4 in fig. 16 the generation of control signals for a pop operation is started: C2 is charged and the demultiplexer MUX2 is switched over. Thus at $\varphi 1$ of the next cycle $\varphi 1$ is routed to the SHR output instead of the TRL output of the controller circuit.

The timing scheme of this LIFO example is somehow tricky. This is the typical price to pay for using dynamic registers with 2-phase clocking. Currently this method yields the most simple data path structure in many cases. A more simple timing scheme, such as e.g. using a single phase clocking, would have required a more complex flipflop circuit. This would have been a substantial waste of chip area, since this larger flipflop would have been placed repeatedly over the entire array. One problem of this type of dynamic circuits is, that it is difficult to map into a RT level descripotion. Such a mapping would be very simple for circuits using static flipflops. Also the use of pass transistors makes mapping into the RT level more difficult, since it is not a directed element. It may be used in either direction of the channel. The direction of data flow depends on the output impedance of the circuits connected to source and drain of the pass transistor.

Fig. 17 shows the map of the entire push down stack including the push and pop drivers. Fig. 18 b shows a RT level representation of a push driver of one word of the flipflop array. This fig b also shows the topology, and how it fits to the topology of the flipflop cells (see fig. 18 a). The push driver is repeatedly placed at the top of the array (see fig. 17), so that each word of the store has its own driver. Fig. 19 b shows the pop driver RT diagram, and how its topology fits to the flipflop cell of the array (fig. 19 b). Note, that the pop driver circuit does not differ from the push driver circuit. The only difference is, that it is rotated by 180°.

So the entire push down store requires mainly the design of only two cells to be used repeatedly: the flipflop cell and the driver cell. One of these cells, namely the flipflop cell has been even more easy to design being snythesized from two identical half cells as we learned above and from fig 13. The only additional circuit to be added consists of four pass transistors to generate the inputs OP & $\varphi 2$ and OP & $\varphi 1$ to the drivers. Also no routing is needed inside the array of these two types of cells, since all connections are generated by just placing the cells. That's why this LIFO array is an example of structured design, saving design time. A minimum number (2) of types of simple cells is used to layout almost 100% of the push down store. Almost no individual wiring has to be layed out.

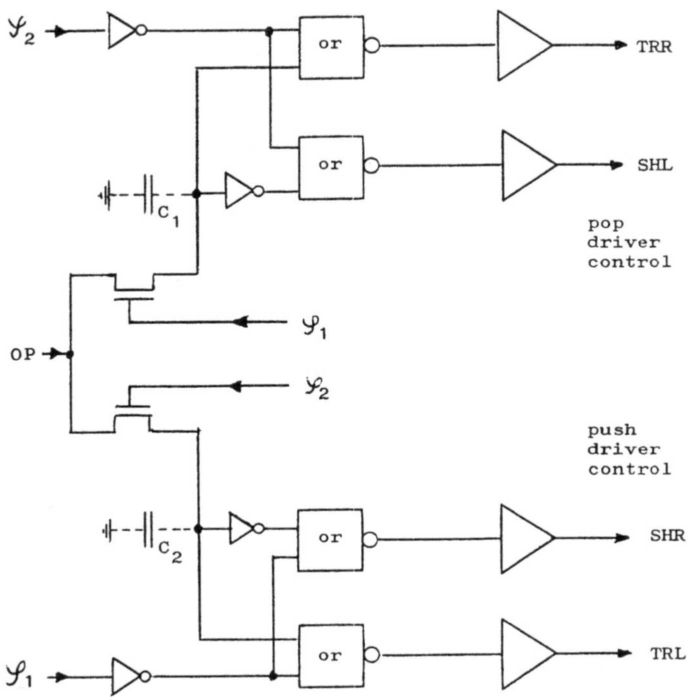
φ_2
or
TRR
or
SHL
C_1
pop
driver
control
φ_1
OP
φ_2
push
driver
control
C_2
or
SHR
or
TRL
φ_1

Fig. 14

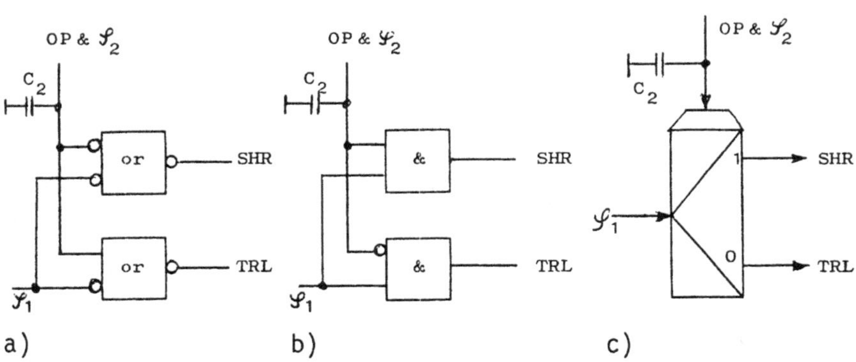
OP & φ_2
C_2
or
SHR
or
TRL
φ_1
a)
OP & φ_2
C_2
&
SHR
&
TRL
φ_1
b)
OP & φ_2
C_2
1
SHR
φ_1
0
TRL
c)

Fig. 15

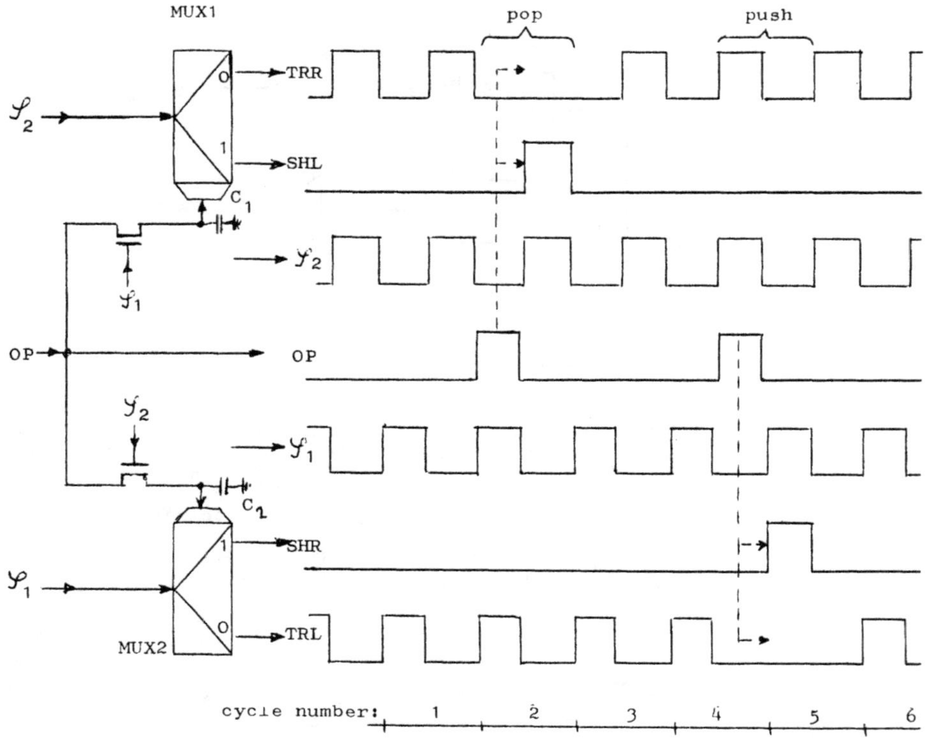

Fig. 16

Finite State Machines

This section refers to chapter 3.11 in the book and only part of it is repeated here. So it is advisable to have the book available when reading this section of this paper here. Just the introduction to finite state machines will be presented in more detail here than it has been done in the book. A second feature of this paper here is the introduction of a second type of flow chart which is more hardware-near: CLARE`s ASM chart. Also the equivalences between state diagrams, flow charts, and ASM charts will be shown and illustrated. Finally RT level control modules will be introduced as an alternative means to implement controllers. Detailed description of PLA use and layout and the application of 2-phase clocking to dynamic registers in nMOS silicon gate depletion load technology is not repeated here in this text. For details the reader should study sections 3.10 and 3.11 of the book.

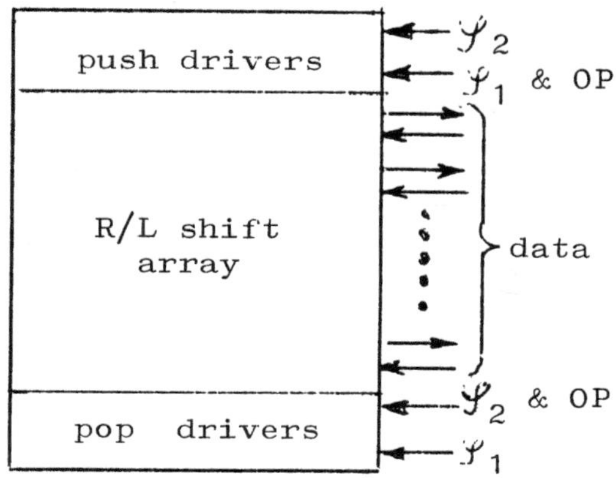

Fig. 17

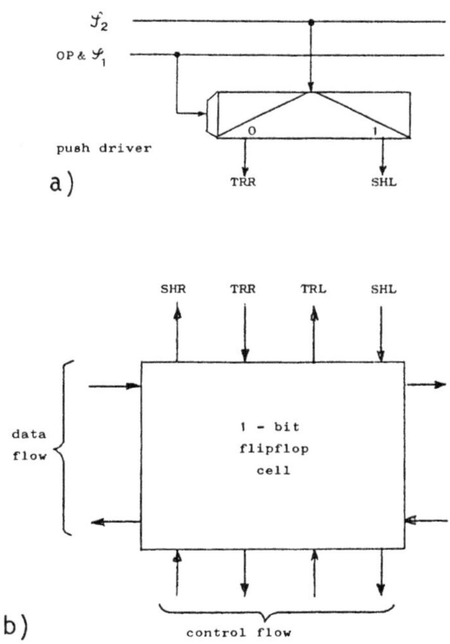

Fig. 18

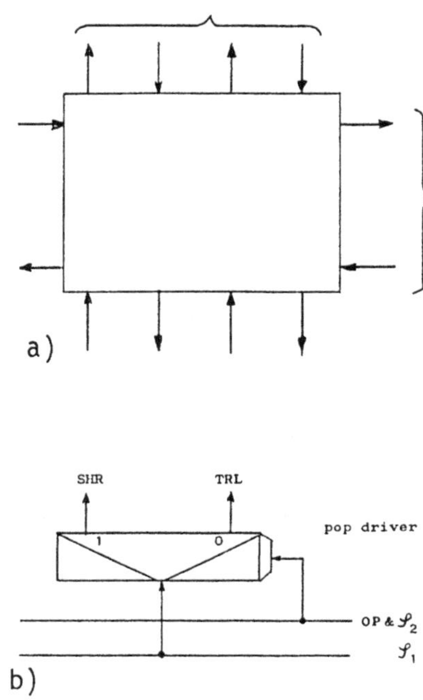

Fig. 19

A finite state machine is one method to model controllers. Also other functions, such as e.g. recognizers may be modelled by a finite state machine (f.s.m.). A controller is a device which generates sequences of control words to operate some data manipulation unit or other operation unit. The above section showed an example, where a shift array register was the operation part. Normally any autonomous unit of a data processing system consists of 3 functionally different parts (fig. 20): the control part, the operation part, and the timing part. To implement some sequence of operations the controller sends control words to the controlled part (the operation part). Whenever the operation required next depends on some results of previous operations, this is signalled back to the controller. These "feed back signals" are the basis of decisions on branches inside the controller

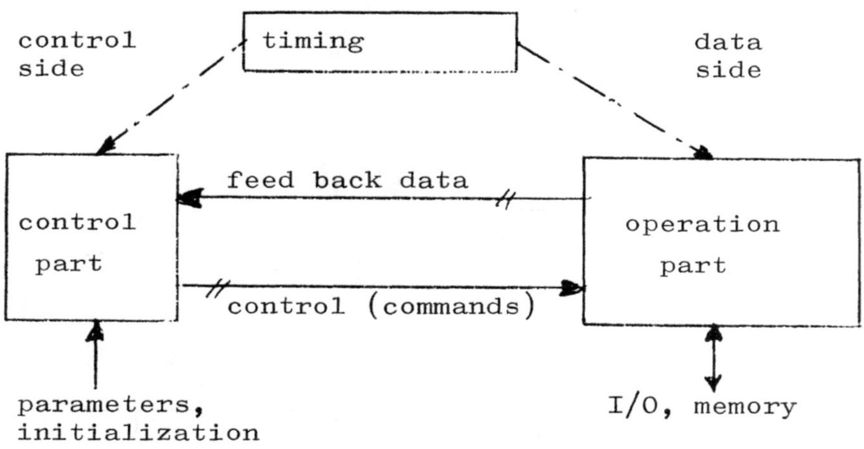

forms of control:

program and sequencer (Moore-type)

microprogram and sequencer (mostly Moore-type, sometimes Mealy-type)

minimized finite state machine (f.s.m.)

PLA-implemented f.s.m.

RT control module network (f.s.m. is subset only)

Fig. 20

specification. To coordinate the actions of both, the controller and the operation part, a timing part is needed (see fig.20), which generates and distributes clock signals to the other parts of the entire processor system. In case of asynchronous cooper-tion of controller and operation part the coordination is done via "handshaking signals" directly generated inside these two units, so that no separate timing part is found. Normally both, controller and operation part are also connected to the outside world: the controller for initialization selection of one out of the repertoir of its control algorithms, the operation part for moving data in and out from peripherals or memories (compare fig. 20).

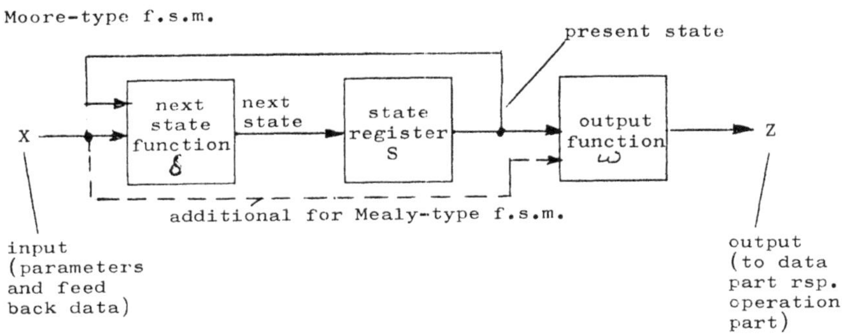

Fig. 21

Forms of Controller Implementation

Various forms of control implementation are known. But all of them may be mapped by a f.s.m.. We distinguish encoded control (centralized control) and distributed control (such as e.g. the RTM control module system to be explained later). In central control we can distinguish programmed control and hard-wired control. In programmed control we have the control algorithms stored in a memory, called program store. Any programmable processor in a computer is of that type. The program store together with its accessing unit, called "sequencer" forms the "controller", comparable to the f.s.m. in a hard-wired control. The sequencer reads in the program store and generates sequences from it. The "program counter" in the sequencer, which is the address register for accessing the program store, is equivalent to the "state register" known from f.s.m. modelling and from hard-wired control.

Hardwired control is traditionally modelled by f.s.m. specification. A f.s.m. consists of 3 parts (see fig. 21): a next state function δ, the output function ω (both are normally implemented

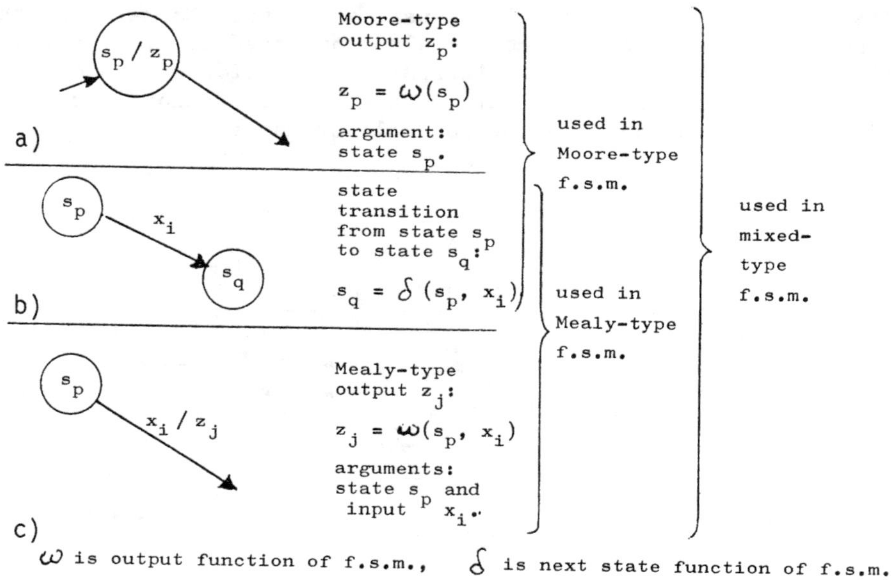

Fig. 22

as combinational logic), and the state register S. The output Z of the function generates sequences of control words to be sent to the operation part. The input X is connected to the source of feed back data from the operations part, and, to sources of initialization parameters from the outside world. The most important models of f.s.m. are the Moore machine and the Mealy machine. The difference is the number of inputs to the output function ω: for a Mealy machine only S is used as an argument to the ω-function, whereas in a Mealy machine both, S and X are used as arguments (the additional path in the Mealy machine is shown in fig. 21).

The function of a f.s.m. is specified by a state diagram or an equivalent tabular representation (transition table). The Mealy-type state diagram differs from the Moore-type state diagram in a very clear way. During operation a f.s.m. goes through a sequence of states being represented as small circles in the state diagram ("nodes" in the "automata graph"). Since a f.s.m.

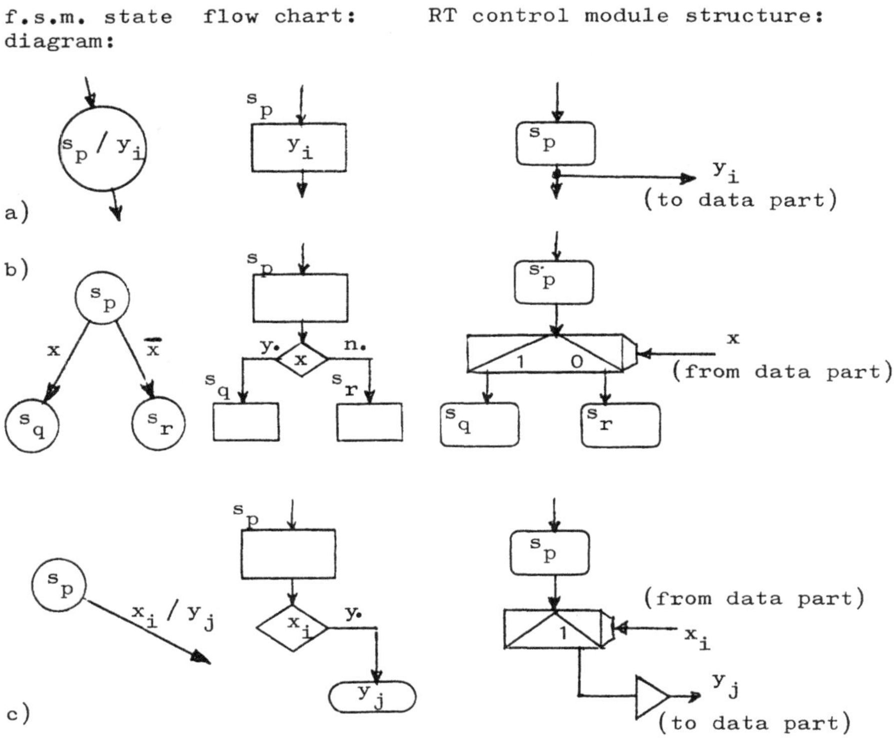

Fig. 23

can be in only one single state at a time it is modelled as a single token flow system. This means that the current state of a f.s.m. is indicated by a token inserted in the state diagram into the particular node which represents this state. Not more than one token must be in a f.s.m. state diagram.

Fig. 22 exemplifies the state diagram formalism. Fig. b shows the representation of the δ-function which is the state transition function: when the f.s.m. is in state sp and the input is xi, then the next state will be sq. This formalism for specification of state transitions is used for both type of machines, for Mealy machines and Moore machines. However, the specification of output functions is different. Fig. 22 a shows the formalism for Moore-type outputs: when the f.s.m. in this example is in state sp, then the output is zp. The output only depends of the state, that is, why it is written into the node and thus associated with the state only. Fig.22 c shows the specification of a Mealy-type output: when the f.s.m. is in state sp and, at the same time the input is xi, then the output is zj. The output is written behind the input at an edge of the state graph. Thus it is associated with both, the edge (and its input), and, with the node (and its state). An f.s.m, may also be a mixed Moore/Mealy-type machine, i.e. it may have both types of outputs: Moore outputs and Mealy outputs. In this case you find both formalisms in one state diagram, that of fig. a and that of fig. c. For completeness also a third type of machine be mentioned; the storage-type f.s.m., where state and (Moore-type) output are identical. The state code is directly fed to the output, and the node in the state diagram does not specify an output. The node looks like that in fig. 22 b: state and output equal sp.

Flow charts are equivalent to Moore-type state diagrams. This is illustrated by figures 23 a and b. The action box of a flow chart is equivalent to the node of a state graph (see fig. 23 a). The test inside the action box specifies the f.s.m. output. The only difference is, that in flow charts the variation of notations is much wider: not only numbers or binary codes, but also names and action descriptions may be used. This means, that instead of the output its desired effect may be inserted into the action box. Also the state notation may be different from that is state graphs. It is written outside the box (see sp in the middle of fig. 23 a) and it may be an identifier (also called "label" or "state label") instead of a number or a code.

Register Transfer Control Modules (RTCMs)

There is also a third notation which is equivalent to state graphs. It is a subset of the RT control module formalism (RTCM formalism) known from the description of a hardware prototyping kit called "pdp-16" by Digital Equipment corporation. This kit allows the direct hardware implementation of control structures. It uses a flipflop as an action box (fig. 23 a, right side), a demultiplexer as a decision box etc. The insertion of a token into a node of a state graph is modelled by writing a 1 into an action box flipflop. A "1" in the sp flipflop means, that the f.s.m. is in state sp. Control flow over edges of the graph is modelled by transmission of the 1 over a wire into action box inputs. A branch is implemented by routing a 1 through a demultiplexer (compare middle and right side of fig. 23 b). Controllers implemented by such a RTCM kit are distributed controllers. Instaed of a state register and a coded state there is a set of flipflops spread over the entire structure. As long as the kit is used only for f.s.m. implementation the state code is a singulary code, since only one token may exist.

An important difference between flow charts and a RTCM implementation is, that there are no control codes or action notations in the action box. Instead, there is a wire going from each action

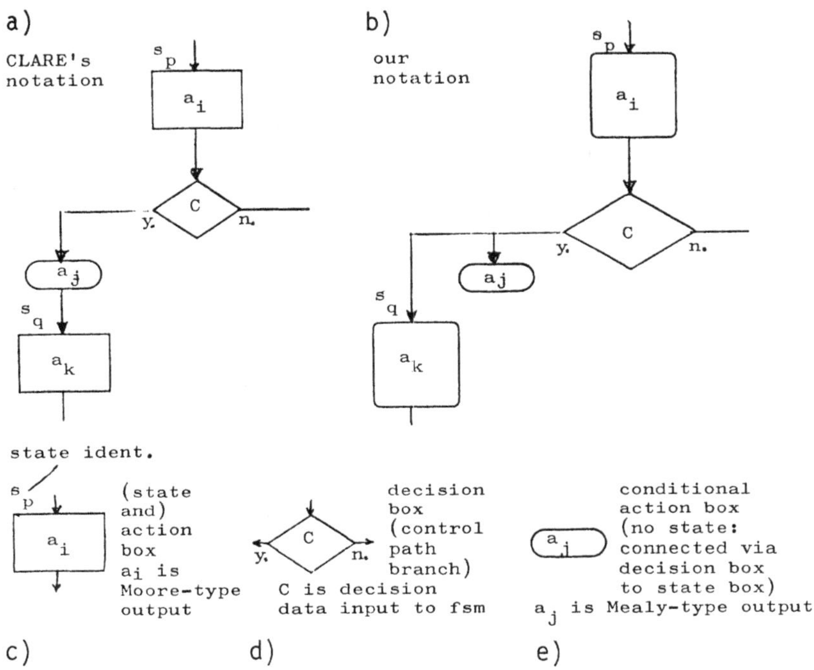

Fig. 24

to the data part. It conveys the evoke signal to evoke the action inside the data part. In right side of fig. 23a the signal yi, which is high when a 1 is in the box sp, is connected to the particular function inside the data part which executes action yi.

Fig. 23 b, left side and middle, maps the decision box of flow charts into a state graph. The boolean variable x (also called "condition" x) is an input to the f.s.m.. The decision box is associated with state sp (see label at upper action box), and, it tells, that the next state is sq, when x=1. Otherwise, when x=0, the next state is sr (compare state labels at the two lower action boxes). The left part of fig. 22 b shows the equivalent state graph representation. The right part shows the equivalent RTCM structure, where a 1-to-2 demultiplexer (1 bit wide) shows the a hardware implementation of the decision box. Controlled by condition x, fed back from the data part (compare "feedback data" in fig. 20), the demultiplexer routes the token flow: it determines, wether a 1 in flipflop "sp" will be written into flipflop "sq" or sr". It is an important part of the RTCMS definition, that the RTCM flipflops are always read destructively, i.e. reset by reading (this will be explained later in more detail).

ASM charts

Flow charts can only be used to model Moore-type f.s.m., however, not for Mealy-type machines. To achieve also Mealy-type representations, CLARE /CL73/ introduced an extended flow chart formalism by introduction of an additional type of box, the "conditional action box" (see fig. 24 e). This box does not represent a state, as the normal (unconditional) action box would do. It is only conditionally associated to some ordinary action box. The middle

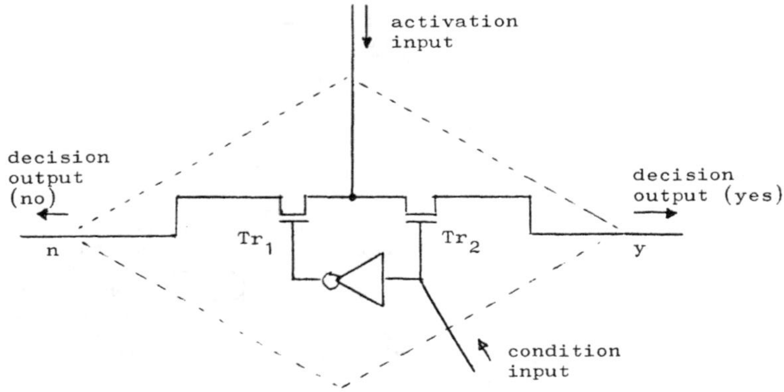

Fig. 25

part of fig. 23 c gives an example: the action is executed, when the f.s.m. is in state sp only if at the same time condition xi is true. This is represented by the fact, that a conditional action box (CAB) must not be connected directly to an action box. It must always be connected to the output of a decicion box, i.e. it is always depending of a condition.

CLARE called his notation "ASM chart" (algorithmic state machine chart). CLARE`s notation (fig. 24a) is somehow confusing, since the CAB`s arc to the input of a state box (state sq in this example) resembles a state transition from aj. To avoid such confusion we use the slightly different but clear notation shown in fig. 24b, called OSAM (our ASM) chart. (equivalent to fig. 24 a). The right part of fig. 23 c shows a RTCM representation of a hardware implementation of a CAB. When the f.s.m. is in state sp, and, at the same time condition xi is true (xi=1), the evoke signal yi is sent to the data part (the triangular box represents an output, since it looks like the C level notation of an output buffer of the f.s.m.). Note the equivalence of the 3 notations in fig. 23 c. Fig. 25 shows a hardware implementation of a decision box (also compare fig. 9 c and d).

Implementing Finite State Machines

Finite state machines may be implemented in a classical way including minimization of both, transision function and output function. However, when a larger number of states is involved, this classical design is difficult, and time consuming. The logic designed by this method is called random logic, since the structure is highly irregular, and much space is wasted for connections. If only a small machine is to be designed, simetimes also the RTCM structure implementation may be considered. For larger f.s.m. implementations a PLA implementation is an optimum method. It uses a highly regular underlying structure which may be programmed by only one rectangle per variable in a term. This programming is done in only one mask level, so that individual layout is simple and only one mask has to be produced in addition to the other masks which remain unchanged.

In chapter 3 of the book the PLA application is explained in detail and using 2 examples. A PLA mainly allows the specification of a number of AND functions (terms) in one ROM ("AND" ROM, or AND plane) and in an "OR" ROM ("OR" plane of the PLA) termes may be ORed to a number of PLA outputs. The AND plane has two input rails per input variable, so that each variable may be used in a term directly or negated. The reader not familiar with PLAs should read chapter 3 of the book, since here is no space to do the complete introduction.

Outputs from fsm:

HL (1 : 0) Hwy lights
FL (1 : 0) Frd lights
ST = start timer

Inputs to fsm:

C = car is waiting
TS = time-out short
TL = time-out long

HL and FL coding:

code	meaning
00	green
01	yellow
10	red
11	-

ST coding:

ST = 1: "start timer"
ST = 0: "do nothing"

abbrev.: Hwy = Highway
Frd = Farm road

controller states (states of fsm):

Q (1 : 0)	mnemonics	meaning	comment
00	HG	Hwy green	Frd.: red
01	HY	Hwy yellow	Frd.: red
11	FG	Frd green	Hwy.: red
10	FY	Frd yelloww	Hwy.: red

Fig. 26

The Traffic Light Controller Example

The following section refers to section 3.11.1 of the book, which describes design and PLA implementation of a f.s.m. to control the traffic lights of a cross road. Fig. 26 summarizes all mnemonics used in this example. A farm road (Frd) is crossing a highway (Hwy). Both sides of the farm road have car contacts (C) to indicate a car waiting in the farm road. A "time-out short" timer (output: TS) determines the duration of "yellow", whereas a

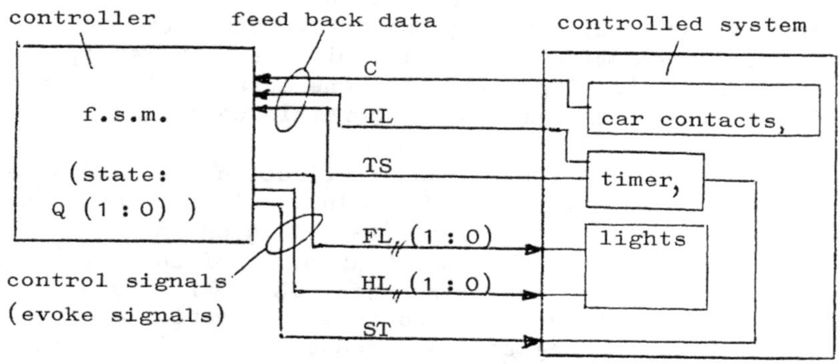

Fig. 27

"time-out long" timer (output is: TL) guarantees some minimum duration of traffic flow between changes of direction. Also study section 3.11.1 of the book concurrently when reading this section. Not all essentials are repeated here in this paper.

Fig. 27 shows the block diagram of the entire design problem: the f.s.m. to be designed, and its environment. Car contacts, timers, and the lights are components of the "data part" (the controlled part). C, TS, and TL are feed back data, whereas ST ("start timer"), FL(1 : 0) (2 bits wide, controls farm road light), and HL(1 : 0) (2 bits wide, controls highway light) are f.s.m. output.

Fig. 28 a is the state graph of the controller to be designed. The graph is derived from table 3.2 in the book. Since there are outputs specified in both, the nodes and at the arcs, the f.s.m. is a mixed Moore/Mealy-type machine. The catenation of HL and FL, 2 bits each, form the Moore-type output. ST (1 bit) forms the Mealy-type output (specified at the arcs in fig. 28a). The simple straight-on PLA solution (see plate 8 in the book for stick diagram) requires 10 terms (also see table 3.2 in the book) where the AND plane has 5 two-rail inputs (f.s.m. inputs C, TL, and TS,

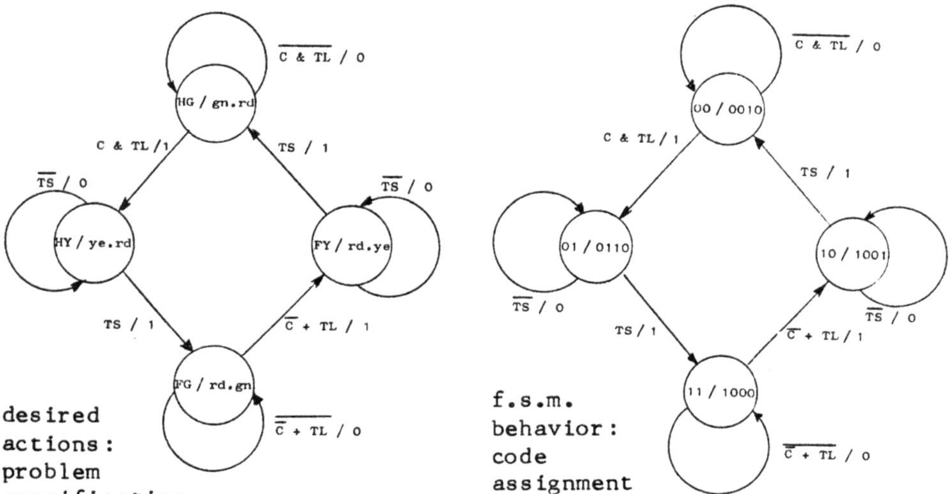

desired actions: problem specification

a)

f.s.m. behavior: code assignment

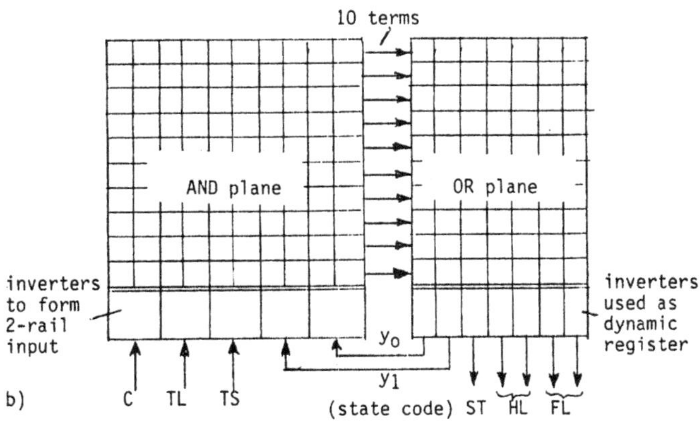

b)

	Q	HL,FL	comment
	00	0010	Moore
	01	0110	output
	11	1000	
	10	1001	
C TL TS		Q+ ST	
1 1 x	00	01 1	next
x x 0	01	01 0	state
x x 1	01	11 0	funct.
1 0 x	11	11 1	and
0 x x	11	10 1	Mealy
x 1 x	11	10 1	output
x x 0	10	10 0	
x x 1	10	00 1	

c) split-up PLA bit pattern

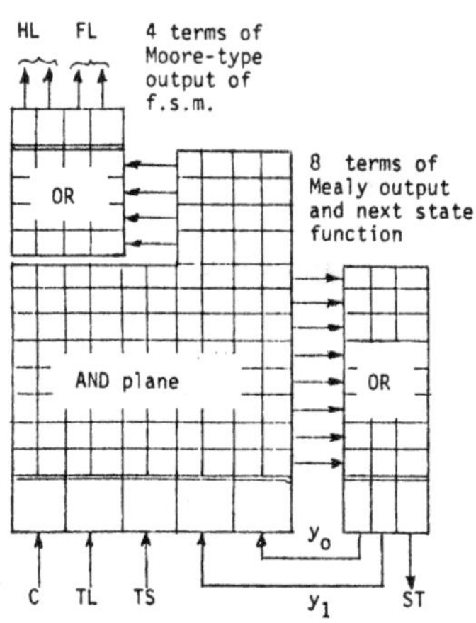

d)

Fig. 28

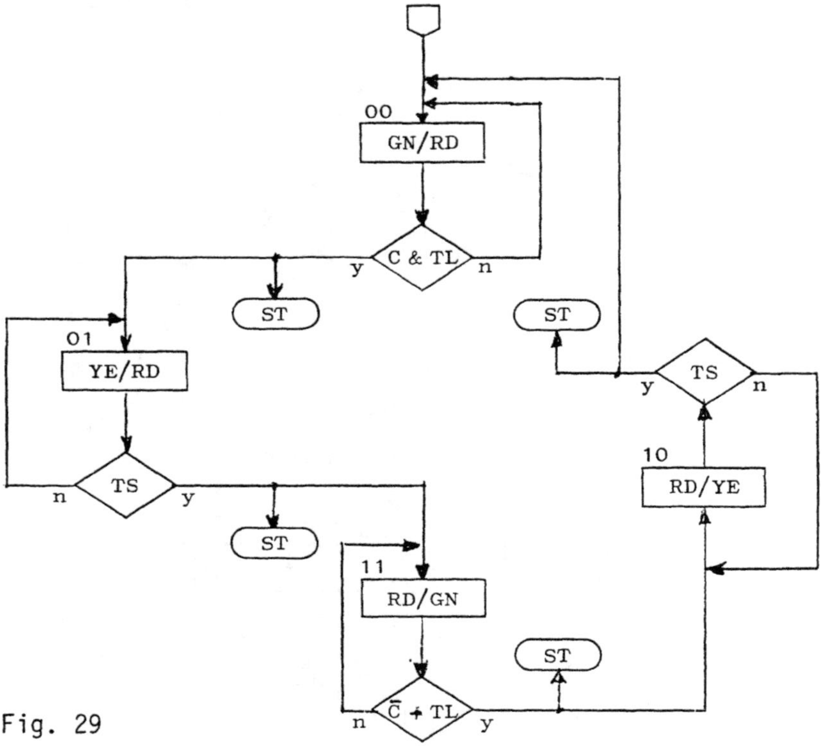

Fig. 29

and the present state variables y0 and yl). The OR plane has 7 outputs: HL, FL (2 bits each), ST, and the state code (2 bits: y0 and yl). Fig. 28 b shows the block diagram of this PLA solution.

In this example a little space may be saved, if the Moore-type output and the state transition function together with the Mealy-type output would be implemented in separate PLAs. Fig. 28 c shows these 2 different funtions, where the Moore output has 4 terms with only the state as an input, and with 4 output bits. The other function has 8 terms with only 3 output bits. Fig. 28 d shows the split up PLA implementation, where the 5 input AND plane is slimmed down to only 2 inputs for those terms which are needed to generate the Moore-type output. A split up of PLAs into a system of smaller PLAs may lead to substantial reduction of space needed, when the original PLA planes have been sparsely programmed. Such splitting up is not difficult, since most solutions may already found by mere inspection of the PLA programming bit pattern.

Fig. 29 shows the ASM chart of the controller specification. It has been derived by one-to-one mapping from the state graph in fig. 28 a. A flow chart representation would not be feasible, since flow charts cannot model Mealy output. Fig. 30 shows the RTCM implementation of this chart. Fig. 30 d shows the state flipflop unit and its pins for token input/output: the extension is used, if more than one token source is to be connected to the flipflop. The circuit diagram of an nMOS implementation of this

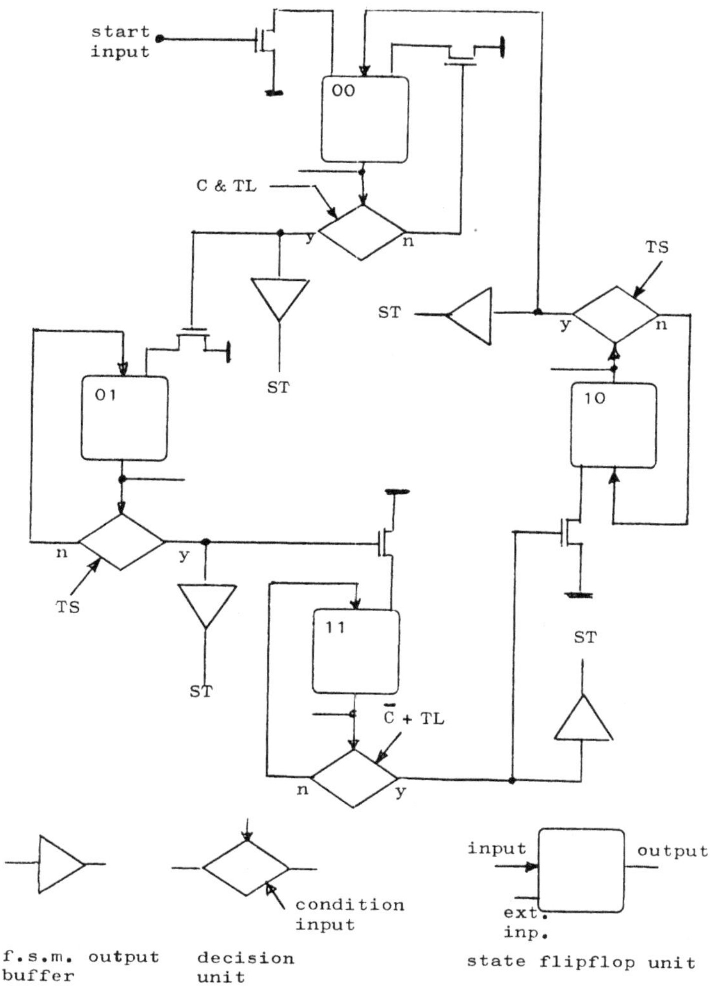

Fig. 30

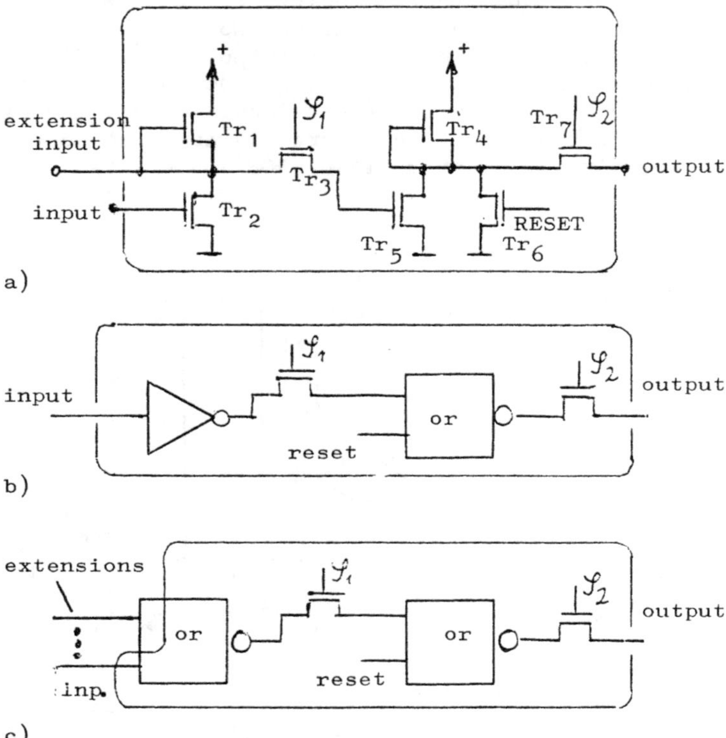

Fig. 31

flipflop is given by fig. 31 c (the reset input of this flipflop is not shown in the RTCM diagram in fig. 30a). For each input extension an additional transistor in parallel to Tr2 is needed, to be connected from outside to the extension input (compare fig. 31 a and fig. 30). The logic diagrams of a single input and an extended input flipflop are shown by figures 31 b resp. c.

It is very easy to construct the complete C level diagram of this distributed f.s.m. implementation from the RTCM chart in fig. 30. The decision box circuit had been shown in fig. May be, that the traffic controller example used here is a little bit too complex to demonstrate the usefulness of RTCM implementation of control. However, for a number of sequencing problems, such as e. g. the timing sequence generation, it may replace random logic, often used in addition to PLAs. In many cases an RTCM structure with e.g. around 3 states may be more easy to design than classical random logic. Especially, when a singulary coded f.s.m. output is required, a RTCM structure saves the output function logic, since the state code may be directly used as an output.

Summary

This paper looked at chapter 3 in "the book" /MC80/ from a different point of view. It demonstrated that the design and understanding of nMOS dynamic digital circuits could be made even more easy by also using register transfer level descriptions in addition to logic, circuit and topology level notations. A few of the examples from chapter 3 have been revisited. It has been shown that a push down stack together with its driver circuits may be designed almost 100% from one and one half very simple cells. Also the topic of finite state machines has been covered, where also the equivalence between state diagrams, flow charts, extended flow charts (ASM charts), and the RTCM structure implementation has been shown. It was explained, that the traffic controller from chapter 3 of the book is a mixed Moore/Mealy machine. Also its distributed implementation has been introduced as an alternative to PLA implementation. The demonstration of basics of structured design methodology and its comprehensibility has been improved.

Literature

/CL73/ Clare, Ch. R., "Designing Logic Systems Using State Machines", McGraww-Hill, New Zork, 1973.

/HA73/ Hartenstein, R., W., "Hierarchy of Interpreters for Modelling Complex Digital Systems", Lecture Notes in Computer Science, Springer-Verlag, Heidelberg/New York, 1973.

/HA77/ Hartenstein, R. W., "Fundamentals of Structured Hardware Design", North Holland Publ. Co., Amsterdam/New York, 1977.

/HE80/ Hartenstein, R., "VLSI-Bausteine in geringen Stückzahlen" Elektronische Rechenanlagen, Vol 22, No. 4 (Aug. 1980).

/HR80/ Hartenstein, R., (see in this volume)

/HZ75/ Hartenstein, R., Zaks, R., "Microarchitecture of Computer Systems", North Holland Publ. Co., Amsterdam/ New York, 1975.

/KF79/ Foster, M. J., Kung, H. T., "The Design of Special VLSI Chips , COMPUTER 13, 1 (January 1980).

/MC80/ Mead, C., Conway, L., "Introduction to VLSI Systems", Addison-Wesley, Reading, Mass., 1980.

/SE80/ Sequin, C. H., "Instruction in MOS LSI Systems Design", COMPUTER 13, 3 (March 1980).

Shared Cultures:

CIF Library, Starting Frames, and Scalable Design Rules

Reiner W. Hartenstein

Kaiserslautern University, D-675 Kaiserslautern, F.R.G.

Preface

This paper describes a new form of cooperation between designer, user, mask maker, and, manufacturer of digital LSI circuits. This innovative organization may be the beginning of a second breakthrough in the application of integrated circuits. The lack of designers may be fixed within 2 or 3 years. At low cost the fabrication of integrated circuits in very small quantities could be made possible for a wide community of designers having only limited background in technology. A computer science professional may learn the design within 2 or 3 months. This paper gives detailed information on the new organization, called MPC' (Multi-University Multiproject Chip Set Project), currently running in the U.S.A. A growing community shares in the "culture" of MPC, held together by the compatibility of a common design language, and in the utilities of a central facility.

1. Introduction

Programming and LSI design have some surprising parallels w. r. to organization problems /HA79/. Number of bugs compared to complexity, design cost, debugging problems, etc., are comparable. The well known term "software crisis" stimulates to use the term "chipware crisis" in comparing both sides. The well known curves on increasing design cost by Gordon MOORE, often copied, are not repeated here (for copies e.g. see /HA80/; Gordon MOORE's curves are very likely to be found in at least one other paper within this volume here). The summary is: the current bottleneck in VLSI design is design cost, the designer's ability, and the designer gap. Technology is not the bottleneck. This paper describes the philosophy of a beginning "shared culture" with the goal to overcome this bottleneck.

1.1 Trends in Custom LSI

From 1973 to 1980 the percentage of completed custom LSI designs, compared to the total of designs increased from a few percent to more than 50 percent. This is only an indication of the urgent demand. However, it is still very difficult and costly for a user to get his custom designs implemented. Systems houses want to have their own LSI design and fab facilities for two reasons: to become independent, and, to achieve designs which fit better to their requirements. Currently this is only possible by multi-million dollar efforts. That's why custom design is still urgently needed. By the way, there is also the opposite development on the way: not only systems houses start their own fab facilities, but also semiconductor houses go into the systems and equipment business. This development has also been called "vertical migration".

Not only for large houses, but also for smaller companies there is a demand for custom LSI circuits. Experienced microcomputer users found out, that interface and program development is much more expensive than had been expected before. That`s why in many cases custom LSI may have lower development cost than microcomputer application development. Another reason of custom LSI demand is the high fabrication cost of printed circuit boards, especially in applications other than consumer electronics, when also testing is expensive. In one case it has been reported, that in quantities of 200 custom LSI was cheaper than PCB fabrication. The comparizon had been very conservative, since LSI cost also included development, whereas PCB cost did not. In another case a break-even point was reported at a quantity of 78.

One remedy is, that the user has his own design facilities. Especially when the intended production outputs only chips in small quantities, it is not economic to have own production facilities. The fab line would be used only at an extremely low percentage of its capacity. However, the investment needed to have in-house design facilities is quite substantial (almost 1 million US dollars, or even more). That's why the only reasonable interface to the manufacturer is the "mask tape", used to control the manufacturer`s pattern generator (this will be explained in detail later in this paper). Let us call this form of user/manufacturer cooperation "custom development" to distinguish it from custom design. One disadvantage of this form of cooperation is lack of independence of the manufacturer.

Users of off-catalogue integrated circuits normally design their printed circuit boards with such circuits, which also may be purchased from a "second source". Such a back-up would also be desirable for buyers of their custom-developped LSI circuits. However, normally second sourcing is impossible, since the mask layout patterns produced by design facility are closely adapted to the process parameters of a particular fab line. Also when using the same technology, such as e.g. nMOS depletion load silicon gate technology, the design rules normally differ from fab line to fab line because of individual process parameters.

So the demand is, to have a design facility, the "mask tape" output of which could be submitted to several manufacturers to guarantee independance by second sourcing. However, it is even difficult to find just one manufacturer. This is an even stronger argument to strive for availability of second sourcing. What is needed to make feasible a second sourcing for custom-developped LSI circuits is a new designer/manufacturer interface. This new interface should feature compatibility of layout design files.

A standard data format is needed to transfer design files to the manufacturer. Currently the different pattern generator (PG) data formats for different mask pattern generators are used. What is needed is a standard pseudo PG language which could be converted conveniently into a particular PG input language by using a a computer. During such a conversion run also design rule adaptation could be performed automatically, if needed.

2. The Traditional LSI Design Organizsation

The design process for LSI digital circuits includes a hierarchy of methodological levels, with their specific notations and methods. In each level structures are implemented from elements, which have been implemented at the next lower level. In LSI design we may distinguish the following levels (for a table and for details see/HR80/): device physics level, geometry level, topology level, circuit level, gate level, register transfer level, state machine level, and sometimes even higher levels, such as e.g. the installation level. The role of the device physics level is in most cases that of a service level, which delivers ready-to-use recepees for the design process. However, all the other levels are involved in the "systems" design process more or less closely coupled and interactively. And, as a matter of fact, most LSI or VLSI design is or should be a "systems" design problem to achieve better design solutions.

Normally all the persons involved in the design process are specialized in one, or sometimes two of these levels. Fig. 1 shows the the organizational structure of cooperation in LSI chip

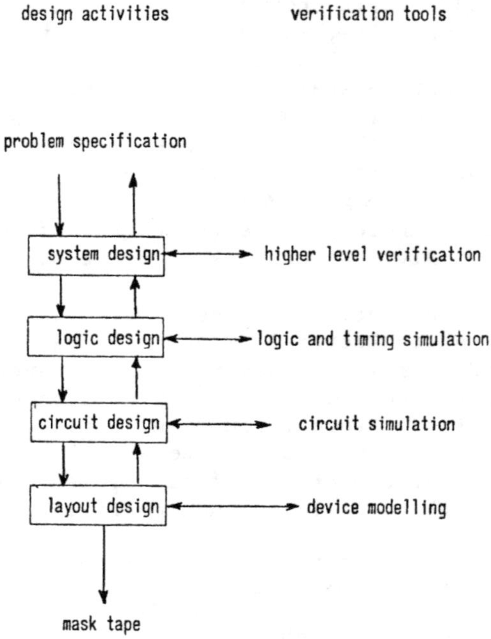

Fig. 1

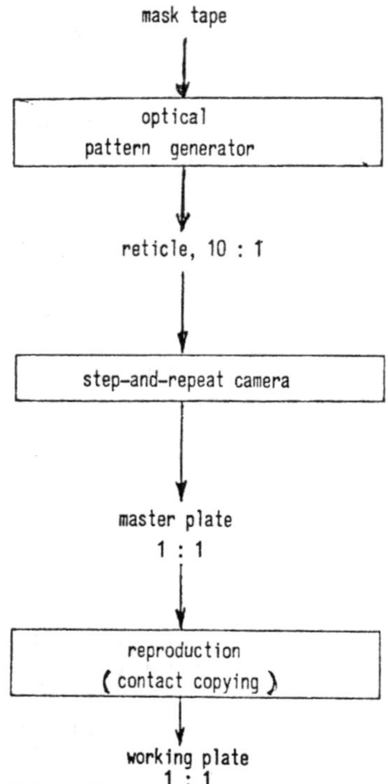

Fig. 2

design. The left side shows its pattern of interaction, and the right side shows the verification and service tools used by the particular level. There are feed back loops in the organization. A circuit specialist may reject a logical designers submission and may request changes to meet circuit level restrictions. Also a layout specialist may request modifications of circuit level specifications because of layout level restrictions.

All this illustrates, that the cooperation is very complex. The various specialists have communication problems because of their different backgrounds. This and the large extent of interaction required is the reason, why bugs are introduced by interfacing between these levels. The large extent of interaction (also the communication is time consuming) and the existence of iteration loops because of rejections result in a long design time and thus in high design cost. One of the most important cost factors is the fact, that for completion of one design a number of different persons with different background is required. It would be desirable, that just one person could complete the entire design from problem specification down to layout level.

Another important cost factor stems from the design objectives for off-catalogue circuits for mass production and marketing. The objectives are: low price and high performance to have advantages over competitors in the market. Sales in very high quantities bring back the high development cost: The design objectives derived from these goals are: high yield and high performance. The high yield reduces production cost and high performance makes the circuit attractive to buyers. To achieve high yield the chip area has to be optimized: the smaller the chip, the higher the yield.

A price has to be payed for this. The design rules are taylored to squeeze chip area required for the circuit. Such squeezing design rules increase design time. These rules are difficult to handle and skilled persons are needed. The optimization of area adds more loops to the iteration in the design process. Also the optimization with respect to performance of the chip needs skills and time. Verification is more difficult as a result of optimization. That's why excessive use of computer aids for verification is needed: simulators specialized on different levels, and other verification software packages, such as e.g. design rule checkers (also see fig. 1). All these computer aids are costly and sometimes one simulation run needs hours of CPU time on very powerful computers. Debugging needs a number of iterations of such expensive runs.

2.1 Conventional Mask Making

Also conventional mask making and its interface to the design department is very costly. The design optimization illustrated above and the squeezing design rules make mask making from layout specifications a very critical process. Because process parameters differ from fab line to fab line, the layout has to be adapted to the input required to the mask pattern generator (PG). There are two solutions: the design rules may be adapted, or the layout may be adapted after the design process is finished (geometries of e.g."wires" may be shrinked or blown up a little bit). The latter method using computer algorithms is advanced and not yet widely spread, but it is advantageous, since it yields in design compatibility, at least to some extent.

Conventional mask making requires a high degree of expertize from the designer. He has to know, wether positive or negative masks are used. He has to add alignment marks, test pattern, test circuits and scribe lines to the layout specification. He has to consider process parameters when completing the design. Also Other knowledge on mask making and fabrication is required. One problem is, that this interface is not documented and it is not static. That's why much interaction between design department and fab line is required which has to be done by a very much skilled

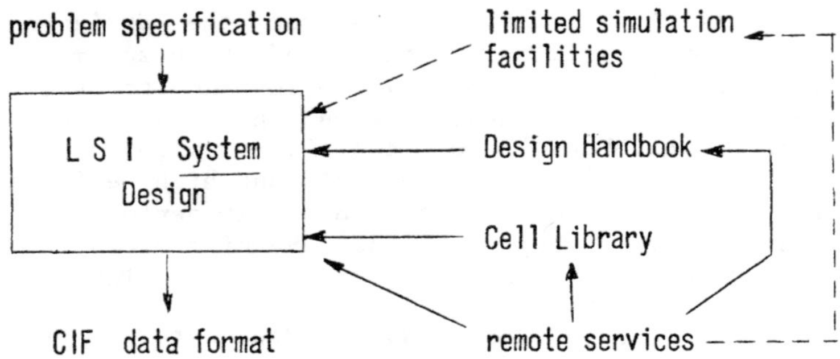

Fig. 3

layout specialist of the design team. This specialist has full responsiblilty for the success of a submission to the fab line. One error may cause, that a complete charge of processed wafers has to be discarded. This means a waste of thousands of dollars and a delay of 1 to 2 months, which also means a waste of design capacity.

Conventional mask making is taylored to mass production. It could be done with less steps and thus less expensive for production of only small quantities of chips. Let us first explain the method currently used. Normally the mask maker or manufacturer refuses to organize mask making in a different way. Fig. 2 illustrates the traditional organization. Three steps are required per layout level. At the first step the PG machine generates a 10:1 reticle by means of a sequence of rectangular flashes (i.e. it has 10 times the size of the chip; for PG process also see /HS80/, or chapter 4 in /MC80/).

The other two steps are performed by projection or contact print. The second step it the production of a master plate by means of a step-and-repeat camera. By repetitive exposure a matrix of 1:1 patterns is generated, so that the entire wafer size is coverd with copies of the same pattern. One mask normally contains the patterns of several hundred chips. Since a minimum production charge is about 10 to 20 wafers this way of mask making fits to production quantities of around 10 000 chips or more.

The third step is the generation of working plates, which are 1:1 copies of the master plate. Since in production the mask will loose its quality by damage or wear out, the master plate is a back-up to generate new working plates for mass production. The working plate is manufactured by means of contact printing, or, for smaller geometries, by projection methods. The above three steps in mask production are required for each layout level. At least five layers are included, depending on the particular technology used.

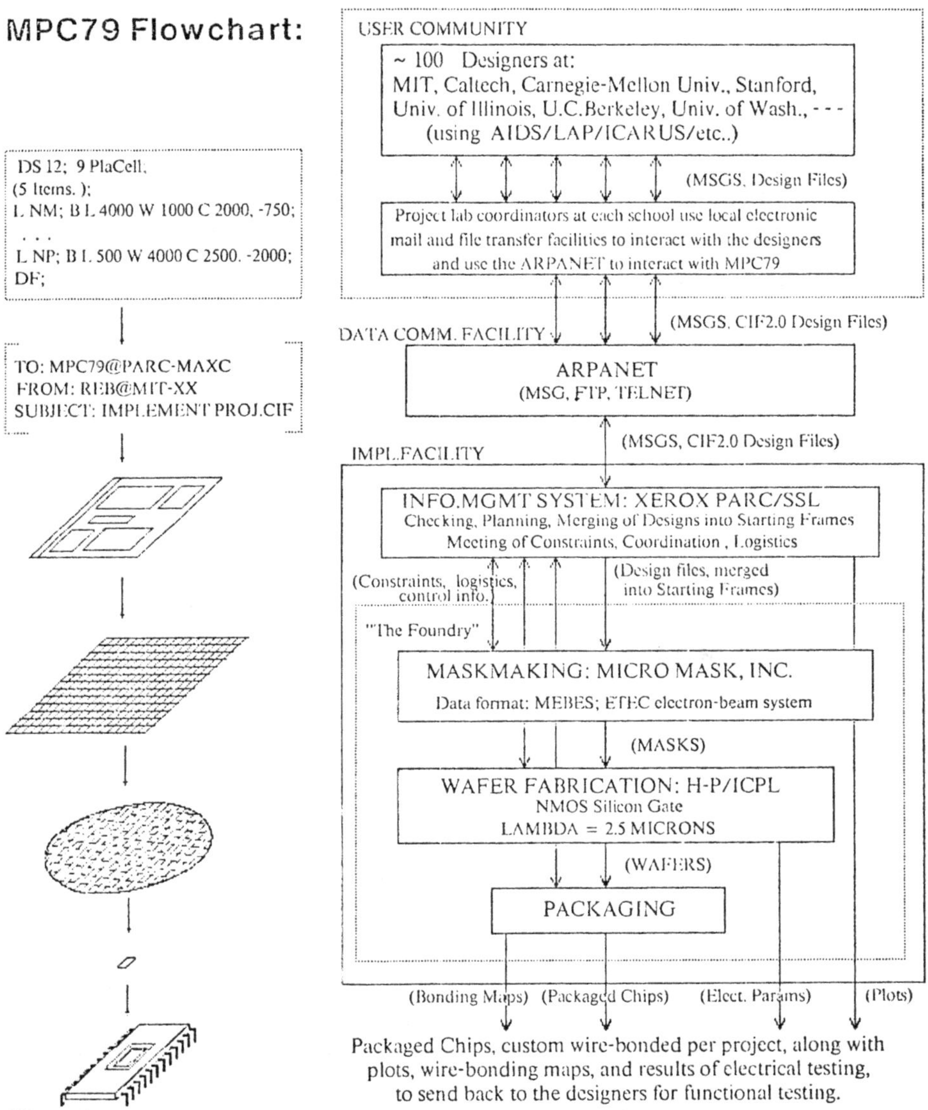

Fig. 4

We noticed, that conventional mask making is expensive (currently ca. 10 000 US dollars or more are reported for a mask set). It does not fit to chip production in small quantities, because the step-and-repeat camera produces too many copies of the same chip layout. Of course, the master plate could be used directly to save the cost to produce working plates. However, it is risky, not to have a backup. So what is needed for custom-developped chips manufacturing, is a new form of mask production, which is less expensive. A way to decrease the cost of mask making is used in the MPC' approach, which will be explained later.

2.2 Problems of Traditional VLSI Development

To characterize the current state of the art in LSI design a German term would fit nicely: it is a "Buch mit 7 Siegeln" (a book with seven seals). It is considered to be a subject of only very few highly skilled persons. It is considered to be a such difficult topic, that only a few students dare an attempt to enter this field. This designer image is not the only obstackle on the way to fix the designer gap, and, to make LSI design "culture" accessible to systems people.

Another problem is the high cost of chip design and production because it is taylored to mass production. This is a problem, because the demand for custom designed and custom developped LSI circuits requires an economic way to produce chip samples in low quantities. Let us summarize the most important problems on the way to go.

- expensive and time consuming designer training
- expensive equipment needed also for training
- expensive mask making
- designer/manufacturer interface requires to much expertise
- this interface lacks of compatibility features
- no second sourcing for custom designs/developments
- communication overhead in the design process
- too much specialisation within the design team

A number of restrictions which are tolerable for a wide variety of existing of future design environments may open new ways to cope with all these problems. For instance, in a number of design environments it is possible to design only digital circuits, to use only one technology (preferrably the most comprehensive one), and, to tolerate a limited yield and a performance not quite optimal. Also a somewhat higher chip count on a PCB could be still less expensive than PCB design from off-catalogue chips. Also the use of a process with less advanced geometry resolution (say 5 or 6 microns or so) may be still profitable in many cases. All these limitations could be tolerable in design environments different from those directed towards mass production of chips. A detailed description and explanation the MPC' project and the underlying instructional philosophy will illustrate the advantages possible by these limitations.

3. An Experimental Inferface Organization

It has been explained above, that the traditional designer/manufacturer interface is one major source of expense. A pilot project funded by the DOD demonstrates an innovative interface. A service center called "remote entry fast turn-around VLSI implementation service" (IZ = Implementierungs-Zentrum) is inserted between designer and manufacturer (fig. 3). It centralizes the expertise and adaptation needed to submit mask tapes for chip production. It also employs methods to decrease mask production cost by at least one order of magnitude for small quantity production of a class of chips. An experimental version of such an IZ is running at Xerox PARC (Palo Alto Research Center). It is a center of "shared culture", shared by a growing community of users. More details (also see /BC80/) will be given in the following sections.

3.1 Instruction on LSI Design

Innovative curricular efforts at a number of universities in the US, based on the so-called "Conway/Mead course" /MC80/ generate test input to the experimental IZ at Xerox PARC. At the same time these courses, currently held at one and a half douzen of US universities, are a means to fix the designer gap within the US, probably within a few years. These courses and their success highly depend on the IZ use. A further availability of such an IZ to the universities would mean a revolutionary second "breakthrough" of integrated circuit technology. A spreading of such a culture could considerably influence competition on the world market of microelectronics. It would not only fix the designer gap within a short time (rescue of silicon valley`s leading role ?), it would also be a major push in "vertical migration".

The new university course smoothly fits into the ACM curricular recommendations /SE80/, and may be given at the end of the undergraduate program of universities to students of Computer Science and Electrical Engineering. Using an innovative text book /MC80/ it is tought in 2 hours lectures and 2 hours exercises per week within one quarter. After 6 weeks the students start with their own design efforts. After the end of such a course many students report their surprise about the success. Many students say, that they never believed before, that they ever would dare to enter the field of LSI design. After the course they report their surprise, how easy it is to design integrated circuits. A number of students, after completion of such a course, start a small design project. After about 2 to 3 months they test samples chips of their design, prívided by the IZ at Xerox PARC.

The secrets of success of these courses are based on the new type of text book and on the availability of IZ services at low cost. The main features of the text book are: only a limited background in technology is required, only one technology is used (nMOS silicon gate depletion load, which is easy to understand), only digital circuits are subject of the text, and, very simple device models are used. A very important point is the fact, that structured design is stressed, which also enhances productivity in design exercises (for introduction to basics of structured design also see /HR80/, in this volume here). This is the main objective of the book, and all other parts of it serve to minimize the prerequisites.

The new curricular approach makes structured design of digital integrated circuits accessible also to systems-type people. These people are more likely to contribute to the underdevelopped fields in VLSI design, such as systems design methodologies, systematic design to cope with complexity, higly parallel system architecture, good algorithms and architectures suitable for VLSI implementation. To achieve further progress in VLSI design is is required, that these people have direct access to the real world of silicon without being converted into technologists. Fig. 1 of the contribution /HR80/ in this volume illustrates, how different the two worlds are, and, why VLSI "systems" design is not a topic for technology.

An important feature, with this goal in mind, is the use of conservative design rules, easy to learn, and somehow compatible to a number of fab lines. Also the use of CIF, a standard data format for mask layout specification is tought and exercised. CIF (Caltech Intermediate Form) is a pseudo PG language, which could be easily converted into any PG input format (a PS is a numerically controlled device). These design rules, together with other curricular minimization features, require less technology background than usual. The students of this culture also share in the IZ's cell library, which includes ready-to-use circuits or circuit kits, such as various input pads and output pads, together with their pad drivers, buffers, a PLA kit, etc.

3.2 The Impact on Digital Systems Research

Also researchers attended the new course at universities. Most of them reported, they never believed before, that they would enter the field of IC design. However, less than 1 year later, they already had sample chips of their own designs in hands. Topics are covered, such as a LISP machine, a geometry machine for computer graphics, multipliers, floating point processors, digital signal processing hardware, interfaces and line protocol circuits for computer networks, base architectures of local micro computer networks (also see /DH80/, /DK80/), data aquisition, and process control applications and many other applications. This is an indication that accessibility of IC design to systems people opens new fields of research and development. Another indication is the ICCC (Int'l Conference on Circuits and Computers) started October 1980 in New York) which brings together systems-type professionals and technologists.

4. The New Interface and Compatibility Concepts

The term "interface" also rises the question for compatibility. The cooperation of the IZ at Xerox PARC with its designer community is based on an interface standard. There are, speaking more exactly, two standards, one directed towards the IZ user, and one to interface mask making and packaging. The user interface includes the specification of a standard layout data format, a hand shaking protocol for remote entry of design files, and, design rules for compatibility. The manufacturer interface of the IZ is the so-called "starting frame" which specifies standard scribe lines, alignment marks, test patterns, and, test circuits, as well as a packaging and pinning standard. The following parts of this paper give some details on both interfaces. The flow chart in fig. 4 illustrates the interface functions of the IZ at Xerox PARC (also see /BC80/, or, /HA80/).

4.1 The Implementation Guide

A part of the cleaned-up designer/manufacturer interface is the "implementation guide" (e.g. /HS80/), or, "design handbook". It is a documentation of the designer/manufacturer interface, so that no more personal interaction is required to prepare the layout design data format. It may also include some instructional parts, such as e.g. the "VLSI implementation guide" /HS80/ from the IZ at Xerox PARC. This guide includes a cell library, a few design examples together with documentation examples, and, also some introduction on mask making and fabrication. All this is organized in a way, that also users with limited background in technology are enabled to start design practise. The replacement of personal designer/manufacturer interaction by a design guide requires the use of a fabrication process which is frozen, or, conservative design rules, to achieve some insensibility towards minor process parameter modifications. The guide from Xerox PARC also includes the description of a layout CAD algorithm. (Its listing in a portable language, such as e. g. PASCAL, would be desirable.) Its purpose is described later.

4.2 Design Rules for Compatibility

In a design environment for production of chips in small quantities, such as in research and training, the optimization of chip area is not always a primary goal. In such cases it is not necessary to use to the critical design rules which are optimized for the individual parameters of a particular fabrication process. That's why a second set of more relaxed "conservative" design rules may be tolerated. For example, critical distances between areas could be avoided by adding some safety factor to it. Such more relaxed design rules could be elaborated in a way, that the designs could be submitted successfully to several fab line with different process parameters. This means compatibility of the design. This form of insensitivity is also a protection from parameter modifications of the same fab line, so that updating of the design handbook can be avoided.

Another form of compatibility of layout designs is achieved by technology-relative design rule specification. This is done by specification not in absolute dimension, such as in microns, but relative to some scaling factor. The scaling factor used in /MC80/ is called "lambda", and each dimension is expressed in multiples of lambda. This method allows to use the same design file for differently scaled technologies. For example, when lambda is 2.5, a five micron technology is produced, or, if it is 2.0, a four micron minimum channel length and line width is assumed. However, for such a scalability of design rules it is required, that the bonding pads are somehow larger than usual, so that after reduction of the scaling factor still the minimum pad size for reliable bonding is reached.

4.3 Intermediate Data Format for Compatibility

Pattern generators are numerically controlled devices which accept a particular layout data format on the mask tape. The problem is, that there are several PG manufacturers using different data formats. That's why often for submission of design files to different manufacturers, or, to different mask makers, also different data formats have to be generated because of incompatibility. To solve this problem, a standard input language is required to submit designs to the IZ at Xerox PARC. It is CIF (Caltech Intermediate Form) serving as a standard intermediate data format. It is finally converted by the IZ into the PG format needed. The IZ user does not have to bother about this.

This CIF language is similar to a PG input language, so that conversion into various PG formats is easily programmed. However, it has some additional features usually not found in PG input forms. One is a "symbol" definition and call mechanism similar to macro

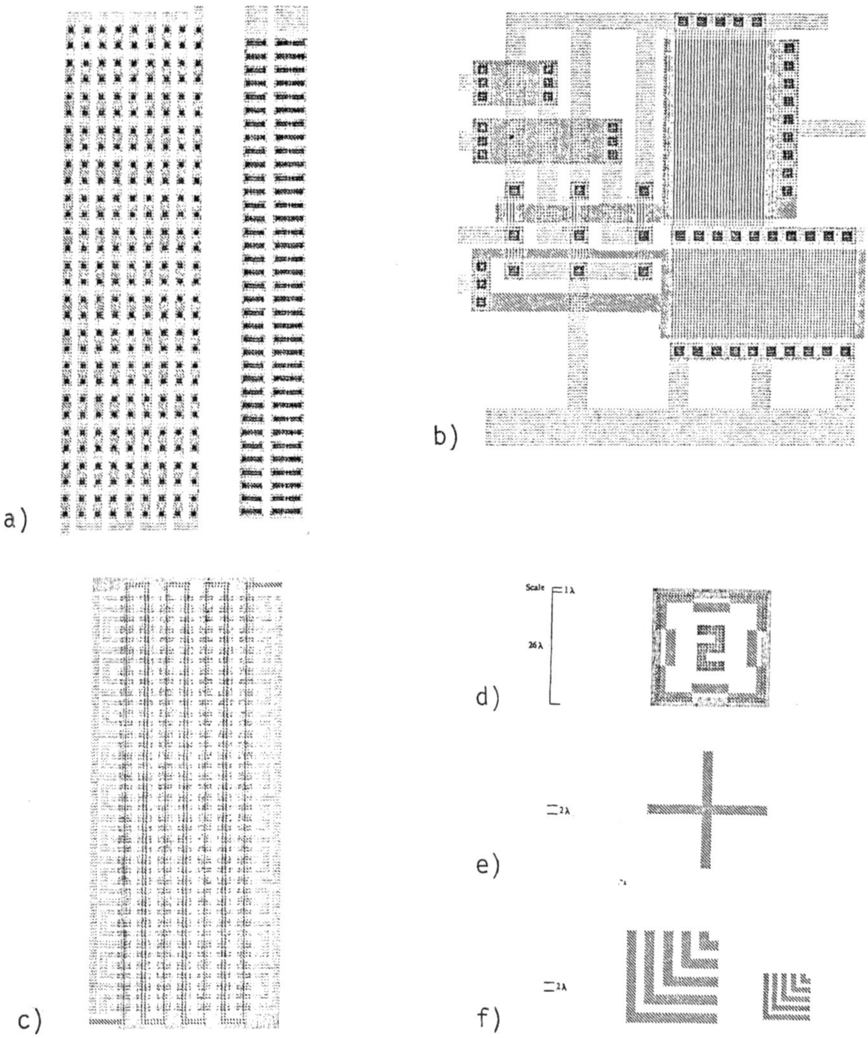

Fig. 5

concepts which allows repetitive placement of layout modules. This mechanism causes a high degree of data compaction, when the designs are higly regular. Also rotation and tranposition mechanisms are included in CIF. For a more detailed CIF description see /SQ80/ in this volume, or /MC80/.

Let us also look in the future, to see how to overcome limits. Since anyway a conversion from CIF into a PG form is required, at the same time also other adaptations could be computed on a CIF file. For instance, a compaction algorithm could create a more

Fig. 6

critical layout to save chip area. The compaction algorithm could be adapted to a particular fab line and its individual process parameters. When all this processing is done by the central IZ, the IZ user does not have to bother about it.

It is interesting to know, that meanwhile in the US a standards committee on layout languages has been set up. CIF is a candidate to become standard, since it is already widely used by about two douzens of Universities. Also one mask maker company in silicon valley announced, that also CIF code will be accepted soon from their customers.

4.4 The "Starting Frame" to Support Compatibility

Also compatibility on the manufacturer side of an IZ is desirable to reduce expenses. One feature is, that a standard set of test circuits and test patterns and a standard placement of uniform alignment marks is used by the IZ. The advantages are, that the IZ user has no responsibility and thus no problems with these important parts of the chip. Since this confectioning of the chip layout is centralized, a standard format is guaranteed. So the

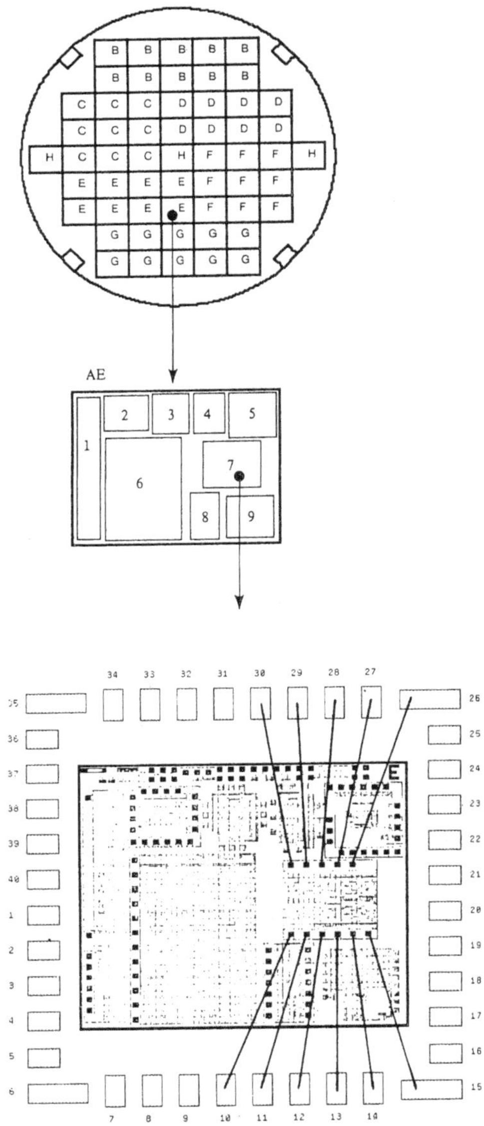

Fig. 7

Fig. 8

manufacturer may always use the same set of needles to test the various circuits. Also the alignment set-up is much more economic because of uniform placement of marks. More details on the confectioning used by the IZ at Xerox PARC are described in /HS80/.

The test circuits and their extra pads, and, alignment marks and test patterns are visible near the upper edge of the chip in fig. 6. Fig. 5 shows a number of test circuits and other patterns of the starting frame, such as (a) contact hole tester, (b) test transistor, (c) layer tester, (d) alignment mark, (e) critical dimension cross, (f) etch test patterns, (not shown:) shift register and ring oscillator.

4.5 Shared Culture by Compatibility

The compatibility which is given by the common CIF format, the scalability of design rules, and, by the use of a common cell library, is the basis of a "shared culture". In a similar way, as programming languages (in computer user clubs also assembly languages) are the basis of some "software subculture", the compatibility keeps the IZ user community together. Designs or cells may be exchanged between IZ users. The same circuit may be plotted on different equipment at different places which join the culture. The designs may be processed on different software. Also the exchange of software is made possible by this compatibility. All this facilitates, that many places and persons may share in this culture.

5. Reduction of Design Cost

A number of compatibility features mentioned above have at the same time the effect of reducing fabication cost. This has been achieved, partly by the mask making process, and, partly by the services of an implementation center. As a result of these measures also the fabrication of very small quantities of chips has become feasible. This fact, and the availability of a number of services by the IT could be utilized to reduce design cost considerably. These means for design cost reduction will be explained in the following sections. Also some proposals for future research on IZ services and their influence on design cost will be discussed.

5.1 Inexpensive CAD Equipment for Start-Up

The implementation guide by Xerox PARC (see above) includes the description of a simple layout CAD algorithm. It is a help to the beginner to set up his own small CAD software on a computer in his environment. At least two members of the MPC' community have implemented this algorithm in PASCAL and have given it away to other members for free. Due to its portability this algorithm is running on the computers which are available anyway in universities. The only additional equipment required for a start-up of check plot facilities for the Mead/Conway course was a small four color plotter, and, maybe also a controller for it.

This means,that in a university environment the start-up investment for exercising LSI design has the order of magnitude of 10 000 to 20 000 US dollars, provided, the services of an IZ are available. At the University of Kaiserslautern, for example, a 9872A Hewlett Packard Plotter has been purchased for less than 10 000 Dollars. A Hewlett Packard 13290A intelligent terminal, which was already available from a microprocessor interfacing lab is used as a controller. For plotting larger layout designs, such as e.g. for research, the plot is patched together.

Another proposal, made by Prof. Francois ANCEAU, University of Grenoble, is, to run the CAD package on a PASCAL micro engine. This would mean, that with a start-up investment of roughly 20 000 Dollars teaching the LSI design courss may be started in an environment, where no computer center is accessible. Since the cost of CAD equipment adds somehow to design cost, the use of the equipment described here yields a substantial reduction of design cost.

5.2 Sharing in, and Reduction of, Cost of Mask Making

The production cost of chip prototype samples adds to design cost. As has been said above the cost of traditional mask making is too high for fabrication of sample chips in small quantities. One method is the use of the master plate directly which reduces the number of steps per layout layer from 3 to 2. A better way is used for submissions of the MPC' project to the IZ at Xerox PARC. By means of an electron beam pattern generator ("E-beam") the working plate is produced directly, i. e. in a single step, since its high resolving power does not require the 10:1 intermediate reticle. This reduces the number of steps from three to one. Sometimes the pattern is even written directly on the wafer.

Since the step-and-repeat operation is done by the E-beam directly, also different patterns may be written on a plate. So there may be the masks of several different types of chips on one maks, so that different types of chips may be produced on the same wafer. For example, 10 different types of chips, with 16 mask pattern repetitions of each, may be placed on one wafer's mask. The upper parts of figures 7 and 8 show an example of such a multichip wafer, and the placement of some copies of each chip. So 10 designers could share in the mask and fabrication cost of one production charge. Such "multi chip wafers" can only be produced by using an E-beam, because on a step-and-repeat camera the alignment problem would be to difficult.

Especially, when student exercises are submitted, the required chip size is rather small with less than 4 square millimeters in most cases. Since an E-beam may process chip sizes up to 8 by 8 millimeters or more, several small circuits could be placed on one chip. Such "multiproject chips" are prepared by the IZ at Xerox PARC, where sometimes more than 10 circuits are placed on one chip. Fig. 6 shows an example (from Stanford University) of such a multiproject chip. Each project circuit has its own bonding pads, and the bonding process decides, whose designer's circuit from the chip will be finally connected (e.g. see middle and lower part of figures 7 and 8). If, for example, chips with an average of 10 circuits are placed on a 10-chip wafer. The share of one designer would be one hundredth of the mask and fabrication cost.

So the combination of multiproject chips on multichip wafers is a considerable reduction of fabrication cost, although an E-beam operation is more expensive than that of optical PG equipment. Another advantage is given by the use of multiproject chips. By careful selection of circuits a standard chip size can be kept without substantial waste of chip area. So it is no problem to use a standard scribe line arrangment, which is a further rationalization factor. The multiproject chip assembly is done on a computer of the IZ. For more details on multiproject chips on multichip wafers see /BC80/.

5.3 Cheap Samples and Fast Turn-around Time

It has been shown above, that a different way of mask making reduces fabrication cost by more than an order of magnitude. It has been reported, that the fabrication cost for a few samples of a project is less than 100 US dollars. Because of fully automatic acceptance of user design files by the IZ and the standardized starting frame the turn-around time of fabrication could be reduced. Turn-around times of 4 weeks from design file submission via ARPAnet to mailing and delivery of the packaged chips have been reported.

These two features, cheap sample fabrication, and fast turn-around may influence the methods of design verification. For instance, a cell to be used iteratively, could be tested by separate submission of a test circuit which contains only that cell, and, may be, some additional circuitry just for testing purposes. This may have advantages over simulation of that cell. However, there is almost no experience about the design success of this verification method. The question is, wether good designs could also be worked out without using any simulator. This would mean, that no simulator experience is needed by the designer.

5.4 Centralized Design File Checking

Another service by the IZ currently running is the checking of the design files submitted. Currently the design file, to be submitted in CIF format, is checked for format errors. In case of errors, a message is given back via the ARPAnet to the submittee. Of course, submissions for a multiproject chip are also checked with respect to their size and number of pads etc.

A further service possible by the computer facilities of an IZ, could be a design rule checking service, the output of which could be a check plot with inserted diagnostic indications. In case of remote entry facilities it could be transmitted back via the computer network. This would be of help to the designer and could possibly reduce design cost.

5.5 Efficient Check Plot Inspection

By experience within the MPC' project a number of effective methods of visual check plot inspection have been found out. Visual check plot inspection can be very efficient due to the superior pattern recognition abilities of the human being. A surprisingly high percentage of design rule violations and other errors may be found by visual inspection of check plots. This is an advantage of chipware debugging over software debugging, where no visual pattern recognition methods are used. The efficiency of visual inspection can be further improved by the use of optimal check plot representations.

Very often contiguous areas on a layout are synthesized from a number of rectangles. This is a reason for difficulties to recognize the real shape of these areas. By Carlo SEQUIN it has been found out, that check plots are much better readable, when the entire area has a contiguous boundary line and the area is not interrupted by lines dividing it up. If a special algorithm converts the mask layout into a special plot layout of this kind the plot's readability for inspection is improved substantially.

Besides it is very efficient for visual inspection, not to plot all layout layers on the same plot. Inspection is more easy, when only a subset of layers is on the plot, and, several separate plots are used (e.g. see /HS80/ for detailed proposal). For example, the detection of bad metal coverage over contact holes a plot with just only the metal and cut layers would be more effective than the entire plot of all layers. To check critical distances a plot with only diffusion and poly layer (and, may be, also the ion implant layer) is better than the plot of all layers.

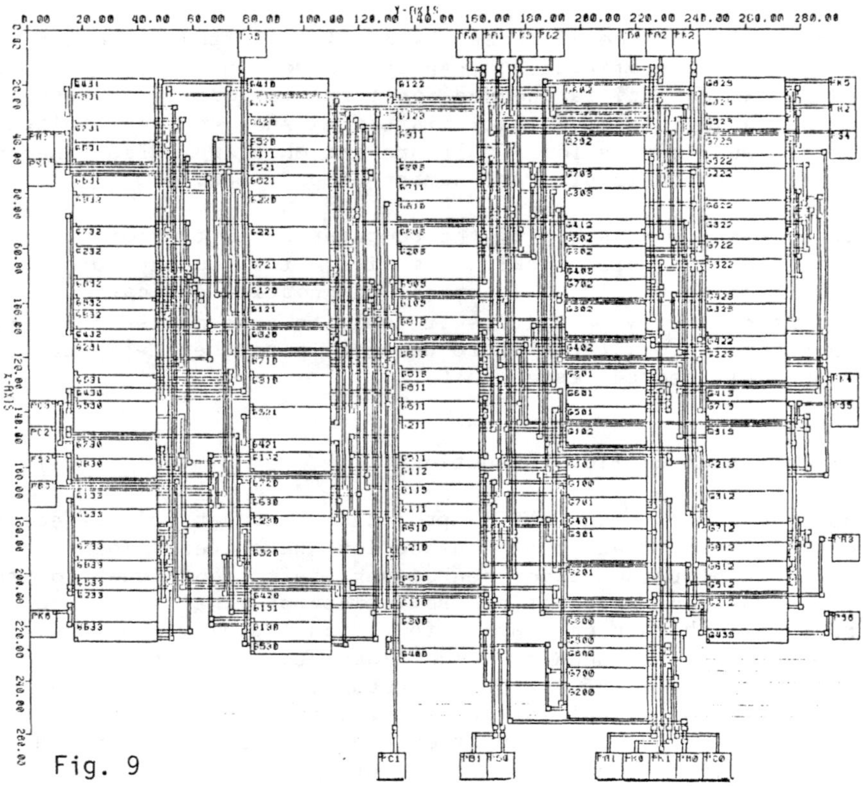

Fig. 9

This illustrates, that improved check plots enhance design error detection. This is a means to reduce the number of bugs in a design substantially, without using any computer aids, such as design rule check algorithms. By the way, design rule checkers are (due to the current state of the art in practice) not that effective, as would be desirable. It is reported that hours of CPU time are needed, or, no 100% detection is achieved, depending on the particular implementation. As a consequence, visual detection of design rule violations is a means to save design cost.

5.6 The Influence of Structured Design on Design Quality

Structured hardware design (also see /HR80/) is a means, not only to reduce design time (see above), but also to reduce the error rate. Let us look at an example of an array of "snap-in" cells. ("snap-in" means, that the connections to neighbour cells are bound automatically by mere placement of the cells.) If, for instance, an array of 16 by 16 cells is designed, only one out of the 256 copies of the cell has to be verified. When this cell has

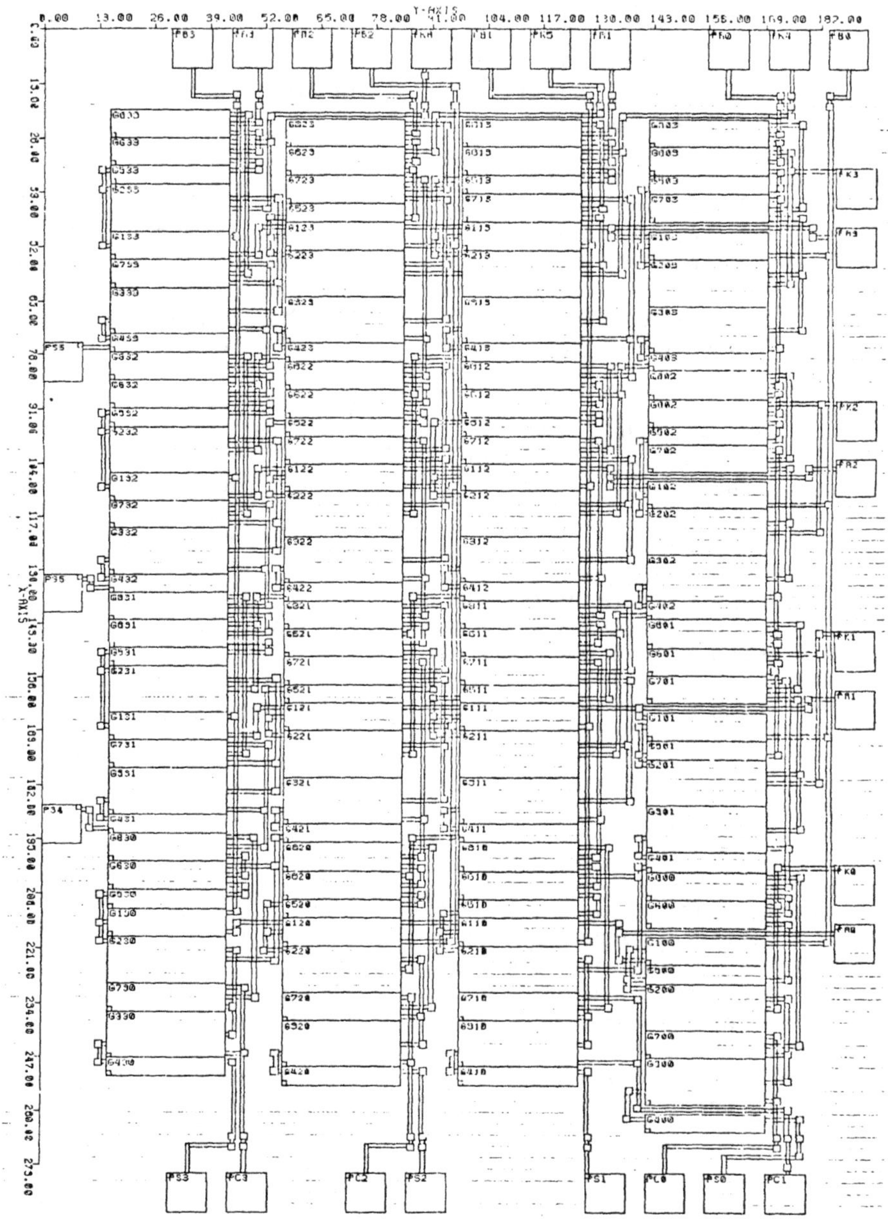

Fig. 10

been very carefully checked and simulated, also with respect to the topology and geometry to its connection to neighbours within the array, the correctness of the entire array is very likely.

This has demonstrated that structured design is a means to reduce the error rate, and at the same time, a means to reduce simulation efforts. By the way, structured design manually can produce designs of higher quality, than routing algorithms can do. This will be demonstrated by the following example of a 4-by-4 bit integer multiplier design exercise. Fig. 9 shows the result plot of a well-known routing program (commercially available). (The cell library includes 2-input NAND, inverter, and EXNOR.) At this first attempt the placement has been generated by the algorithm. The routing channels needed are up to 13 lines wide. Fig. 10 shows the result of a second run, where the placement had been specified by a human being who knows the structuredness of circuit. Chip size has been reduced to about 60% of the original size.

Fig. 11 shows a manual layout using structured design methods. This design is an output of the exercises of a course on "Introduction to VLSI systems", held last winter in Kaiserslautern. Instead of 3 logic level cells a 1-by-1 bit multiplier snap-in cell has been designed and optimized (design time: two days). The middle of fig. 11 shows the 4-by-4 array, where 16 copies of one special cell replace 64 standard cells in the routing program apprach (Fig. 9 and 10). The peripheral wiring layout, such as for connection of the pads and placement of buffers (taken from the cell library in /HS80/), has been carried out by hand. The result is a chip with only about 10% the size of the original attempt (fig. 9).

6. Other Centralization Proposals

It has been discussed above, how an IZ as an interface between designer and manufacturer can decrease design cost considerably. The interfacing is organized as a centralized service. Also the organization of other services, which have no interfacing functions, or, which are not primarily interface functions, can be implemented economically. Since the IZ with its powerful computer facilities are needed anyway in such a "shared culture", also other centralized services could be included without adding too much IZ operating cost. Plotting and design rule checking had been proposed above.

To keep the IZ subscriber's equipment cost and man power requirements low, also a remote entry simulation service would be an interesting feature. In Switzerland, for example, the organization of a governement-supported simulation center is discussed. This is an indication of demand for such centralized simulation facilities. An IZ, like that of the MPC' project, would allow to attach such a service to the existing organization. This would be probably much more economic, than having a separate simulation service.

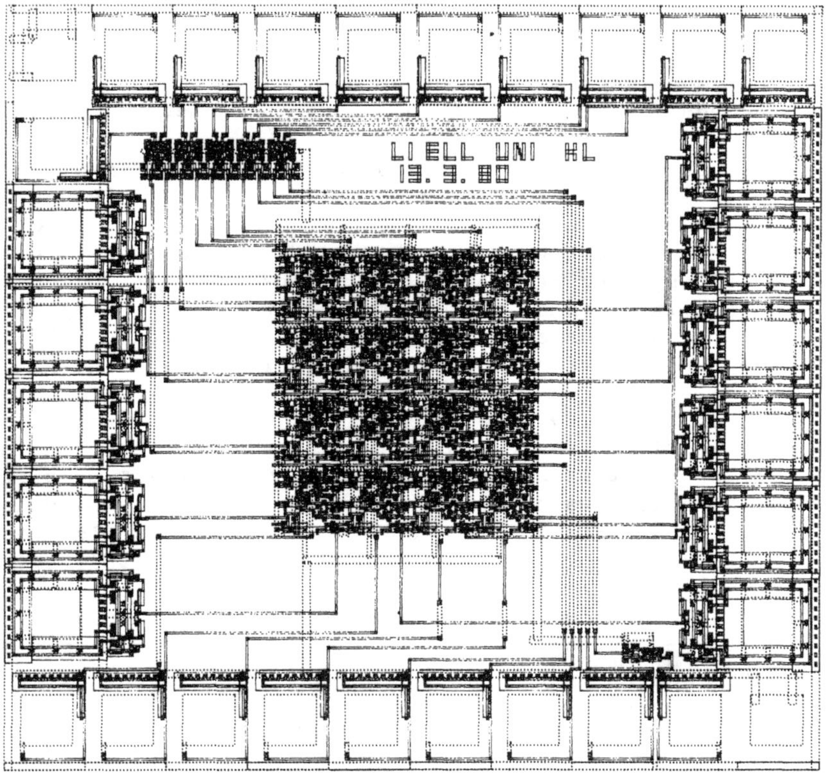

Fig. 11

6.1 A Circuit Level Simulation Service Proposal

A further advantage of the combination of IZ and simulation service stems from compatibility considerations. The standard input format for the IZ could also be the input to the simulator. The layout data input format (such as e.g. formatted in CIF) is checked anyway. A design rule check should be added as a second step of preprocessing. After these 2 preprocessing steps the probability of simulator input errors would be decreased.

The third preprocessing step in such a combination should be the automatic conversion of the layout specifications to simulator input. There are existing experimental examples of such converters for circuit level simulation. It is devisable to run such conversion algorithms on the IZ, since much processing power is needed for analysis of the inputted geometry.

The process parameters and transistor models required for a circuit level simulation should be provided by the implementation service. These data could be stored in the computer of the IZ. This would be an effective help to the "technologically naive" systems designer using such an IZ, since to get access to realistic transistor model parameters requires a high degree of expertise. Thus a "shared simulation culture" would allow a further substantial effort on the IZ subscriber's side.

Another interesting service would be plotting the circuit diagram of the submitted design. Since geometry data are included in the CIF input, it is not too difficult to implement a circuit level plot algorithm as an extension of the CIF-to-simulator-input converter program. This form of feedback would be an additional debugging aid for inspection by the IZ subscriber.

One problem is, that the stimulation input of the simulator has to be generated by the designer. However, in industry many designers have to provide test patterns anyway (such as e.g. for production quality control). For a IZ subscriber this requirement to provide test input would be also a stimulation to the designer to produce a good design. Since design for testability is very important for highly complex designs, such a requirement would be a good basis to extend the Mead/Conway course concept.

6.2 RT Level Specifications and IZ Use

The use of RT level languages, taken from the field of CHDLs (Computer Hardware Description Languages), as notations for the specification of LSI design problems may reduce the number of bugs in a design. If such a language, such as e.g. CASSANDRE, DDL, ISP, KARL (see /HR80/, /HA77/), or RTS, is implemented, it may be used as an RT level simulator input. Already at syntax analysis time of such a simulation run a number of diagnostics is edited. The consequence is, that a design problem specified in such a language in a syntactically correct way, contains a reduced number of bugs.

A comparizon between two different design approaches has been reported /HA78/, where the same processor has been designed from a narrative specification, and, also - by other designers -

from a KARL specification. In the latter case much less bugs had been introduced, and the design, debugging and prototyping was completed in less than half the time, compared to the other case.

CHDLs allow a very comprehensible, concise, user-oriented LSI design specification. Such descriptions, as e.g. in KARL which is a PASCAL dialect, are also conveniently readable by software-type specialists. That's why such a language should be used for design problem specification. Such a specification could be maintained throughout the design process, to be finally available as part of the product description.

If such a CHDL is used for design problem submission in a custom design environment, it could be used by systems level customers without any LSI design experience for submission of their design problems. The design then has to be processed by the IZ. The 1st step, the conversion into a logic diagram, for example, could be done by qualified employees of the IZ. When the customer has a RT level simulator available, or, when the submitted design is checked by a RT simulator running at the IZ, the amount of manual efforts and the extent of customer/designer interaction will be considerably reduced. So the use of CHDL input may be a means to increase the number of subscribers of a IZ, by adding another type of user to this community.

An automatic LSI synthesis from CHDL input is currently not yet feasible in an IZ environment. Although a number of experimental algorithms are running, which can handle a certain type of digital architectures (such as e.g.data path structures of CPUs), their maturity is not sufficient, to reach the robustness needed for an IZ service or for IZ users.

7. Experiences

The use of a "shared culture", supported by an IZ, has been experienced in a series of MPC' actions: MPC78, MPC79, and MPC580, together with the implementation service at Xerox PARC (also see /BC80/). The next action MPC1280 is planned for the end of 1980. Fig. 12 gives a survey of participating universities, or, at least of those intending to participate. The actions are synchronized with the structure of the academic year. That's why students of LSI design courses produce most of the designs submitted to the IZ. MPC78 was carried out only in connection with Lynn CONWAY's course given at MIT in the winter quarter 1978. MPC79 was carried out in the winter quarter 1979 with participation of 6 US universities. The next action was called MPC580 (which means, that the deadline for submission was in May 1980). Meanwhile the number of universities increased to a douzen (see fig. 12). For the winter quarter 1980 (deadline: december 1980) the MPC1280 action is planned, with probably one and a half douzen US universities.

7.1 Experiences with MPC' Actions

More detailed reports are found in /BC80/ about the MPC79 action, where 82 designs of more than 100 students have been implemented. Experiences /CS80,MF80/ say, that about 20% of the students of a LSI design course finally complete a design project and get a few working chip samples. These students' part time design activities during the subsequent quarter of the academic year take about 2 to 3 months. The design of the regularly structured parts of the projects takes the order of magnitude of one week, whereas the layout of the random parts and the chip peripheral wiring takes most of the student's time.

From one university it has been reported /MF80/, that from 4 circuits (first implementation attempts) one had not yet been tested, two worked satisfactorily, and one had been disappointing with respect to performance. It was the only one which had been simulated before. It turned out, that, in reaction to simulation results, the circuit had been changed. After the chip had been tested, it was found out, that the first version was better. The wrong parameters had been used for simulation.

7.2 Inertial Forces in Europe

Originally also MPC admission of universities outside the US was planned. Two universities in UK and the Kaiserslautern University planned a participation. The CIF data files on magnetic tape rsp. on Hewlett Packard mini cartridges were mailed to a participating university in the US, which was linked to the ARPAnet. However, finally in May 1980 the DOD closed MPC580 for foreign participation, since the IZ and its accessibility has become a political issue.

No official statements about the motivation of this "implementation embargo" have been made. Sure, some small student circuits are not of any strategic significance. The question is, wether protectionistic motivations contributed to this governement decision. Wether or not this was a motivation, this decision stimulates attention within Europe. Also politicians recognize, that such an implementation service may have a heavy impact on future innovations in techology application. It may initialize major structural changes in the electronics industry. It allows to make technology accessible also to very small companies. It has been mentioned above, that the designer gap could be fixed. Also this is an important issue in international competition.

M P C - Multi-University Multiproject-Chip Set Project

participation since:

MPC78	MPC79	MPC580	MPC1280
M.I.T.			
	Caltech		
	Carnegie-Mellon-Univ.		
	Stanford University		
	U. C. Berkeley		
	U. o. Rochester		
		G.W.U. Washington, DC	
		Kaiserslautern Univ.	
		U. o. Bristol	
		U. o. Colorado Springs	
		U. o. Edinburgh	
		U. o. Washington, Seattle	
		Yale University	
			Syracuse University
			U.C. Los Angeles
			U.C. San Diego
			U.o..Illinois
			U. o. Oslo, Norwegen
			U.o. Santa Barbara
			U.o. Santa Clara
			U.o. Southern California
			Washington U., St. Louis

Fig. 12

However, the degree of attention payed to this new innovative development, is different in the various states in Western Europe. A fast reaction of the French governement on the embargo news was a grant to purchase an E-beam PG for IZ purposes to a lab in the CNRS national scientific research center. In Italy an IZ project is discussed between a semiconductor house and a number of universities. In the Netherlands a subcommittee of the parlament proposed an IZ-based design center. In July it has been discussed by the plenum of the parlament. Two technical universities are interested in organizing the center. Also in UK and Denmark similar plans are discussed, also at political level. In Sweden, a grant has been given to the University of Linkoeping to organize some sort of design center. In Norway, the defense department in cooperation with the University of Trondheim is planning an effort. Two compatibility-oriented design handbooks (nMOS and CMOS) have been ordered from a consulting firm in the US. In the Federal Republic of Germany there are discussions going on with the Federal Ministery of Technology. However, unfortunately, the research funding budget of this ministery had been cut down recently.

One problem is, that the driving force of these innovative concepts comes out of a community of systems professionals, apart from the well-established community of technology-oriented professionals. For part of these professionals the proposal is a conflicting topic. Also the communication gap between these two communities is an obstackle to a possible promotion of the idea of "shared cultures". However, the NATO ASI, the proceedings of which you have in hands when reading this paper, has been the place, where the "large scale" dialogue between these two communities has been started.

8. Summary and Conclusions

A survey and out look has been given on principles and impacts of VLSI implementation services, which play the role of centers of "shared cultures", shared by a major community of systems specialists practicing LSI design. Such centers are an important medium to bridge the technology/informatics communications gap and technology transfer gap. An experimental example of such a center and experience by its users has been reported. Also its impact on LSI design instruction and on computer science curricula has been described. A number of proposals for the extension of services of such centers illustrates their significance even further.

Acknowledgements

Figures 4 thru 8 are courtesies of Xerox PARC, thanks to Dr. Lynn CONWAY, manager LSI systems area at Xerox PARC. Let me appreciate the encouragement by Prof. Carlo SEQUIN, University of California at Berkeley, which helped in my decision to extend my activities over to LSI structures. We also appreciate having received nice checkplots of our MPC-submitted designs from Carlo SEQUIN (instead of packaged chips, see section 7.2), which are a demonstration of compatibility.

Literature

/CB80/ Conway, L., Bell, A., Newell, E., "MPC79 - a Large-Scale Demonstration of a New Way to Create Systems in Silicon", LAMBDA, Second Quarter 1980.

/CS80/ Carlo Sequin (University of California at Berkeley): personal communication.

/DH80/ Dieckmann, J., Hartenstein, R. W., "A Local Micro Network to Support Software Modularity", EUROMICRO-Symposium, London, Sept. 1980.

/DK80/ Dieckmann, J., Hartenstein, R. W., Konrad, W., "The ESRA Network Kit for Hardware-Aided Support of Software-Reliability", ICCC, Int'l. Conf. on Computers and Circuits, New York, Oct. 1980.

/HA77/ Hartenstein, R., W., "Fundamentals of Structured Hardware Design", North Holland Publishing Co., Amsterdam/New York, 1977.

/HA78/ Anderson, P. A., Nygard, M., "A Study of Different Languages and Related Tools for Description and Simulation of Digital Systems", Report 6/78, University of Trondheim, Div. o. Computer Sc., Trondheim, Norway, 1978.

/HA79/ Hartenstein, R. W., "LSI Chip Design: From Evolution to Revolution ?", (invited paper: Int'l Symp. on Microcomp. and Microproc. Application, Budapest, Oct. 17-19, 1979),. Report 12/79, FB Informatik, Universitaet Kaiserslautern, 1979.

/HA80/ Hartenstein, R. W., "VLSI-Bausteine in geringen Stückzahlen", Elektronische Rechenanlagen 22, 4 (Aug. 1980)

/HK75/ Hartenstein R. W., Koch, G., "The Universal Bus Considered Harmful", in /HZ75/.

/HP79/ Hartenstein, R. W., von Puttkamer, E., "KARL - A Hardware Description Language as Part of a CAD Tool for VLSI", Proc. Int'l. Conf. on Computer Hardware Description Languages and their Applications, Palo Alto, Oct. 1979.

/HR80/ Hartenstein, R., "Basics of Structured Design Methodologies: Data Path and Finite State Machines", (see in this volume).

/HS80/ Hon, R., Sequin, C., "VLSI implementation guide", Xerox PARC, Palo Alto, 1980.

/HZ75/ Hartenstein, R. W., Zaks, R., "Microarchitecture of Computer Systems", North Holland Publ. Co., Amsterdam/ New York, 1977

/ME79/ Mead, C. A., "VLSI and Technological Innovation", Proc. Caltech Conf.on VLSI, California Institute of Technology, Pasadena, Cal., 1979.

/MC80/ Mead, C. A., Conway, L., "Introduction to VLSI Systems", Addison Wesley, Reading, Mass., 1980.

/MF80/ Mike Foster (Carnegie-Mellon University): personal communication.

/SE80/ Sequin, C., "Instruction in MOS LSI Systems Design", COMPUTER 13, 3 (march 1980).

/SQ80/ Sequin, C., "Layout, the Caltech Intermediate Form, and sticks formats" (in this volume)

STANDARD INTERCHANGE FORMATS FOR INTEGRATED CIRCUIT DESIGN

Carlo H. Séquin

Computer Science Division
Electrical Engineering and Computer Sciences
University of California, Berkeley, CA 94720

ABSTRACT. *Designing very large scale integrated circuits requires interaction with many software tools. These tools can only be employed efficiently if they work from one and the same representation of the partially completed design. Standard formats for such an intermediate representation are discussed.*

1. INTRODUCTION

Babylonian confusion reigns in the area of design and verification tools. Layout languages, simulators, design rule checkers and so on are developed at many institutions. Often these programs are strongly influenced by the local computing equipment and by the habits of the programmer. This results in a hodgepodge of software modules, all using different languages and different input and output formats. In the process of designing an IC, it is thus often necessary to translate the description from one representation into another, in order to make use of different tools. Since nobody has the time to develop the many possible translation programs between all these formats, this translation must often be done by hand. This leads to errors, or at least to uncertainties whether two different representations describe the same physical circuit. Clearly, the existence of a standard format is not just a luxury, but an urgently required necessity to permit designers to cope with VLSI designs.

2. MASK GEOMETRY DESCRIPTION

Some progress has recently been made in the standardization of the formats used to describe the geometry of the masks needed in the fabrication of the IC's. Based on some preliminary work at Caltech in 1977, Sproull, Lyon and others [Sproull 80] have defined an intermediate format to link together the many possible input and output devices. This language, called CIF, or Caltech Intermediate Form, serves as the standard intermediate representation between the input devices used by designers to express their layouts and the machines producing masks or checkplots (Fig. 1). The existence of such a standard guarantees that only one interface module per device need to be developed, rather than n*m translation programs between n input and m output devices.

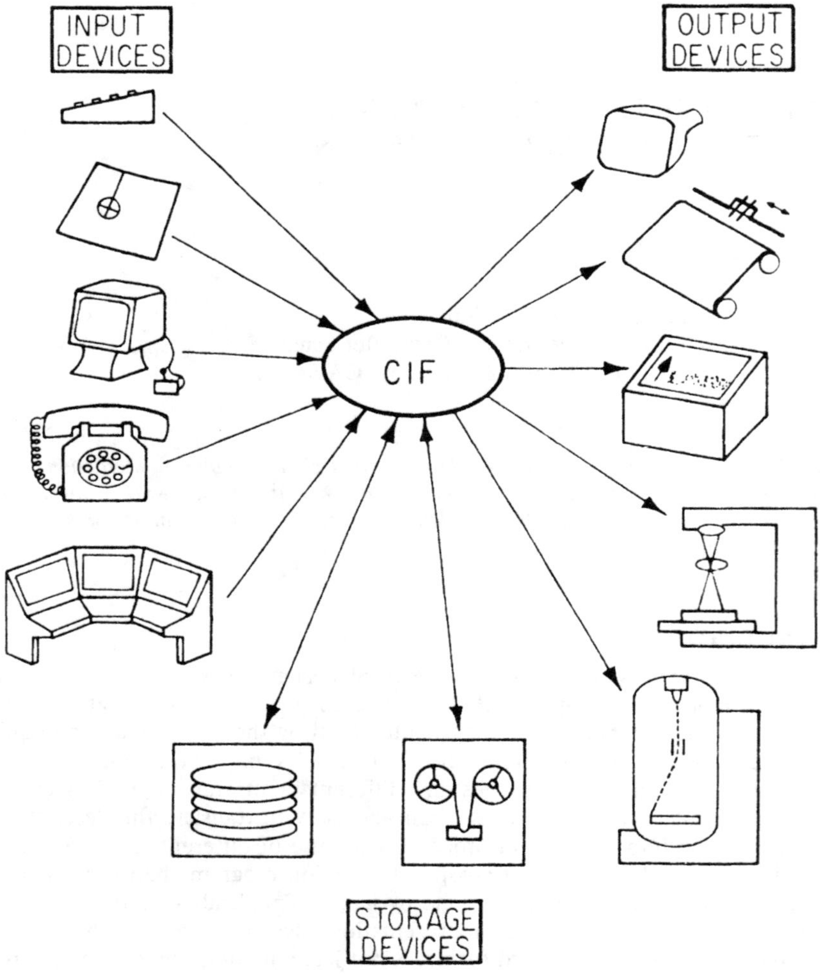

Figure 1. The role of CIF2.0 as the intermediate link between input and output devices.

2.1. The Caltech Intermediate Form

A definition of CIF2.0, and a description of its syntax and of some semantic aspects can be found in [Sproull 80], and no attempt will thus be made to repeat such a description at this point. Suffice it then to say, that CIF is an ASCII file describing the graphic items from which masks for IC fabrication are composed. Literally every geometric object in the layout is specified with a precision of 1/100 micron. To reduce the length of the description, congruent groupings of geometric shapes can be lumped together into macros, called *symbols*, which can then be called to appear in different locations, possibly mirrored and rotated. The geometric primitives of the language are: rectangular *boxes*, *round flashes*, general *polygons*, and *wires* of uniform width. The CIF file also specifies on which mask level each geometric item is to appear.

While fraught with some flaws, to be discussed later, CIF has already proven its worth to the design communities at many universities. In particular, CIF was crucial to the success of the multi-project, multi-designer chip sets coordinated by Xerox PARC during 1979 and 1980. Several designs, done by many contributors at different universities, were submitted as CIF files to the coordinators at Xerox, either via the ARPAnet or even over ordinary telephone lines. These files were then properly merged and converted to the format required by the mask house. Conversion programs from CIF to many output and plotting devices do exist and are available from university sources. Masks generated optically or with an electron beam exposure system have been made from CIF files.

Since CIF was not intended to be a design language, it is rather cumbersome, even though possible, to input a particular design by directly writing the corresponding CIF code. Many different front ends have thus been developed to make this task less menial, and to produce a more humane interface to the computer. For instance, some students faced with the prospect of having to type hundred of lines of CIF code, have written a simple front end macro processor to accept a LISP based file that describes the layout in a higher-level and much more flexible format. Another approach resulted in the interactive input program MKCIF [Krause 79], which accepts mnemonic names, relative coordinates, and has many of the desirable features that are missing from CIF2.0. It also prompts the user for the next entry, in case a line contains an incomplete CIF command. Another approach is to construct a higher-level layout language by extending an existing programming language; an example is ICL [Stone 78] built around Simula. None of these programs are of course a substitute for a proper design system based on an interactive graphics editor.

2.2. Limitations of CIF2.0

Some limitations in CIF show up when many projects from different places need to be merged. The lack of nested symbol definition prohibits clean scoping for the names used by different designers. Collision of the same symbol numbers (CIF does not permit real names) referring to different subcircuits can thus result. Therefore the "DD"-command has been introduced. It permits to remove a whole set of symbol definitions, so that the same numbers can be used again by a different designer. Individual symbol numbers can also be reused by simply overwriting the old definitions.

Unfortunately, this way of handling assignments of symbol numbers leads to many problems when CIF is not strictly interpreted in real time, as its original intended usage would prescribe. Rather than outputting each feature immediately, as soon as the corresponding command is encountered, CIF is often used with a raster-scanned output, where all features need to be sorted with respect to their x-coordinate (in the direction of the slow scan) before anything can be put out to the display, printer or mask maker. In this case all references to "old names" must be remembered, and all calls must specify to which version of a particular name they refer. These problems can be alleviated to a large extent if the redefinition capability offered in CIF is not used at all. When different designs are merged, all definitions and calls could be modified with an identifier of the designer, in order to end up with unique names (numbers).

Another shortcoming of CIF is the lack of an iterative command that permits one to specify in a single statement a one or two dimensional array of identical cells. An explicit statement of such iteration is important for structured design rule checking and for improved efficiency in the preparation of electron beam masks, where explicitly repeated patterns could be duplicated more efficiently. The problems with precision and rounding of the coordinates in array specifications, which historically have led to the omission of such an array command, are not more serious than those encountered with nested symbol calls and rotation.

The use of symbol numbers, rather than mnemonic names, is another unnecessary restriction. It is a consequence of the original goal to make the programs working on CIF as simple as possible. Symbol numbers were intended to be used directly as indexes into the array containing the symbol table. However the overhead of using a hashing scheme on symbol *names* is well worth the additional flexibility it provides for the usage of mnemonic or hierarchically structured names. Some users have introduced a way around this particular limitation by adding the names in comment form after the symbol definition command. A special user command in the CIF software can make use of this additional information.

In spite of its flaws, CIF has had a strong beneficial impact on the university community with respect to sharing design ideas and software tools. Certainly CIF2.0 is better than no standardized description at all. Some mask houses have recently started to show interest in such a standard, and may, in the not too distant future, actually provide their machines with interfaces capable of accepting CIF files.

3. STICKS-LEVEL STANDARD FORM

For the actual design process a description of a circuit at the CIF level is inadequate; too much of the intent of the designer is lost, once the representation is reduced to a description of the mask geometry. A higher level representation is needed which also provides a good interface to all the design tools that operate at that level and that are frequently used during the design phase. Such a Structured Interchange Format, called STIF, is currently being developed at University of California, Berkeley. The basic objectives and main features of this language will be discussed in the following. As an appendix a first tentative description of this language, STIF 1.0, is presented. It is being used at U.C. Berkeley during the summer of 1980 to gain experience with the implementation of such a standardized database for IC's and with the development of a cohesive set of tools working from such a format.

STIF 1.0 represents a first step towards a standardized, portable database containing all relevant information about a fully or partially designed integrated circuit. The parts of the language that will be defined in a first iteration address primarily issues related to layout geometry, logical and physical interconnectivity, and hierarchical representation and protection of cells. Hooks are left for future extensions of the language permitting to include with each cell suitable models for simulation, information on history and status of the cell, and, possibly, even relevant test patterns.

3.1. Need for a Standard Format

The exponential growth of single-chip circuit complexity forces us to rely more and more on the computer to keep track of all aspects of a design of an integrated circuit or system. Many tools already exist, and a whole host of new ones is under development, that read or operate upon descriptions of integrated circuits. While these tools internally may require significantly different data structures for efficient manipulation and perusal of the information, the input and output formats of these programs need not be equally varied. Just as CIF2.0 provides an interchange format among a large number of input, output and storage devices, that deal with geometrical mask information, a standardized portable database is required to capture all relevant information about an IC design.

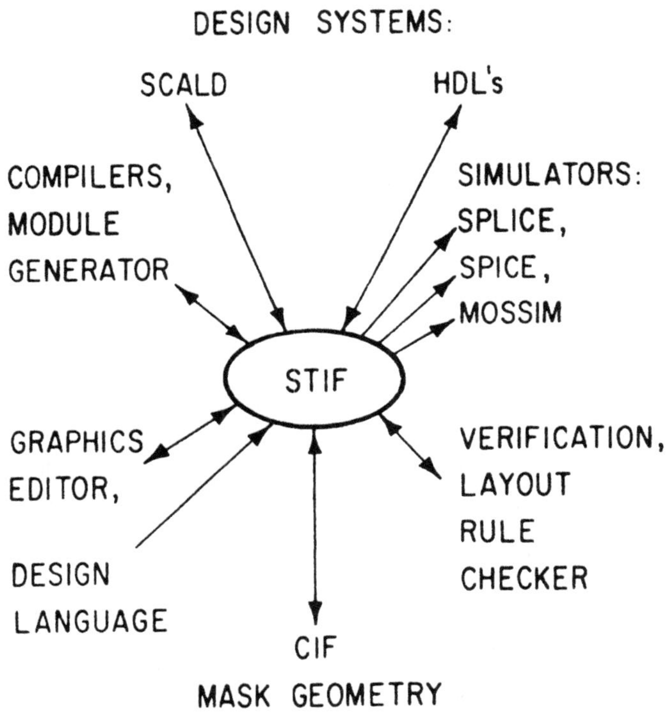

Figure 2. The role of STIF as a link between various software tools working on the description of an integrated circuit.

As shown in Figure 2, such a database can then act as a STRUCTURED INTERCHANGE FORMAT (STIF) between tools such as interactive graphic "sticks" [Williams 77] editors, PLA generators, layout compactors [Hsueh 79], logic verifier, and circuit simulators [Nagel 75]. A common, agreed-upon standard would not only permit the joint designs of very large systems and sharing of cell libraries, but would also facilitate sharing the software of the various new tools being currently developed in many institutions.

By necessity STIF 1.0 will be strongly influenced by the assumptions we make about the kind of tools that will operate upon the designs of integrated circuits; as we gain more experience with such tools, the demands on STIF will probably change. On the other hand, some interchange format is required in order to obtain a cohesive set of tools that can work on the various aspects of an IC design. Obviously a few iterations will be necessary before a format is found that supports the required tools most effectively.

3.2. Intended usage of STIF

Syntax and semantics of STIF must be defined in such a manner that a STIF description can be used to specify completely and unambiguously the layout of an integrated circuit or system, containing the exact routing of all interconnects. At the same time it must be possible to extract a higher-level, more abstract view of a design, to permit the designer to work in an uncluttered representation, e.g. a "sticks" format, that allows him to focus on the important topological issues. This sticks representation is probably also the most suitable format for the original creation of a STIF file with an interactive graphics editor. During input the actual geometry stored with the file would normally be loose enough, or would be expanded if necessary, so that no design rule violations result. This layout geometry can then be compacted in a layout compactor such as CABBAGE [Hsueh 79], given a suitable set of layout rules, in order to produce the geometry of a viable layout for implementation. The representation of the compacted geometry is formally the same as the original one, and it can thus be displayed again either in the "sticks"-like representation or "fleshed-out" as it would appear on the mask set. The choice depends entirely on the taste of the user and the purpose of the display.

The binding of suitable simulation models to each cell definition permits a ready interface to simulation tools of various degrees of accuracy or abstraction. As a minimum, this information has to classify the cell (e.g. mosfet, norgate, flipflop) in a way that the chosen simulator can understand, and must specify how the logical connections of the model correspond to the formal terminals of the defined cell. Since all interconnections in STIF are explicit, simulation of the circuit is possible without explicit instantiation of the layout geometry.

STIF may prove particularly useful in the design phase, since it is quite possible to supply files that are only partly specified to various tools that work on STIF files. For instance, logic simulation at the higher levels could proceed long before all subcircuits have been specified in terms of actual transistors, provided the designer defines some tentative models about the desired behavior of the lower-level functional blocks. Similarly, circuit simulation could be carried out before the layout has been fully defined, if one is willing to consider the formal interconnects between nodes as ideal equipotential regions. Alternatively, one can certainly do checkplots and inspect for layout rule violations without having complete simulation models pinned to all cells.

With these various representations combined into a single data-base (layout, logical connectivity, circuit schematics ...), consistency is not automatically guaranteed. Consistency can be achieved in various ways: Only structured access to the database through a well defined interface is permitted, and all redundant information is automatically generated by proven tools. Alternatively, individual representations can be entered individually, but each cell definition needs to pass through a *verification tool* in order to get *certified.* If any change is made to that cell, or one of its components, the cell gets automatically de-certified and needs to be verified again.

While this may appear cumbersome, it is certainly more efficient to do this certification and to store redundant information with each cell than to regenerate this information (e.g. a circuit model by circuit extraction) every time a simulation is run.

In summary, STIF may be used to describe an evolving design through various steps of successive refinement. A register-transfer-level or circuit-level description can be turned into a "sticks"-like form with the addition of explicit routing information for the interconnects. This loose, "topological" description can then be converted to an actual layout through the addition of information about the design rules of a particular process and the use of a layout compactor.

3.3. Specific Objectives for STIF 1.0

The ultimate goal for STIF is to become a standardized portable database for IC design. While this goal cannot be achieved in one step, it should constantly be kept in mind so that all preliminary steps are aimed in the proper direction. STIF 1.0 is a first tentative subset of such a database, and, as such, deliberately omits some aspects of an all-encompassing IC design system; notably there is only a token statement for information aimed at various simulators, and information for testing is not even given such a place holder. Some of the specific considerations that lead to the present format of STIF 1.0 are listed in the following paragraphs:

STIF 1.0 contains geometrical primitives (rectangles, polygons ...) so that the details of an actual mask layout can be specified unambiguously.

A simple syntax and a terse and compact format have been chosen to reduce both the complexity of the parsing process as well as storage space and file transmission time.

A textfile format based on the ASCII character set is being used for the purpose of portability. This property also makes it possible for humans to read and debug a STIF file directly, which will be quite necessary in a University environment, at least in the near future, while more sophisticated tools for interaction with STIF are still under development.

STIF 1.0 has a hierarchical structure with proper scoping for definitions, names, and parameters, which permits cooperative designs by several groups. Different designers can share the same cells where necessary, but their individual designs can be shielded from interference by other designs.

As much as possible of the intent of the designer is preserved in a STIF description, in particular the hierarchical structure, the iterative nature of regular arrays, logical connectivity, and models specifying the behavior of each particular cell. This information is hierarchically integrated with each cell definition, rather than distributed through many disjoint files.

A strong orientation towards hardware description systems such as SCALD [McWilliams 78], simulators such as SPICE2 [Nagel 75] or SPLICE [Newton 78], and the mask description format CIF2.0 [Sproull 80] has been built into the language to facilitate interfacing to these already existing and widely accepted tools.

4. CONCLUSIONS

A tentative proposal for a first step towards a standardized database for integrated circuits is presented in the Appendix. Obviously STIF 1.0 can only constitute a "working standard", since more experience with its implementation and behavior in an actual design environment needs to be gained. The language definition is thus likely to change significantly over the next year or two. Nevertheless, the development of such a representation standard and the interfacing of all design tools to its format appears to be crucial for the development of a cohesive VLSI design environment.

REFERENCES

[Hsueh 79]

M.-Y. Hsueh and D.O. Pederson: "Computer-Aided Layout of LSI Circuit Building-Blocks", *Proc. 1979 IEEE International Symposium on Circuits and Systems,* pp 474-477.

[Krause 79]

J.Krause: "CIF Interface", *Project Report,* E.R.L., U.C.Berkeley, Dec. 1979.

[McWilliams 78]

T.M. McWilliams and L.C. Widdoes, Jr.: "SCALD: Structured Computer-Aided Logic Design", *Technical Report No 152,* Digital Systems Laboratory, Stanford University, 1978.

[Nagel 75]

L. Nagel, "SPICE": A Computer Program to Simulate Semiconductor Circuits", *Electronics Research Laboratory Report* No. ERL-M520, U.C. Berkeley, May 1975.

[Newton 78]

A.R. Newton, "The Simulation of Large-Scale Integrated Circuits", *Electronics Research Laboratory Report* No. ERL-M78/52 U.C. Berkeley, July 1978.

[Sproull 80]

R.F. Sproull and R.F. Lyon "The Caltech Intermediate Form for LSI Layout Description", in: C.A. Mead and L.A. Conway, *Introduction to VLSI Systems,* Addison Wesley, 1980, pp 115-127.

[Stone 78]

M. Stone: "IC Design under ICL, Version 1.0", *SSP File No 1336,* Computer Science Department, Caltech, 1978.

[Williams 77]

J. Williams: "Sticks - A New Approach to LSI Design", *Master Thesis,* Massachusetts Institute of Technology, 1977.

APPENDIX: DESCRIPTION OF STIF 1.0

Carlo H. Séquin and A. Richard Newton

Computer Science Division
Electrical Engineering and Computer Sciences
University of California, Berkeley, CA 94720

ABSTRACT. *STIF 1.0 represents a first step towards a standardized, portable database containing all relevant information about a fully or partially designed integrated circuit. The parts of the language that have been defined so far address primarily issues related to layout geometry, logical and physical interconnectivity, and hierarchical representation and protection of cells. Hooks are left for future extension of the language to include with each cell suitable models for simulation, history and status of its development, and possibly even information on testing.*

1. STRUCTURE OF A STIF DESCRIPTION

Cells, often called *symbols* or *subcircuits* by other languages, are the main building block of a STIF description. As a matter of fact, the whole STIF description is jûst a *cell definition*, which may contain further nested cell definitions and other components. This overall syntactical equivalence to a single cell definition has been introduced deliberately, so that individual descriptions of parts of a larger project can easily be merged. The names of the parts must be different; the descriptions can then simply be concatenated. A call statement for each part is provided, that places the pieces in the proper relative positions, and the whole thing is bracketed again with the statement pair **DC ... EC**, which encapsulates each cell definition.

1.1. Definition Hierarchy

Figure A1 gives an example of a STIF definition hierarchy associated with a hypothetical multi-project chip (MPC81). Figure A2 shows the nesting of cells in the STIF description associated with one of the subtrees. Three schools (CMU, UCB, MIT) have contributions on this particular chip, and each school may actually supply several independent designs (D1, D2, ...). A design supervised by any one person (say SEQUIN) can still involve

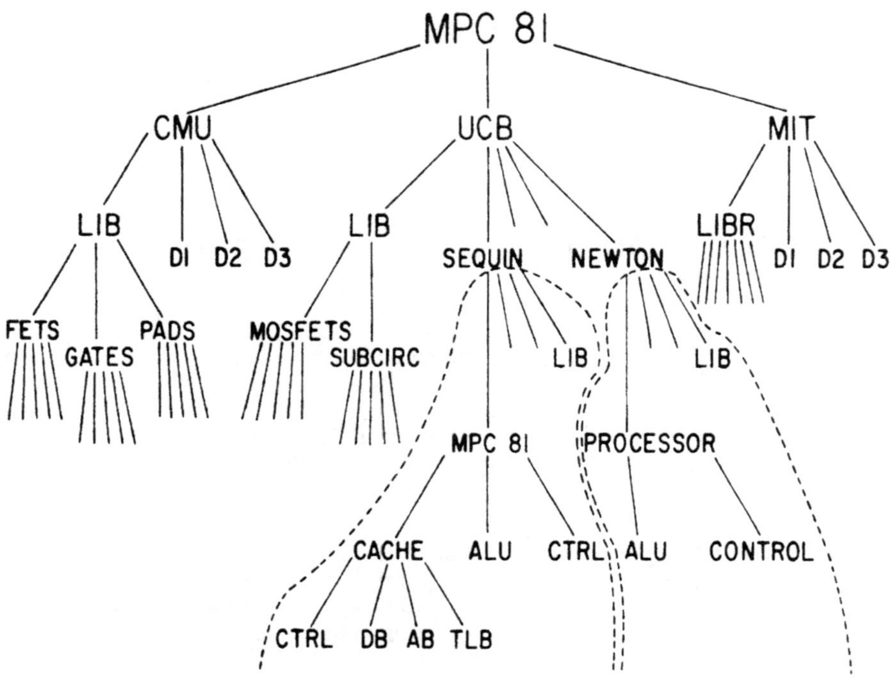

Figure A1. Tree of a hypothetical definition hierarchy of a STIF description of a large multi-designer project.

several designers doing parts of the project. As an example a processor design called MPC81 is shown split into its CACHE, ALU, and CTRL, each one handled by a separate designer. While such a design is evolving, everything in the subtree SEQUIN forms its own self-contained STIF description, probably with its own specialized cell library (LIB). Frequently used, general purpose cells, such as bonding pads, parameterized transistors, and contacts, will be stored in only one place in each institution and are shared by all designers at that school; this would be the function of LIB under UCB.

When individual designs are completed, the corresponding STIF files are combined in a hierarchical manner into larger files, first at the institutional level and finally into one MPC81 multi-project chip. It is mandatory, that in this step all references to names keep their proper meaning, and that no conflict result from the appearance of the same name in several places of the definition tree. Thus an ALGOL or PASCAL-like binding has been chosen: Defined names, parameters, or cells are known within the cell in which they are defined, i.e. in the whole subtree below their definition. A redefinition of a name further down in the subtree will override the higher definition below that particular branch. Thus no conflict arises from the reuse of the same name at two different levels

```
DC SEQUIN
        DC LIB
                DC MOSFET
                ---
                EC
        EC LIB
        DC OTHERS
        EC
        DC MPC81
                DC CACHE
                        DC CTRL
                        EC
                        DC DB
                        EC
                        DC AB
                        EC
                        DC TLB
                        EC
                EC CACHE
                DC ALU
                EC
                DC CTRL
                EC
        EC MPC81
EC SEQUIN
```

Figure A2. Example of the nesting structure of cell definitions in a STIF description corresponding to part of Figure A1.

(e.g. MPC81). It is currently a subject of debate whether all names defined higher up in the tree should be imported *explicitly* or whether some of the tools operating on STIF should at least generate a warning for each reuse of a name, so that accidental reference to the wrong definitions can be caught. Reuse of the same name at the same node of the definition tree is forbidden. Thus all names to which reference is made must be defined within one of the direct ancestor cells. The search for a particular name progresses upward in the definition tree, at each level checking all siblings, until the proper name is found.

For special needs or to permit hierarchically structured libraries, a mechanism to address any particular node in the definition tree may be desirable. In a UNIX-like manner pathnames downward in the definition tree can be specified with the corresponding cell names separated by "|". Thus, in principle, a designer in the subtree SEQUIN could make use of a particular cell defined in LIB under NEWTON. Such uncontrolled access is dangerous and violates certain principles of data encapsulation and protection. Whether this is a recommended feature, whether it is more advisable to copy the desired cell definition into ones own library or to move it up to a level where it can be reached directly, depends a lot on the particular environment and on the organization of the design team.

1.2. The Calling Hierarchy

A cell or subcircuit definition is, of course, not restricted to contain only other cell *definitions* as outlined in the previous section, but it can make *calls* upon other subcircuits in the definition tree. The only restriction is that no recursion, implicit or explicit, must result from the calling structure.

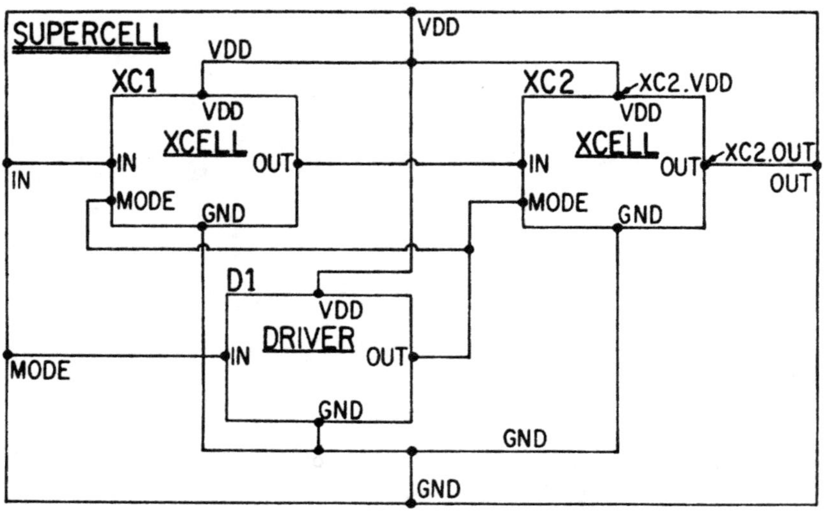

Figure A3. Example of the calling structure within one cell of a STIF description.

Figure A3 graphically illustrates the definition of a subcircuit SUPERCELL which makes calls upon two other cells: XCELL and DRIVER. The cell definition specifies the placement of the subcircuits and the interconnections in between them. It also introduces formal terminals to the outside of SUPERCELL, which act as connection points when SUPERCELL is called by another cell. Since a particular cell (XCELL in this case) can be called more than once inside another cell, instance ID's may be assigned, so that specific terminal points can be unambiguously identified. The chosen instance ID's inside SUPERCELL are XC1, XC2, and D1. These instance ID's have only meaning in the context of SUPERCELL. Terminal points are specified by concatenating an instance ID and a formal terminal name, separated by a dot: e.g. **XC1.VDD** or **D1.IN.** Logical interconnections are then specified by referring to these terminals (see section on interconnects for details).

2. OVERVIEW OF STATEMENTS IN STIF 1.0

Definitions

DN 'short name' = 'hierarchical symbol name' [(['parameters & values'], ['transform'
{text substitution macro to give short names for frequently used cells}.

DV 'parameter name' = 'parameter value' ;
{to give short names to expressions, and for implicit constraints}.

DC 'cell name' [('list of formal parameters and their defaults')] ;
{opening statement for cell definition}.

EC ['cell name'] ;
{closing statement for cell definition}.

Component Calls

C [<'instance ID'>] 'cell name' [(['parameters & values'], ['transform'])] 'location' ;
{cell call specifying parameter values for this particular instance}.

AC [<'instance ID'>] 'cell name' [['version suffix array']] [(['param&values'], ['transform'])] 'nx','ny','location','dx','dy' ;
{call for an array of cells; version suffix to distinguish "1" and "0" cells in a ROM}

Geometrical Primitives

P [<'instance ID'>] 'layer' 'bounding path' :'bounding path' ... ;
{multi-boundary, general polygon; inside is given by winding number convention}.

M [<'instance ID'>] 'layer' 'corner point', 'delta-X, delta-Y', 'delta-X, delta-Y'... [:'other incrementally specified bounding paths'] ;
{multi-boundary manhattan polygon for encoding efficiency}.

R['NSIDE'] [<'instance ID'>] 'layer' 'diameter', 'center' ;
{roundflash approximation, must be between inscribed and circumscribing polygon with 'NSIDE' sides}.

B [<'instance ID'>] 'layer' 'length', 'width', 'turn', 'center' ;
{rectangular box with arbitrary orientation}.

Cell Representation

G 'layer' 'vector path' ;
{for line drawings composed of straight line segments}.

L 'Arbitrary text' 'layer' 'point' ;
{labels in "no-fill" mode, location specified by upper left corner of bounding box}.

Formal Interconnections

XX [<'net ID'>] 'terminal' 'terminal' 'layer' 'width' , 'path' ;
{logical and physical interconnect; implementation is with CIF wire}.

T <'terminal ID'> 'layer' 'width' , 'path' ;
{formal connection segment of framed cell}.

FM 'frame type' 'corner point', 'incrementally specified bounding path' ;
{formal manhattan frame around cell, for protection and representation}.

Constraints

K 'relation' ;
{express constraints explicitly}.

Hooks and Escapes
SIM 'simulator type' 'corresponding information' ;
{place holder for future extension to standardized simulation interface}.
ATT 'type of attribute' 'corresponding information' ;
{place holder to permit inclusion of information on cell status and histor
0 ... 9 'user text' ;
{structured escape to experiment with language extensions}.
... { 'comment text' } ...
{unstructured escape to include arbitrary information, can be nested}.

Parts of a more formal definition of the syntax are given in Table 1. The standard notation proposed by Niklaus Wirth [77] is used: production rules use equals = to relate to identifiers to expressions, vertical bar | for or, and double quotes " " around terminal characters; square brackets [] indicate optional factors (i.e., zero or one repetition); curly brackets {} indicate repetition any number of times including zero; parenthesis () are used for grouping; rules are terminated by a period. The description is limited to the token level.

3. GENERAL COMMENTS ON STIF STATEMENTS

As discussed in Section 4, a STIF description consists of a set of hierarchically structured cell definitions. These cells can contain other definitions, components, and statements describing properties and representations of the cell.

Formally a STIF file is a sequence of *statements*, terminated by semicolons. STIF uses a free format, and statements can thus be of any length. In the following sections we will present the syntactical as well as some semantic aspects of each individual STIF statement, but first we will make some general remarks. All STIF statements start with a keyword. For component calls this is followed with an optional *instance ID* in angle brackets, followed by the formal name of the called component and its parameters, and finally placement information.

Individual *name* or *ID* fields are separated by white-space, i.e. blanks, tabs, and newlines, or by commas. Within expressions, individual numbers or parameters are separated by operators; white-space adjacent to operators or names are ignored. Optional fields will typically be enclosed in some type of brackets: cell parameters appear in round parentheses (), instance names are contained in angle brackets <>, and bounds for values are given in square brackets [].

3.1. Names

Cell names, parameter names, instance IDs, and terminal IDs are unrestricted in length and can be composed of any alphanumeric characters; however they must start with an alpha-character. Alpha-characters are a-z, A-Z, and "_", "#", "$", and "@"; the extra characters have been added to permit a rich mnemonic set of names. The characters "." and "|" may also appear in names or instance ID's, but they are used for the formation of hierarchical names or ID's and thus have a predefined

```
StifFile          = defCell {stifStmnt} endCell.
stifStmnt         = StifFile | DN_Stmnt | DV_Stmnt |
                    M_Stmnt | P_Stmnt | R_Stmnt | B_Stmnt | G_Stmnt | L_Stmnt |
                    FM_Stmnt | T_Stmnt | XX_Stmnt | C_Stmnt | AC_Stmnt |
                    K_Stmnt | SIM_Stmnt | ATT_Stmnt | userStmnt.
defCell           = DC name [parDecList] semi.
endCell           = EC [name] semi.
DN_Stmnt          = DN name eql pathName [parDefList] semi.
DV_Stmnt          = DV name eql value semi.
P_Stmnt           = P [ident] type path {colon path} semi.
M_Stmnt           = M [ident] type path {colon path} semi.
B_Stmnt           = B [ident] type value cma value cma value cma point semi.
R_Stmnt           = R [ident] type value cma point semi.
FM_Stmnt          = F M_Stmnt.
T_Stmnt           = T name type value path semi.
XX_Stmnt          = XX [ident] instName instName type value path semi.
C_Stmnt           = C [ident] pathName [parDefList] point semi.
AC_Stmnt          = AC [ident] name [versionSpec] [paramDefin]
                        arraySize cma point cma point semi.
K_Stmnt           = K relation semi.
SIM_Stmnt         = SIM type simParamDefin semi.
ATT_Stmnt         = ATT type attParamDefin semi.
comment           = "{" commentText "}".
commentText       = {commentChar} {comment} {commentChar}
commentChar       = any ASCII character except "{" or "}".
userStmnt         = digit userText semi.
userText          = {userChar} {comment} {userChar}.
userChar          = any ASCII character except ";" or "{" or "}".
parDecList        = "(" paramDeclar {cma paramDeclar} ")".
paramDeclar       = paramID [eql value] [ "[" value cma value "]" ].
parDefList        = "(" paramDefin {cma paramDefin} ")".
paramDefin        = paramID eql value | transform.
versionSpec       = "[" name {name} "]".
ident             = "<" name ">".
transform         = R eql value | MX | MY .
arraySize         = digit {digit} cma digit {digit}.
instName          = { name "." } name.
pathName          = { name "|" } name.
path              = point {cma point}.
point             = value cma value.
relation          = value rel value.

value             = "(" value ")" | wsp | number | name | expression | no content.
expression        = ~~Expressions in C-style using basic 4 operators.
number            = ~~Floating point numbers defined in the usual way.
```

Table 1. Tentative formal syntax of STIF 1.0 at the lexical-token level.

meaning. STIF uses the ASCII character set, but as a convenience to users with a more restricted keyboard, it DOES NOT distinguish between upper and lower case letters in names. While the STIF parser should accept characters of either kind, it may internally convert them all to the same kind.

All cell *definitions* form a tree structure. At each node of this tree all cell and parameter definitions as well as all instance IDs must have unique names, and redefinition of names is not permitted. Explicit or implicit recursion in the calling structure is forbidden.

STIF is *descriptive* rather than procedural. Thus names cannot be redefined nor can parameters be reassigned at the node at which they are defined. Thus there is no general purpose assignment mechanism. In particular, parameters cannot be incremented and cannot depend on each other recursively. Thus the ordering of call statements and definitions within a cell definition is irrelevant! In summary, the places where names can be introduced and which must be searched when going up the tree, are the three types of *definition statements:* define name **DN**, define value **DV**, and define cell **DC**. These statements will be discussed later.

3.2. Parameters

STIF is *not* a programming language, nor even a language to describe programs that generate layouts. It is a portable format that permits one to describe an integrated circuit during the evolution of its design. There are several good reasons for having parameters in STIF :

Conciseness. The use of parameters permits a reduction in the number of low level cells that must be specified. For example one generic, parameterized transistor cell can cover a wide range of, say, all NMOS enhancement-mode pull-down transistors.

Commonality. Reference to the same cell specifies most clearly the designers intent to use the same type of device or subcircuit in different instances.

Changeability. If it is necessary to make a systematic modification in a whole class of cells, the appropriate change in a properly parameterized cell can often take care of it.

Adaptability. The use of a proper set of parameters in the bottom level cells permits one to express explicitly the hooks for a simple set of design rules. A redefinition of these parameters will often be sufficient to adjust a layout to a new process if compaction programs such as CABBAGE [Hsueh 79] are used.

A simple parameter mechanism has thus been built into STIF. All *value* fields can contain simple arithmetic expressions, which themselves can contain formal parameter names.

To make a STIF file a self-contained description of a circuit, the specification of the values of all parameters must be part of the STIF file itself. Parameters can be specified in two ways. They can be introduced with a **DV** statement if they serve the role of *global constants*, and are thus passed down the definition hierarchy. The scope of these parameters is then the subtree below their introduction. Alternatively, they can be handed down the calling hierarchy from level to level as *formal parameters*.

3.3. Values and Expressions

Value fields may contain integers, floating point numbers, parameter-identifiers, simple arithmetic expressions, white-spaces, or no characters at all between separating commas; the latter two leave the value field undefined, and it would thus resort to the default value, if one has been defined. All *units* used are understood to be the basic units of the MKS system. Thus the unit *length* is the meter. Convenient multiplication factors can be defined with the **DV** - statement.

It would be desirable to have a more efficient encoding of an overall grid unit applied to the layout related values. As in CIF2.0 [Sproull 80], it would be conceivable to declare a unit length in the cell definition that only affects the geometrical values within that cell definition, but does not affect the values in cells called from outside this subtree. A terse and easily readable representation could then be obtained by expressing all dimensions in lambda-units, and introducing the lambda-unit-length at the top of the hierarchy. However, some difficulties arise because of the possibility to parameterize the cell definitions.

We will tentatively introduce a reserved parameter **F** *in each cell definition. All linear geometrical dimensions are scaled with this factor after all parameter substitutions have been made at face value. Further cell definitions contained within this cell assume the same scale factor unless they have their own factor F explicitly set to a different value. Dimensions within called cells defined outside a cell with a scale factor are not affected.*

Values do not have an inherent *absolute* precision. For STIF 1.0 it is specified that the accuracy of numbers should be equal or better to that obtained from 32-bit floating point numbers as defined in the emerging IEEE floating point standard [Coonen 79].

Simple arithmetic expressions involving declared or formally passed parameters and the operators **+**, **–**, *****, and **/** are allowed in all places where a numerical *value* is expected. Parentheses () can be used to specify ordering within these expressions. Number or parameter fields within an expression are separated by operators.

4. DEFINITION STATEMENTS

For brevity and for readability of the STIF file, aliases can be assigned to lengthy cell specification in the definition tree; if the cells possess long library names, or have a long list of parameters that need not be changed, a text substitution macro can be used to introduce an abbreviated name. In these path names hierarchical levels are separated by "|".

The **define name** statement has the format:

DN 'reference name' 'path name'[**(**'parameters & values'**)**] ;

Example

DN TE4 componentLibrary|MOStransistors|EnhType(L=Lmin,W=4*Lmin);

This would permit to reference this particular transistor model simply as: TE4.

Parameters that serve the role of constants throughout a subtree of the definition hierarchy can be specified with the **define value** statement, which has the format:

DV 'parameter name' = 'parameter value' **;**

Example

DV AS = de * W ; {Source area, diffusion extension, gate width}

On the other hand, formally passed parameters are declared in the cell definition statement. At the same time they can also be assigned default values, as well as optional bounds. Upper and lower bounds for the value of a parameter are introduced after the default value field. These bounds are enclosed in square brackets []. The full **parameter declaration** thus takes on the form:

... 'param name' = 'default value' **[** 'max value'**,** 'min value' **]** **,** ...

Actual values of these parameters are specified in the call statements. If a value is not specified, the default value is used; if the latter is not available, or outside specified bounds, it is an error. Parameters that are formally handed down from definition to definition are the proper mechanism for circuit-related parameters. As an example consider a value WW which could specify the drive power of an inverter, and internally would be used to define the width of the component transistors of the inverter.

Cells are the main building block of the hierarchy of a STIF description. They preserve the modular aspects of a large design and make it possible to expedite operations such as design rule checking and simulation. These objectives are accomplished through the use of an abstract frame around the cell and formal connection points on this frame. At each level in the hierarchy, the cells at the next lower level can be considered abstractions to which connections can only be made at specific connection sites, called *terminals*. Information for the representation of a cell in a circuit or logic simulator, and other relevant information, can be stored with the cell definition. Cells can call other cells, formal interconnects and geometrical primitives.

Cell definitions have the format:

DC 'cell name' [**(**'formal parameters, optional defaults and bounds'**)**] **;**
  definitions
  attributes
  representations
  constraints
  components
EC ['cell name'] **;**

Example

DC norgate(in=2, k=4, F=2.5e-6) ;

EC;

In this cell definition, F is a reserved parameter and corresponds to the scale factor on all linear geometrical values. In the above case all linear dimensions are meant to be in units of 2.5 microns. This scale factor also applies to all cells defined within the above cell carrying this scale factor.

5. GEOMETRICAL PRIMITIVES

To be able to specify an actual layout unambiguously, STIF must contain geometrical primitives. These primitives in STIF 1.0 are different from the ones in CIF2.0; the goal was to end up with a more compact encoding for the most frequently used features and to avoid ambiguities in interpretation. STIF 1.0 currently contains: General polygons **(P)**, incrementally specified manhattan polygons **(M)**, rectangular boxes of general orientation **(B)**, and approximation to roundflashes **(R)**. Wires have been deliberately omitted as *geometrical* building block, since they should only be used as interconnection primitives. In STIF, all geometrical items can be called with parameters and their layer is part of the call rather than a "mode" as in CIF2.0. These primitives can also be provided with instance ID's so that they can act as *terminals* for logical interconnects.

Polygons have the format:

P [<'instance ID'>] 'layer' 'bounding path' [:'bounding path':'bounding path'...];

Example

P window 3.5,7.5, 2,5.-2, -7.1,-3;

The polygon is specified by its boundary. The start and end points of the describing path are automatically closed by a straight line. The polygon boundary can be self-intersecting. The *inside* of the polygon is defined with the *non-zero winding number* convention [Hon 80, Newell 80]; points are *inside* if they are truly encircled in either direction by the bounding path. The polygon can also have multiple boundaries; individual bounding paths are separated by colons. In order to generate a hole in a larger feature, the two boundaries have to be specified with opposite orientation - then the winding number in the hole becomes zero.

Manhattan polygons have the format:

M [<'instance ID'>] 'layer' 'corner point', 'delta-X, delta-Y', 'delta-X, delta-Y'...
[:'other incrementally specified bounding paths'] ;

Example

M metal 434,725.5, 12,16 :436,727.5, 8,12;

The manhattan polygon is defined by one or more incrementally specified bounding paths. Each such bounding path consists of a starting point and pairs of increments delta-X, delta-Y. Each bounding path is assumed to be closed with another delta-X, delta-Y pair (in that sequence) necessary to bring the bounding path back to the starting point. Thus for the simplest case of a manhattan box the format becomes:

M [<'instance ID'>] 'layer' 'corner point', 'delta-X, delta-Y' ;

Since manhattan features are the most frequently used constructs in IC layout, it is appropriate to have a separate efficient representation. The definition of inside and outside is handled as in the case of the general polygon.

Boxes have the format:

B [<'instance ID'>] 'layer' 'length', 'width', 'angle', 'center' ;

Example

B polySi 42, 6, 45, 78.3, 54.3;

The box is specified by its length, width, angle of rotation around its center, and by its center coordinates. Rotation of the box is specified in degrees in the positive (ccw) or negative (cw) direction. To be consistent with the principle that transforms are executed in the sequence in which they are specified, rotation has been placed ahead of the specification of the center coordinates.

Roundflashes have the format:

R['NSIDE'] [<'instance ID'>] 'layer' 'diameter', 'center' ;

Examples

R diff 4, 23,45;
R8 poly 4, -17.5,22;

The roundflash is specified by its diameter and the location of its center. Since most devices can only generate an approximation to a roundflash, the specification can become rather ambiguous. Attaching an integer number after the **R** can specify a minimum quality of the approximation. The number 'NSIDE' specifies the minimum number of sides to be used in a polygon that approximates the circle. More specifically, this definition conveys the intent to approximate a circle as closely as possible and specifies that the approximation has to lie between the inscribed and circumscribing polygons with 'NSIDE' sides onto the desired circle. If the circle to be approximated is assumed to be centered at the origin, then the inner and outer polygons acting as bounds for the approximation are assumed to cross the positive x-axis at a right angle (Fig. A4).

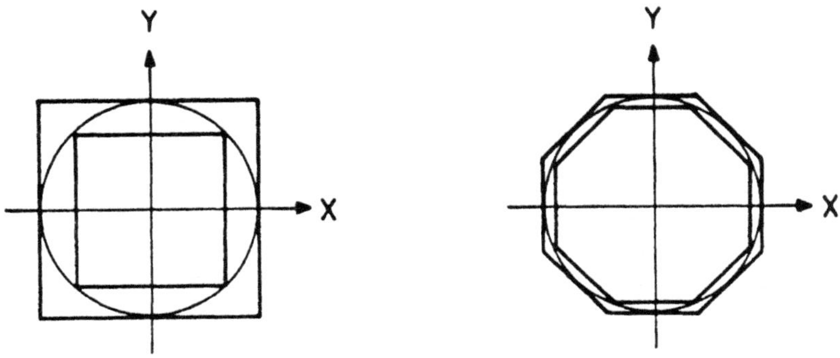

Figure A4. Approximation bounds for R4 and for R8.

6. CELL AND ARRAY CALLS

Instances of defined cells can be called and placed individually or in orthogonal arrays. At the time of the cell *definition*, a set of formal parameters can be specified, and optional default values and bounds can be supplied. At the time of the cell *call*, some of these parameters may be respecified, while others are simply not mentioned. If actual parameter values are not specified for all formal parameters, the default values are used; if they are not supplied in the cell definition, an error results. The call statements also provides information about possible transformations, to be applied to the cell, and about the cell placement.

6.1. Cell Calls

The minimum information that must be contained in a cell call is the name of the cell and the coordinates of its placement. Additional optional information concerns parameters and transformations of the cell, as well as an instance ID.

A full **cell call** thus has the format:

C [<'instance'>] 'cell name' [(['param & values'], ['transform'])] 'location' ;

Example

C <nor21> norgate(in=4, mx, R=-90) 22.3, 7.6 ;

An instance ID can be specified in the call statement, so that this particular use of the cell and the corresponding connection points obtain unambiguous logical names. All instance ID's within one definition must thus be unique. Internal to the programs working from STIF, it may often be necessary to catenate many such instance ID's (separated by dots) to represent the pathname in the hierarchy of definitions, the length of these instance IDs should thus be kept short, although there is no formal syntactical restriction.

6.2. Transforms

Rotation and mirroring of a cell are specified through reserved parameters R, MX and MY, and are considered an additional parameterization of every cell. These "local" transforms are thus specified inside the parentheses which also contain the values of the user defined parameters; they can thus be included in the **DN** - statement. Since mirroring and rotation will be relatively unfrequently used, while x and y coordinates, indicating the placement of the cell, are practically always needed, the translation part has been separated from the "local" transform specifications, and is a separate, *required* field.

As in CIF2.0, individual transformations are applied to the cell in the order in which they are listed. However, since the translation is not part of the transform, it will always be done after the local cell transformations, i.e. mirroring and rotation around the cell origin. The flags MX and MY indicate complementation of the x and y coordinates of the cell, respectively. Rotation is specified in degrees in the positive (ccw) or negative (cw) sense. Thus the transform (MX, R=90) would first mirror all x-coordinates in the cell definition and then rotate the cell one quarter of a full turn, i.e. 90 degrees in the counter-clockwise direction.

6.3. Arrays of Cells

Future VLSI chips will rely even more heavily upon regular array constructs. The definition of an array makes some implicit statements about the layout, e.g regularity, use of identical cells with identical parasitics, tight interconnects... . It is important to preserve this information in the file describing the integrated circuit. The replication statement has been restricted to orthogonal arrays. This should be quite acceptable for practical IC designs. Hexagonal arrays can always be converted to orthogonal arrays by combining two adjacent cells into one cell for 2-dimensional iteration. More complicated geometries can still be assembled from two levels of cell calls, by first specifying one row and then calling these rows in a second iterative call.

An array call contains the same information as a cell call plus the additional specifications about number of columns and rows and the corresponding pitch in both dimensions.

The format for an **array of cells** then becomes:

AC [<'instance ID'>] 'cell name' [(['param'], ['transform'])] 'nx', 'ny', 'location'
'dx', 'dy' ;

Example

AC <P2> pattern(param=value, R=90, My) 4, 4, 87, 44.7, 2.5, 3.6 ;

6.4. Version Suffix Array

A mechanism is needed to introduce a piece of code describing the bit pattern in a ROM or in a PLA. One approach is to define all necessary versions of cells that make up the array, i.e. for a ROM a "1"-cell and a "0"-cell could be defined. The information that spells out which cell type needs to be called in a particular place of the array is contained in a special *version suffix array* included in the array statement. For generality, entries in this suffix array are short ASCII strings which are used as suffixes to the 'cell root name'. Thus each element[i,j] in the array is the cell whose name is formed by the concatenation of the 'cell root name' and 'version suffix array[i,j]'. If no suffix array is specified, then the default is an array of blanks; hence, the called elements are all 'cell root name'. All cells called in that context need to have the same external dimensions and connection points and must accept the same calling parameters.

The full format for the specification of an array of cells thus may contain **a version suffix array** which is included in square brackets:

AC [<'instance ID'>] 'cell root name' [['version suffix array']]
[(['cell parameters'], ['transform'])] 'nx', 'ny', 'location', 'dx', 'dy' ;

Example

AC PLAand [1 1 0 x 1 x x 0 x 0 0 1 1 1 0 x] (R=90, My) 4, 4, 23,79.5 7.2,4.8;

This array would be based upon the definition of three cells: PLAand1, PLAand0, and PLAandx. Since a STIF statement can extend over several lines, large version suffix arrays could also be formatted in a more readable form, which may be particularly important for large ROMs.

AC PLAand [
1 1 0 x
1 x x 0
x 0 0 1
1 1 0 x
] (R=1/4, My) 4, 4, 23,79.5 7.2,4.8;

To minimize errors, the specification of the version matrix has to be full, and thus give exactly one suffix for each component of the cell array. Specifications are in row major order, thus specifying cells for subsequent positions incremented by dx, for nx positions, then specifying the next row at a y-position incremented by dy, and so on, until all ny rows have been described.

7. FORMAL INTERCONNECTIONS

A powerful concept associated with the notion of a cell is the construct of a *frame* that presents an abstraction of the cell at the next higher level of the hierarchy. It carries all the connection sites for external contact to the cell. Formal interconnections can then be specified using these connection points.

7.1. Frames

Frames around cells can serve a many different purposes. In the simplest case they are simply representative for the cell at the next level of the hierarchy and thus provide an abstract view of the cell. In addition frames can be used to protect the contents of a cell from being disturbed by infringing features from the outside. If the frame around the cell is kept loose enough, and if frames are not allowed to be overlapped by other frames, by geometrical primitives, or by formal interconnections, then the innards of the cell can be truly protected. This has the advantage, that cells can be checked individually for layout rule violations and can be simulated independently. This is the basis for the development of hierarchical layout rule checkers and simulators, which can bring orders of magnitude shorter run times compared to programs running on the fully instantiated designs.

In STIF 1.0 all frames must be manhattan polygons, until some experience has been gained with the actual use of frames. A **manhattan frame** is introduced with the statement:

FM 'frame type' 'corner point', 'delta-X, delta-Y', 'delta-X, delta-Y'...
[:'other incrementally specified bounding paths'] ;

Example

FM protect 722,366, 14,21 -7,-11 ;

A *frame type* field has been introduced, corresponding to the *layer* field in the statements calling geometrical primitives. It permits expansion to more than one type of frame. It may be useful, i.e. for denser packing, to have separate frames, indicating closest possible spacing on the metal layer and on the polysilicon layer. The technology package that personalizes STIF 1.0 to a particular process will also specify what frame layers are being recognized. In the simplest case, a single frame can serve as a protective boundary as well as a carrier for the formal interconnection points.

Obviously, layout density will suffer as a consequence of using these frames too far down in the hierarchy. However, in the realm of VLSI systems, the highest priority may have to be assigned to getting the system to work at all, rather than to squeeze out the last square micron of layout area. The analogy with the trade-offs between well-structured high-level programming and memory space requirements should be noted.

There is no requirement that all cells must include a frame statement. There are some structures, e.g. transistors and contacts, which are used so often that there must be some facility for referring to them symbolically. However, it would be inappropriate to demand that such cells possess a protective frame with formal terminals on it. If the protective frame is left out, cells can be packed much more tightly, or can even fall right on top of one another. On the other hand, cells can then no longer be certified individually by a design rule checker, nor can a circuit description be considered reliable. Such unframed cells correspond fairly closely to the symbols in CIF2.0 [Sproull 80].

7.2. Terminals

In STIF all interconnections must be explicit. For that purpose, cells carry so called *terminals* to which all external logical connections are made. Each *interconnect* between two cells corresponds to an explicit statement, linking two terminals physically and the corresponding two nodes logically. All terminals must thus carry logical names, so that they can be properly identified.

Terminals are formally specified connection sites on cell frames. This abstraction implies two things about the layout of the internal parts of the cell. No design rule violations must result when a wire of specified width on the proper layer is attached to the defined terminal region from the outside of the cell; and, secondly, there is a connection path to this terminal of at least the specified width on the inside of the cell frame.

In general a **terminal** is specified with the following format:

T <'terminal ID'> 'layer' 'width' **,** 'path' ;

Example

T <carry1> diff 4, 2,4, 2,12, 30,12, 30,08 ;

The path describing the extended terminal must lie completely on the frame of the cell, and, of course, must have the corresponding internal connectivity to back up that whole region. It is quite acceptable to have a terminal that covers the whole frame of the cell; e.g. a metal-polysilicon contact could have metal and polysilicon terminals all around the peripheries of the corresponding rectangles. If the path field consists of a single point, the terminal statement takes on its minimal format, which is useful to contact wires of identical width inside and outside the cell, as will mostly be the case in the higher levels of the systems hierarchy. Thus, if the given path in the T-statement contains only a single point, the segment on the frame to which one can make a connection is equal to a region of the specified width centered around the terminal point.

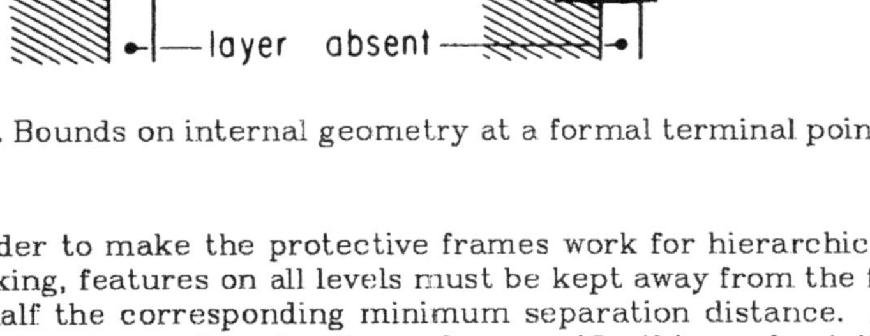

Figure A5. Bounds on internal geometry at a formal terminal point.

In order to make the protective frames work for hierarchical layout rule checking, features on all levels must be kept away from the frame by at least half the corresponding minimum separation distance. A formal terminal on a frame thus implies a few specific things about its layout geometry. Figure A5 illustrate the constraints on the layout in the layer of the terminal.

Unfortunately every terminal is thus associated with a "stub" sticking out to and touching the protective frame. Thus in the neighborhood of terminals, external features can not be run as close to the frame as elsewhere, even if there is no actual connection made to this terminal. Is there a better definition of the terminal geometry that is not so wasteful, perhaps with the actual connection point located a certain distance inside the protective frame ?

7.3. Logical Connections to Geometrical Primitives

At the lower levels, cells will often be used without a protective frame to obtain a dense layout. While one cannot guarantee, that these cells could not be altered by accidental or deliberate overlap with other cells or geometrical primitives, it is still useful to preserve the intended function, i.e. the desired circuit model, with the cell definition. Thus even cell definitions with no protective frame can contain the SIM and the ATT statements to be discussed later.

In order to identify the nodes between which the circuit description holds, and to permit the attachment of logical interconnects, terminals have to be introduced into the geometrical cells. The idea is that one wants to connect to a particular piece of geometry in a flexible manner, without exactly specifying the connection point, and to use logical interconnects between such points. This also helps the circuit compaction program, since it introduces points of explicit stretchability.

Thus, in order to use a box as a logical interconnection point labeled "gate", one would simply provide the corresponding box with the proper instance ID:

B <'instance ID'> 'layer' 'length', 'width', 'turn', 'center' ;

Example

B <gate> polySi W+2*pe, L, 0, 0, 22.5;

Similar constructs can be formed with all geometrical primitives. If a construct of geometrical primitives is built in a hierarchical manner by a nested calling structure, it must still be possible to get to any terminal defined in one of the subcells. Concatenating the corresponding instance ID's (separated by dots) identifies the proper cell instance in the calling structure. The *full terminal name* of a specific formal interconnection point is formed by appending the terminal ID to the instance ID of the cell, separated by a "." . Concatenation of several cell instance ID's is possible, if it is necessary to identify a particular terminal or a labeled geometric feature several levels down in the calling structure.

Examples

... XC.OUT ...
... ALU.OR24.ET45.gate ...

7.4. Logical Interconnections

Interconnects contain explicit information about the logical connectivity between subcircuits and the physical, conductive paths that realize them. Logically the connection is described through the ID's of the two terminals that get connected. These end points can be terminals, named geometrical primitives, and named interconnects. All "loose" ends intended for connection at the next higher level of the design must be tied to a terminal. Since the two terminals may be on different symbols, the *full terminal name* is used.

All interconnections must be specified explicitly. Even connections between separately called, abutting cells with touching terminals must be specified with an explicit interconnect command, if the cells carry protective frames. This is to make it possible to do simulation on large systems solely based on the logical connections without an actual instantiation of the layout.

To specify the exact physical implementation of this interconnect in the layout, the logical definition is extended to include specifications of the layer and of the exact geometrical path taken. Thus, possible intermediate jog points can be precisely specified. The specification of an **interconnect** then takes on the form:

XX [<'net ID'>] 'terminal' 'terminal' 'layer' 'width' **,** 'path' ;

Examples

XX mux.out nor.i2 poly 4, 2.5,3, 7,3, 7,10.2;

XX <VDD> **ALU.Vdd Vdd metal** 6, 4.5,6, 4.5,12, -8,12;

Interconnects physically correspond to contiguous paths in a single "mask" level. The actual physical implementation makes use of the construct of a CIF-wire in the layer specified. The coordinates of the end points of the path description have to coincide with the terminal locations. If both end points have the same coordinates, this may constitute an implicit constraint to the layout compactor stating that the two connected cells should be considered "glued" together.

Incompletely specified interconnections may be acceptable to many programs designed to work from STIF 1.0, since many of them will not look at all the fields in each statement. For instance a block diagram of SUPERCELL can be generated even if the layer and the exact path of an interconnect are not yet defined. Or, a simulation model of SUPERCELL is generated from the respective models of XCELL and DRIVER and their interconnections, in which all parasitics associated with the wiring are ignored. This is useful in early stages of the design, when the layout has not yet been compacted.

7.5. Interconnections between Interconnects - Net ID's

Interconnects can serve themselves as terminals for other interconnects, so that large nets can be built without a need for internal "floating" terminals. For this purpose the existing interconnect has to be provided with a *net ID*, which is then used to specify one of the two logical connections of the new interconnect. Thus, in one sense, the original interconnect now assumes the function of an extended linear terminal. It is permissible to give the new interconnect the same net ID as the one it is connected to, to indicate that they form the same electrical node. However, if different, unconnected pieces are given the same net ID, the simulator may then assume that the two nets are at the same potential.

Similarly, by connecting nets with different ID's into, say, a contact between to layers, one may indicate that the parasitic circuit elements associated with this contact must be taken into account.

7.6. Interconnections in Arrays

An interesting question is how to handle interconnections within these arrays. For the purpose of easy simulation, for proper linkage to higher level hardware descriptions, and for effective error checking, all interconnections should be explicit. But the necessity of specifying explicitly all interconnections between cells in an array seems rather cumbersome, in particular since the cell should be designed to abut as tightly as possible, and the internal wiring can thus be understood as part of the cells that constitute the array.

A suitable convention specifies, that when adjacent symbols in an array call have terminals that fall into the same locations, the logical connection is also implied; if the terminal coordinates do not match up, an error is indicated. The idea behind this convention is to permit omission of explicit interconnection specification only if both terminals involved are (implicitly) defined within a single statement.

Interconnections between to separate arrays, however, must still be explicitly defined. *Some kind of vectorized interconnection statement may be nice, but how do you handle the connections between two deliberately misaligned arrays, in which output n of one array matches up with input n+m of the other array ?*

8. ASSERTIONS AND CONSTRAINTS

A mechanism is required to introduce certain checks on the layout or the connectivity of the circuit, to guarantee that a combination of many different parameters has not lead to some nonsensical values for other fields. It must also be possible to introduce constraints such as: Point A and Point B must not be further apart than 100u, or input and output terminals of a particular cell must have the same y-coordinates. In this document we will restrict ourselves to constraints concerning the layout. Electrical constraints may readily be added once the usage of STIF is extended into simulation and testing.

8.1. Implicit Constraints

A STIF description contains many implicit constraints simply given by the semantics of the STIF statements: formal terminals must lie on frames, frames cannot overlap ... Proper use of parameters makes it possible to introduce many implicit constraints in a much more elegant manner than could be achieved with an explicit listing of all constraints. One such mechanism is the bounds specification for the formal parameters in cell definitions. A verification program, if not the STIF parser itself, will check that the specific parameter values in each cell instantiation fall within the specified bounds.

Another way to introduce geometrical constraints is to use parameterized coordinates. If several cells, A, B, C ..., make use of a declared parameter value **xInv** by specifying their respective locations as xInv,yA - xInv,yB - xInv,yC ... then it is implied that these cells all have the same x-coordinate and thus are lined up on a single vertical line.

8.2. Explicit Constraints

On other occasions one would like to specify constraints which are not so easily introduced in an implicit manner. For this purpose STIF 1.0 contains a **constraints construct** of the form:

K 'relation' ;

Examples

K Y.Acell.out == Y.Bcell.in ;
K WM3 / LM3 >= 16 ; {where LM3, WM3 are length and width of transistor M3}

With such a construct one can specify that the position of two terminals match in the y-direction or that a transistor (M3) have sufficient drive power. The operators recognized in these relations are "<", "<=", "==", ">=", and ">".

In this context additional keywords are used. The identifiers **X.**'terminal or instance ID' and **Y.**'terminal or instance ID' may specify the x and y coordinates of a point-terminal or of the origin of a cell instance. *Additional keywords* **LEFT, RIGHT, TOP** *and* **BOTTOM** *may also be convenient; they could refer to the extreme coordinate values in the x and y direction of the bounding box around the cell. It is also unclear whether means are required to restrict the placement or shape of geometric primitives, e.g. general polygons, and how that can be best expressed. Some experience with the usage of constraints is required to determine the most useful formats, constructs, and keywords.*

9. REPRESENTATIONS

STIF requires the means to represent cells at various levels of abstractions. Labels and symbols that are defined in the language itself would be too restrictive; the users must be able to define their own cell representations. Some capabilities to place textual labels and to define simple line graphics are thus provided.

9.1. Textual Labels

To make cell names or net ID's visible on graphical output, or to list parameters and their values, a mechanism is provided to include alphanumeric labels in the graphical output. Any text using the STIF character set, with carriage returns and spaces as provided in the STIF description, can be included in a particular place of the graphical output. The upper left corner of the bounding box around the text serves as the reference point for placement. The statements to place **labels** thus has the format:

L " 'Arbitrary text' " 'layer' 'point' ;

Example

"DataRAM
16k words
by 8 bits
150ns access" redtext 235,117 ;

Rotation and scaling of text has been excluded until the need for these features is clearly demonstrated.

9.2. Vector Graphics

A simple line drawing capability is included based on paths of concatenated vectors. The statement can specify many separate vector paths which are separated by colons. The construct is closely related to that of the general polygon, with the following exceptions: The vector path is not automatically closed; if this is desired it has to be done explicitly. Such vector paths, even if closed, are not considered to represent an area and are thus never filled in with any shading. Of course any of the geometrical primitives can also be used on a graphics layer and will be represented with the corresponding color or stipple pattern of this layer. The format of the **vector path** is:

G 'layer' 'vector path' ;

Example

G purple 0,2, 3,2, 4,1, 4,-1, 3,-2, 0,-2, 0,2 ; {AND gate}

10. HOOKS AND ESCAPES FOR FUTURE EXTENSIONS

STIF 1.0 represents only the first step towards a standardized portable database for IC design. Many questions are still left open until more experience with new tools and their usage has been gained. One such area is simulation. Separate experiments are going on at Berkeley and elsewhere trying to standardize circuit representation in such a way that many different tools can work from the same representation of the circuit information. The SIM-statement is thus not more than a temporary place holder. When a certain consensus can be reached on the kind of information that various simulation tools require, it will be replaced by one or several more specific statements.

10.1. Simulation

For improved efficiency, simulation models can optionally be stored with each cell definition. Different types of simulation models will be appropriate at different levels of the calling hierarchy and during different phases of the design process. While during an early period, a simple switch simulator could be used to test the basic logical operation of a system, towards the end of the design cycle, full logic and circuit simulation models could be used at the different levels of the calling hierarchy. Since several different simulators may be used, several different sets of parameters for simulation may have to be provided with a cell definition.

Basically the necessary parameter values for circuit simulation would be provided with the cells at the lower levels of the hierarchy. Symbols higher in the calling structure may contain explicit information for logic simulation or for a register transfer simulator. If such information is not explicitly given for a particular cell, the necessary model will have to be derived from the individual models of the lower level cells that make up the cell in question. Specifically, this may mean that if cells containing geometrical primitives are not supplied with simulation information, a circuit-extraction routine has to be invoked that will try to find a suitable representation based on geometrical and process information. For efficiency reasons, once this model has been derived, it should be kept ready for further use with the cell definition.

The spectrum of possible simulation tools is almost unlimited, and it is not possible at this point to define the ultimate generic simulation language to serve all possible simulators. Rather than defining an arbitrary and restricted set of special simulation keywords and parameters, a single keyword **SIM** has been defined, simply as a place holder indicating that this statement contains information for a simulator. A second field then specifies the type of the simulator, and after that the necessary information for this particular simulator is provided in a manner that translates easily into the format of the simulator, while of course remaining within the constraints imposed by the STIF syntax. Separate experiments with different simulators and the specification and translation of their input data are currently being carried out with the Berkeley Language Translator (BLT) [Crawford 79].

The format for **simulation information** is thus:

SIM 'simulator type' 'corresponding information' ;

Example

SIM SPICE NMOS drain gate source bulk L=lambda W=wn AS=value AD=value ... ;

This represents the corresponding information for SPICE [Nagel 75] simulation of an MOS transistor cell.

Of course more than one type of simulator can run from the same **SIM** statement. A simple switch simulator could take the relevant information from the circuit simulation statement, and ignore everything but the logical interconnections to the transistors.

Simulation related parameters should be treated in the same manner as those related to layout geometry. Specific cell parameters, such as the source and drain area of a particular transistor model should be defined in the cell definition itself. Application dependent parameters, such as the width of a pull-down Mosfet, should be passed as formal parameters, whereas processing related parameters, such as the gate oxide thickness, should be introduced near the root of the design hierarchy together with the basic layout rule parameters.

10.2. Attributes

The keyword "ATT" is yet another hook for different classes of information to be tied in with the cell definition. It may be considered a structured place holder for future extensions of the cell definition. Possible uses for this statement are: flags that contain information about the level of simulation that a particular cell can be subjected to, status of design rule checking, average power consumption of the cell. The most important use in the context of a large, hierarchically structured description of a design may concern information concerning the status of the design, e.g. date of last update, name of designer, flags to indicate whether consistency checks or checks for layout rule violations have been performed.

The format of the **attribute statement** is:

ATT 'type of attribute' 'corresponding information' ;

Examples

ATT power 20mA;
ATT designer J.R. Troll, Aloha College;
ATT update 1980 6 1: consistency 1: LRC 0;

11. TECHNOLOGY PACKAGES

STIF 1.0 is general enough to serve different technologies of IC implementation. To make STIF useful, a special package has to be defined for each technology. In this package special keywords and default values can be defined. As a minimum, this package would define all the layers that STIF should know about for a particular process. For silicon n-channel MOS technology, one would introduce the seven layers specified in CIF2.0; in addition at least three more layers will have to be defined:

nf nmos frame

nt nmos text

nr nmos representation

Furthermore, to make it easier to get started with a layout, the most often used primitive cells, i.e. transistors and contacts, could also be provided in such a technology package. A minimal starting set may contain the following cells:

Enhancement mode transistor

Depletion mode transistor

Depletion load

Metal-Poly contact

Metal-Diffusion contact

Poly-Diffusion butting contact

This should be understood only as the startup of a cell library, and, of course, the user may choose to redefine any of these symbols.

REFERENCES

[Coonen 79]
J.T.Coonen: "An Implementation Guide to a Proposed Standard for Floating Point Arithmetic", *Computer,* Vol 13, No.1, Jan. 1980, pp 68-79.

[Crawford 79]
J.D. Crawford: "A Unified Hardware Description Language for CAD Programs", *Electronics Research Laboratory Report* No. ERL-M79/64, U.C. Berkeley, May 1979.

[Hon 80]
B. Hon and C.H. Séquin: *Guide to LSI Implementation,* Second Edition, Xerox PARC, Palo Alto, CA, Jan 1980.

[Hsueh 79]
M.-Y. Hsueh and D.O. Pederson: "Computer-Aided Layout of LSI Circuit Building-Blocks", *Proc. 1979 IEEE International Symposium on Circuits and Systems,* pp 474-477.

[Nagel 75]
L. Nagel, "SPICE": A Computer Program to Simulate Semiconductor Circuits", *Electronics Research Laboratory Report* No. ERL-M520, U.C. Berkeley, May 1975.

[Newell 80]
M.E. Newell and C.H. Séquin: "The Inside Story on Self-intersecting Polygons", *LAMBDA Magazine of VLSI Design, Vol.1, No.2,* pp 20-24, May 1980.

[Sproull 80]
R.F. Sproull and R.F. Lyon "The Caltech Intermediate Form for LSI Layout Description", in: C.A. Mead and L.A. Conway, *Introduction to VLSI Systems,* Addison Wesley, 1980, pp 115-127.

COMPUTER AIDED DESIGN TECHNIQUES FOR VLSI

H. De Man

Katholieke Universiteit Leuven, Dep. Elektrotechniek
Kard. Mercierlaan 94, 3030 Heverlee, Belgium

1. INTRODUCTION

The evolution of IC's from LSI towards VLSI has already been extensively covered in this course. One of the main aspects is a strong evolution from the simple integrated circuit of a few hundred transistors towards integrated systems containing very soon hundred of thousands of transistors.
This fast evolution very definitely leads to the necessity to introduce a new design methodology to master this rapid evolution in complexity. However important this step is, the vast amount of data to be managed also imposes the use of computer aids to reduce the design effort to reasonable proportions.
The purpose of this contribution and the one on simulation is to explore the role of existing and future CAD tools within the scope of the VLSI complexity problem.
In 2. some general reflections about the VLSI design methodology will be given. At the same time the definition and relationship of different CAD tools will be given. In 3. particular attention will be paid to layout styles and their trade-offs. Also new approaches towards VLSI layout such as hierarchical placement and routing, gate-matrix, interconnectivity driven layout, design rule checking, sticks editors and compactive methods will be discussed. In 4. the problem of design verification by simulation as well as the aspect of languages, data-structure and data-base or file management for VLSI design systems is discussed. The influence of VLSI itself on CAD hardware will also be considered. The topic in 4. will be the subject of a lecture under the topic of "Simulation, Modeling and CAD".

2. VLSI DESIGN METHODOLOGY AND CAD TOOLS

2.1. The evolution towards VLSI and its implications on CAD

Traditionally for the design of MSI and even LSI the use of CAD has been limited to the use of circuit analysis programs, such as SPICE, run directly from a circuit diagram at transistor level, often with only estimates of the geometrical data of the devices. For complex digital circuits sometimes logic simulation is used, often only to verify the logic behavior without concern for timing as delay parameters of gates are often unknown or poorly estimated.
The layout is mostly pure manual, often by different people than the logic designers. The circuit schematic is "humanly" transformed into a 500 x or 1000 x drawing, then digitized and completely checked at best by a Design Rule Checking (DRC) program. Test patterns are in almost all cases dreamed up by the "feeling" of the design engineer or at most randomly generated by a Random Test Pattern Generation (RTPG) program and in any case verified by test pattern verification packages (example : TEGAS). As no design for testability is used, Automatic Test Pattern Generation is unfeasable.
This primitive process is still very much daily design practice and clearly has a serious number of drawbacks such as :

- the use of very disparate tools with unrelated input languages and data-base;
- the large number of unverifiable human interventions with error possibility at each step;
- high reductance to try out alternatives;
- poor or impossible debugging after prototype;
- incomplete testability in production;
- etc...etc...

Following this procedure the first round in the design of a tightly packed 1000 transistor digital LSI chip takes roughly 9 manmonths to complete. Nearly 2/3 of the time is spent in the layout phase. Usually at least one redesign is necessary.

Clearly such a situation cannot be tolerated in the future since IC complexity in transistors/chip doubles roughly every two years. A conservative estimate shows the feasability in 1982 of the following typical chip :
- chip size : 1 cm^2
- feature size : 2 μm
- transistors : 350.000 digital : logic & memory.

Translated into some hard facts this means :

- the equivalence of the design of a 25 x 25 km city at street & housing level : streets are 5 meter wide, 350.000 houses;

- map of the city or layout composite drawing : 1 mm features; 100 m^2 mylar;
- number of geometrical units (rectanglar) : appr. 15.000.000 ;
- storage required for layout appr. 250 Mbyte or 4 disks if no data reduction takes place in some way;
- linearly extrapolated first prototype design: appr. 70 manyear;
- probability of working and testable prototype : approx. $2^{-350} \approx 0$.

The above supposes of course no change in design methodology which actually makes those figures even optimistic (!) as illustrated by Fig. 1.

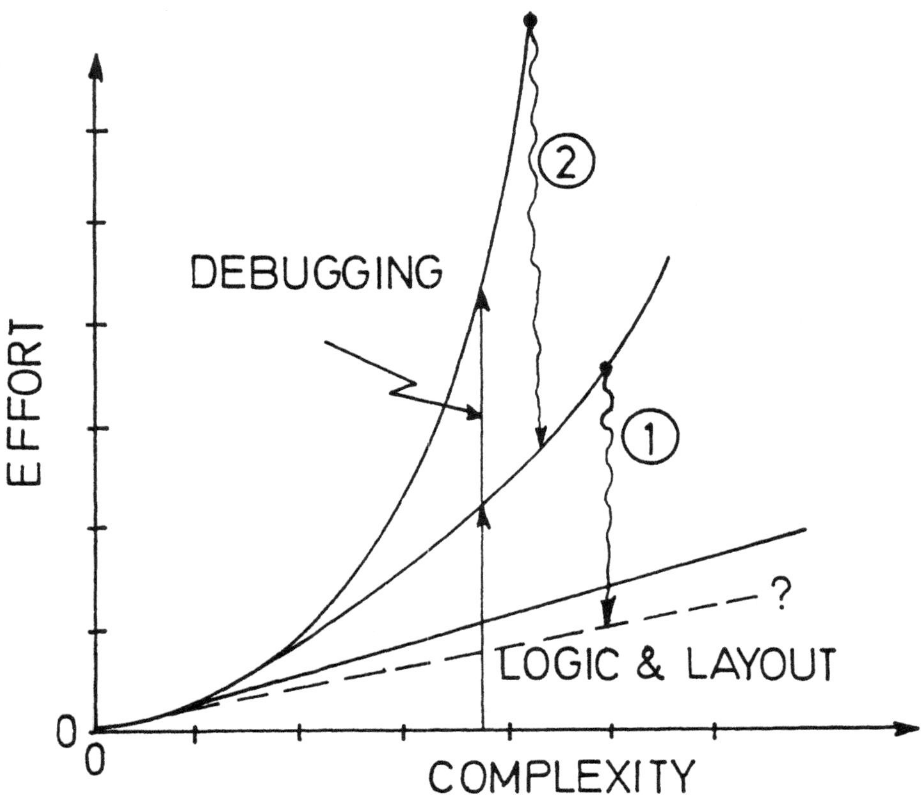

Fig. 1.

Shown there is that, generally, when a human being solves a given problem, then with increasing complexity, two problems occur :

1) If a fixed methodology is used to solve the problem the solution effort increases more than proportional to complexity. Hence there is a need for an other solution methodology which at least reduces the problem back to a linear one... but even this is not sufficient since a VLSI design would be lagging technology by many years ...
We need a design methodology which is sublinear with complexity and it will be shown below that this requires more than "structuring" the design.

2) Of even more importance is the problem of design correctness. It is staggering to notice that if 350.000 transistors would be 350 x (1000 transistor LSI's) and a 1000 transistor prototype chip has 50% chance of working first time then there is 2^{-350} chance for this for the VLSI chip... Furthermore diagnosis of a prototype (related to, but not the same as testability) would be nearly impossible.
As a result without drastical error reduction in all design steps debugging cost would be intolerable.
From the above it follows :

1) the need for design methodology with greatly improved throughput and 100% testability & diagnosability.

2) CAD priority 1 : insure design correctness by checking all human intervention at all stages or/and automate as much as possible.

3) CAD priority 2 : CAD tools should be adapted to/or impose the above design methodology and increase its throughput.

4) The final goal of CAD techniques should be the optimization of the final product cost.
This can be either :

- the design cost for low volume, quick turnaround "custom" VLSI
- the chip size for mass produced IC's.

However the reduction of design cost will certainly become a necessity in all cases since otherwise it will become dominant and unreasonably high volumes would be necessary to write off the development cost. Therefore the emphasis will be shifted from silicon area consumption towards faster and more reliable design making design aids and design automation pure necessity instead of luxury.
Nevertheless in each case where CAD tools need to be conceived, the trade-off between silicon optimization and increased throughput and design correctness should be carefully considered.

5) A very serious problem of CAD tools is that the final users are designers, not computer scientists (could become so in the

future ?).
Therefore special attention to the user interface and interactivity needs to be payed. Many systems look phantastic on paper but work only on the one example published. In this contribution we will try to limit ourselves to these techniques which are either working or likely to work in the near future.

2.2. Design methodology and computer aided design

In this paragraph we will briefly look at the characteristics of complex design problems and look at the way CAD tools should fit into this. We will then study the problems of current LSI design methods and try to derive general properties of VLSI systems likely to alleviate the design problem to hopefully get back in to a "sublinear" problem.
Design in general can be considered as a transformation of a behavioral description into a physical description.
This process is, for complex systems such as VLSI, achieved by hierarchical decomposition of behavior into structure as intermediate form (see Table 1).

- Behavioral description : textual or mathematical description of a system. Table 1 shows how behavior can be considered at different levels of abstraction thereby creating :
  hierarchy in behavior.
  Indicated in Table 1 are the usual levels of hierarchy in electronic systems.
  Shown also are a number of computer languages to describe the behavior of systems. These languages will be discussed in more detail in 4.

- Structural description : behavior is then (not unambiguously) translated into structure i.e. the interconnection of functional blocks. A functional block represents a comprehensive behavioral I/O relationship between signal representation (see Table 1) at the level of behavioral description.
  Clearly structure can also be expanded hierarchically in much the same way as behavior thereby creating :
  hierarchy in structure.
  Table 1 summarizes structural representations. They can clearly be divided into labeled graphical representations (diagrams, schematics, rectangles ...) and textual representations which could be called : structural description languages.
  Well known are for example : circuit simulation input language, logic simulation input language which are single level languages. They are of limited use for a VLSI design since all levels of hierarchy will be necessary.
  This creates the need for :
  hierarchically structured description languages (HSDL) such as SDL [8], MDL [9], BLT [10] which will play in the future a very im-

		BEHAVIOR		STRUCTURE		PHYSICAL
Level	Signal	Abstract form	Computer form	Abstract (Graphics)	Computer (textual)	PCB VLSI
System	Bytes Bits Numbers	Algorithms Flow-charts	GPSS [1] SIMULA [2] ADLIB [3]	Petri-nets Block diagrams	PMS [6] SABLE [7] SDL [8]	
Register	Bytes Bits Clock	Finite state machines State-table State diagrams	ISP [4] RTL [5]	Block diagrams	MDL [5] BLT [10]	
Gate-bit (Opamp)	O,1,X U,D,Z delay	Karnaugh maps Boolean equations p-z transforms	-	Gates-FF-opamps & feedback & interconnection	LOGIC SIM. INPUT	?
Circuit	v(t,w) i(t,w)	Network equations Time or freq. domain	-	R,C,transistor schematics Sticks diagrams	CIRCUIT SIM. INPUT	
Devices	v(t) i(t) N(t,x)	Transport equations of species, electrons, holes ...	-	Geometry (rectangles) and technology	CIF [14] SUPREME	

TOP-DOWN DESIGN

BOTTOM-UP IMPLEMENTATION

Table 1.

portant role in IC design as is to be discussed in 4.

- Physical representation : is the final phase of the design whereby structure, at the appropriate level, is translated into physical layout.

As is clear from Table 1. it is common practice to use a top-down design procedure which means that, starting from a behavioral concept, more and more hierarchical structure is created until a sufficiently low level is obtained which can be translated directly into physical representation.
Finally, the physical entities are then implemented from bottom to top (bottom-up implementation). This procedure is similar to the solution of large software problems and has been traditionally used for the design of large digital PCB systems such as computer systems.
This technique works reasonably well for PCB design, since most functional structural blocks translate directly into well-defigned physical entities (e.g. RAM, ROM, Registers in DIP packages). Design is then partitioning, placement and routing the packages on PCB's. Also electrically the functions are well defined such that the signal level representation can be limited in most cases to the Boolean and delay level.
For IC work the situation is very different since the ultimate design goal is to produce masks i.e. rectangular geometrical representation of the transistor (lowest) structural level whereby the chip size is the optimization criterion.
Historically SSI and MSI has been designed in a top-down fashion down to the transistor schematic structural representation.
This behavior has always been checked by circuit simulation (e.g. using SPICE [1], ASTAP [2], TRSIT [3] etc...).
Then the circuit schematic, which does not contain geometrical or topological information is manually translated into topological and geometrical information by (sometimes different) draftsman using mylar sheets and finally digitizing. The digitizing step creates a geometrical design file. Sometimes the drawing step is done using commercially available graphics editor systems with three or maximum four design stations coupled to a 16 bit minicomputer. We will see in 5. that such systems do not really alleviate the VLSI design problem. They are merely expensive pencil and eraser systems. Alternatives resulting from VLSI solutions are discussed in 4.
The only concern or, methodology here is then to optimize silicon area and we can call it, from the layout standpoint, : random manual layout style. Notice now the following problems :

1. No methodology except density of layout.

2. Circuit description for circuit-simulation from schematic representation before layout i.e. not very realistic simulation (possibility for behavioral errors e.g. timing).

3. Manual translation (error-prone) into physical representation. Probability of two errors :
   - incorrect translation of structured into physical representation
   - violation of layout rules in geometrical layout and therefore :
4. Need for verification of schematic from digitized layout requires :
   - a graphical description language (e.g. CIF [14])
   - schematic extraction from layout [15] [16]

   Need for verification of layout rules or Design Rule Checking (DRC) [17] software for the full design.
5. The above design style is only possible for small circuits or small circuit parts since no human being can handle much more than 30 items at a time and design correctness is an a-posteriori job which is very inefficient as we will see.

The above discussion now leads to the direct conclusion :

use hierarchy in the physical implementation phase.

As we will see, this hierarchy can, but does not have to be, a mapping of structural hierarchy at higher level into physical layout.

Some design automation systems (see 3.) don't do this but in general, when using design aids, it is good practice since :
- it simplifies the design process;
- it enhances design correctness, debugging and testability ;
- it simplifies the use of a procedural CAD language, file management and data-base problems.

However, even if hierarchy in layout is used then still the VLSI design problem has some nasty characteristics on illustrated in Fig. 2.

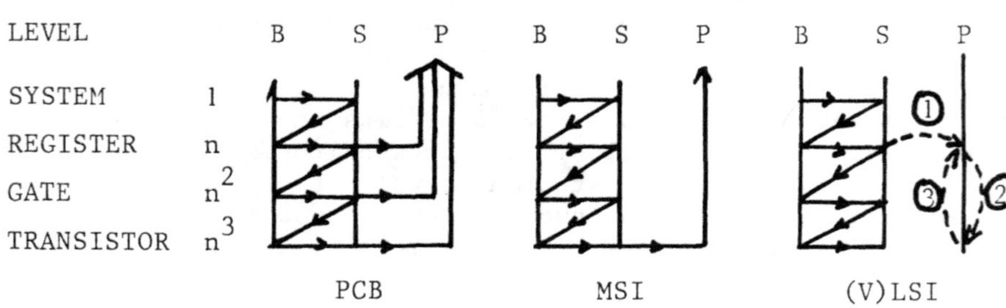

Fig. 2. design styles for PCB, MSI and VLSI.
B: behavior; S: structure; P: physical realization.

Whereas in PCB a RAM, for example, at register level is totally defined in behavior, structure and package, this is not (necessarily) true in VLSI (see dotted line ①).

At the register level only an estimate of the dimensions and interconnections is known. It is only after realization at transistorlevel ② that its geometrical and behavioral characteris-

tics ③ are known unless it was already a block predesigned and stored in a design file or could be built up from elementary predesigned cells stored in a design file.
The above leads to the following property of VLSI design.
Property 1 : in contrast to PCB design in VLSI design the global solution depends very strongly on local information and the global behavior is intimately linked to the local design.
Therefore VLSI design is inherently iterative requiring frequent updates and therefore there is a need for CAD tools capable of :
- intensive data-transport at-and between all levels and representations i.e.:
  horizontal : behavior ↔ structure ↔ physical
  vertical : hierarchy of levels.

- verification for correctness :
  behavioral : simulation, test
  structural : interconnection - structure extraction
  geometrical : design rule check.

- reduction of descriptions, preferrably only one should be used with unambiguous compilation into other representations.

All the above problems are not easy to solve and, to our knowledge, no integrated system exists yet which satisfies all the above requirements. In the following subsection we will try to derive a few other characteristics of VLSI systems which will alleviate the problem. In 3. we will then look at CAD techniques to help layout of VLSI designs and in 4. verification tools will be considered.

2.3. Some particular characteristics of VLSI design.

VLSI integrated circuits will certainly be integrated systems if they are not memory which will probably amount for half the VLSI production. If we exclude memory then what kind of systems will they be ?

1. Predominantly digital with possible small analog interfaces on the same chip. This can be derived from fundamental signal to noise limitations of MOS analog circuits.

2. Digital circuits will need to be synchronous if they are to be 100% testable as no efficient technique is known to insure testability of asynchronous circuits [18] whereas syncrhonous systems can be made 100% testable [19] [20].
Analog circuits in MOS such as A/D and D/A converters as well as filters are sampled, which architecturally can be compared to synchronous behavior in digital circuits.
This leads to property 2 of VLSI systems:

Property 2 : VLSI circuits are time partitioned by non-overlapping multiphase clocks and are therefore amenable to register transfer like [5] descriptions. In analog circuits registers are storage capacitors. This time partitioning enables hierarchical behavior verification or simulation as will be discussed in 4.

3. It is very unlikely that VLSI circuits will be fully random logic circuits since they would be untestable and totally unflexible such that high design cost cannot be written off over the small production volume. Furthermore it is hard to imagine on application requiring a 50.000 gate random logic circuit. Most VLSI circuits therefore will be software programmable computerlike architectures bound to contain much memory potentially with embedded logic. They will be parallel or serial signal processing devices such as speech, music, video processors or number crunching chips. In any case they will have to operate on 8...60 bit wide data and therefore be bus-oriented. Such structures lead automatically to very repetitive structures in space and this leads to property 3.

Property 3 : VLSI systems are space partitioned (repetitive) structures which are memory intensive.
Notice that even analog LSI circuits are very repetitive structures such as ladder filters in which each section is alike except for capacitor values which are built as repetitions of a unit capacitor for matching purposes.
The above property 3 can greatly simplify the VLSI design process of translating structure into physical representation and is beginning to be reflected in CAD layout techniques which, as we will discuss in next section are all cell based hierarchical techniques.

3. COMPUTER AIDED TECHNIQUES FOR LAYOUT.

3.1. Three actual layout styles and their problems.

Fig.3a. shows three existing design styles in use today. Two of them are manual and one of them is a CAD method.
Style I is the random-MOSLSI style. This traditional method has already been discussed before. It is not very well suited for VLSI since it has no real methodology, requires three "hand" interventions (schematic to chip plan; chip-plan to rectangles; rectangles to digitizing). It is thus very error-prone and a disaster to correct the inevitable errors. These are usually only detected after prototyping where debugging becomes exponentially more costly with complexity. If succesful the density of layout is best.
Style II is the so called manual structured or cell matrix design style formally described by Mead and Conway [21] and intensively used in many actual microprocessor designs (e.g. M68000, 8086...). The idea is based on property 3 (space partitioning) by consi-

dering each functional block as a (mxn) matrix of which the elements are identical cells of low complexity embedded in an orthogonal routing of data and control lines. Indeed ROM, RAM, PLA, ALU, MPX, REG, ENC, DEC ... functions are all of this type.

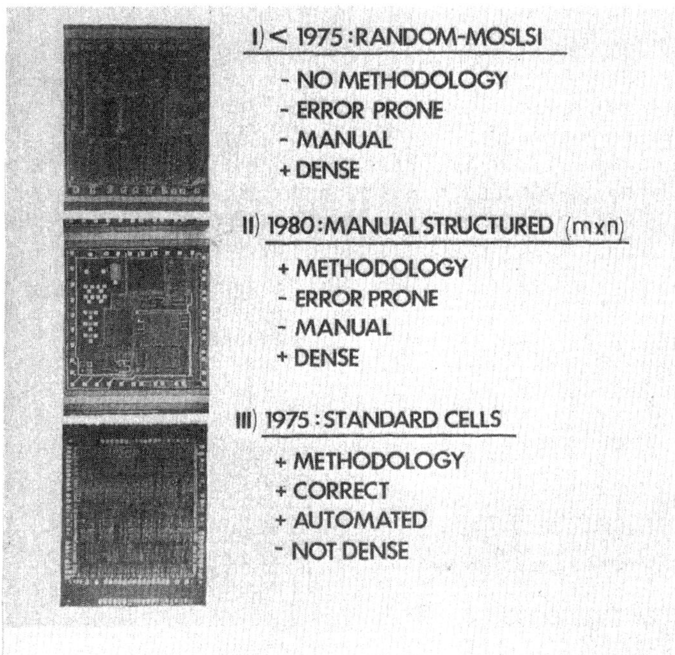

Fig. 3a. Three different layout styles.

In this way the following advantages are achieved :

1. The link between global and local layout is the shape and pinconfiguration of the cell, and the dimension and interconnection of the matrices. This alleviates property 1 of VLSI design.

2. Design at the lowest (rectangle) level is restricted to a small number of matched cells if software for exploiting repetitivity is used both in structural and physical description.

3. Checking can potentially be done hierarchically on a reduced set of data.

4. Data storage can be strongly reduced if again provision is made in the data structure and algorithms to exploit this repetitive nature.

5. Errors are easier to detect since they are local and repetitive too. They are also easier to correct (cfr. microprogramming).

Most of this technique is still manual and, as will be seen below, has kept the layout productivity more or less constant with respect to random design for LSI. Style III is the well known "standard cell" automated CAD layout technique [22] [23] [24] [25]. This technique is the only really automated CAD style (except for master-slice algorithms) which is fully and succesfully used in LSI custom production today.

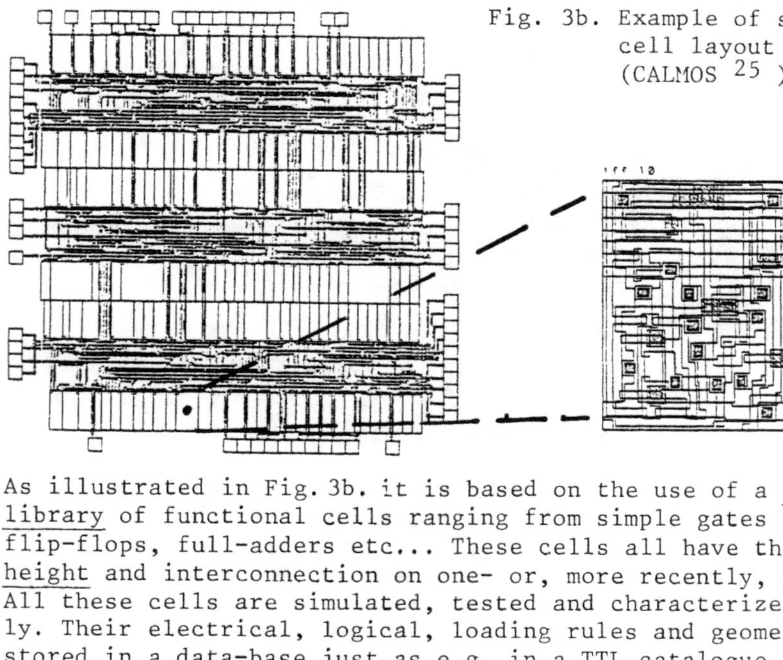

Fig. 3b. Example of standard cell layout (CALMOS 25).

As illustrated in Fig. 3b. it is based on the use of a predesigned library of functional cells ranging from simple gates to complex flip-flops, full-adders etc... These cells all have the same height and interconnection on one- or, more recently, on two sides. All these cells are simulated, tested and characterized separately. Their electrical, logical, loading rules and geometries are stored in a data-base just as e.g. in a TTL catalogue. Clearly this standardization technique reduces the VLSI problem in Fig.2. to the much easier PCB problem.
Starting from one single logic description in terms of the available cells a logic simulation verifies the behavior. When this behavior is correct, the same input is used to :

1. Place the cells in horizontal rows using manual [25] (time consuming) or efficient elaborate clustering and quadratic placement [24] algorithms (CPU time : $n^{1+\epsilon}$ with $0 < \epsilon \leq 1$ and n the number of cells).

2. Interconnect the cells in horizontal routing channels using two orthogonal levels of interconnect and very fast channel routing algorithms. The router algorithm used in CALMOS [24] is linear with complexity. Channel routers for LSI can be guaranteed for 100% routing [24].

3. From the chip layout interconnection parasitics are computed and added to the simulator input. This allows for a load check and/or timing simulation before actual mask production. If the result is unsatisfactory a new design can directly be started by prearranging cells in the placement procedure. The advantages of such a system clearly are :

a) A clearly defined, unambiguous methodology exists starting from a single description.

b) No human intervention at the geometrical phase. This increases correctness of layout rules (in principle no design rule check necessary). Higher than usual yiels for the same chip size are reported since no hidden errors exist.

c) Simulated circuit coïncides with built circuit (no layout to structure translation necessary).

d) Throughput is very high.

e) Computer time is very small e.g. the example shown in Fig.3b. using the CALMOS program contains 234 cells and takes 40 seconds CPU on an IBM 370/158. The design was completed in 4 more interactions with the user for a total layout CPU cost of only 4 minutes.

This first turnaround makes it possible to do LSI runs interactively on any 32 bit minicomputer (simulation and I/O takes much longer than actual layout). The "hit rate" of first right designs is very high. In ref. [25] only one redesign on 75 is reported. It should be noted that a redesign is extremely fast since it only involves new layout runs.

The price paid for this method is illustrated in Fig. 4. It shows the layout productivity in transistors/manday (not including redesign !) versus device density for 6 μm features in layout ($\lambda = 3\,\mu m$). Shaded area I, II corresponds to manual random and recently structured layout of logic circuitry whereas III corresponds to standard cell layout based on data of LTX [22], LOCMOS [25] and CALMOS [24].
Data for I, II are obtained from data on 8080, 8086, 68000, Z80, Z8000 and layouts studied from the last 5 years of ISSCC.
Clearly the following can be concluded :

1. Layout productivity from standard cells is appr. 1 order of magnitude higher than manual layout. Since however the hit rate is much larger than for manual design, the ratio of productivity to working prototype of both styles is much larger than shown in Fig. 4.

The typical design time for a 1000 gate LSI from concept to working prototype for LOCMOS [25] is only 3 months on average.

2. The drawback is clearly in the poorer density (ca. 50...100%) so that the chip size optimization is very poor.

3. Over the last five years when going from random to more structured manual design (I, II) the layout productivity has stayed remarkably constant at ca. 15 transistors/day (Lattin's number is 10/day [26]). Referring back to Fig. 1. this indicates that going to structural design as a methodology just has kept the problem linear but as printed out in [26] that is not enough.
Clearly from Fig. 1 and 4 it follows that from LSI to VLSI the productivity of style II will decrease [26] and in 3.2. we will show that for III the density decreases if nothing is done. Therefore CAD should either have standard cell design towards higher density for the same production rate or CAD methods should be found to increase the productivity and correctness of structured style II.

In the rest of 3. such techniques are considered. In 3.2. methods to improve and extend the standard cell approach towards hierarchical generalized placement and routing techniques (3.2.1.) and, gate matrix techniques (3.2.2) are considered. In 3.3. some new techniques for automation of structured layout are discussed and finally in 3.4. the problem of all design as a structure to layout translation problem will be considered.

3.2. Improvements to the standard cell technique.

Fig. 5. shows the major problems with a classical standard cell approach when applied to VLSI (i.e. more than 1000 cells).
The following problems arise :

1. When the row length increases and cells are randomly interconnected it is clear that routing space grows quadratically. At best, growth would be linear if only local interconnection occurs. Practice indicates that routing area grows as $n^{1.5}$; n being the cell count. Therefore VLSI will decrease density even further as indicated in Fig. 4.

2. When the routing channel length grows, there is a tendency for routing congestion in the middle, creating empty silicon spaces i.e., another factor in density losses.

3. Standard cells are nice for random logic (controllers, control logic etc...) but are less suited for bus oriented regular structures such as ROM, RAM, PLA, MPX etc... which are often used in VLSI.

4. VLSI comes about by scaling technologies which is contradictory to "established" libraries i.e. fast, reliable cell up-

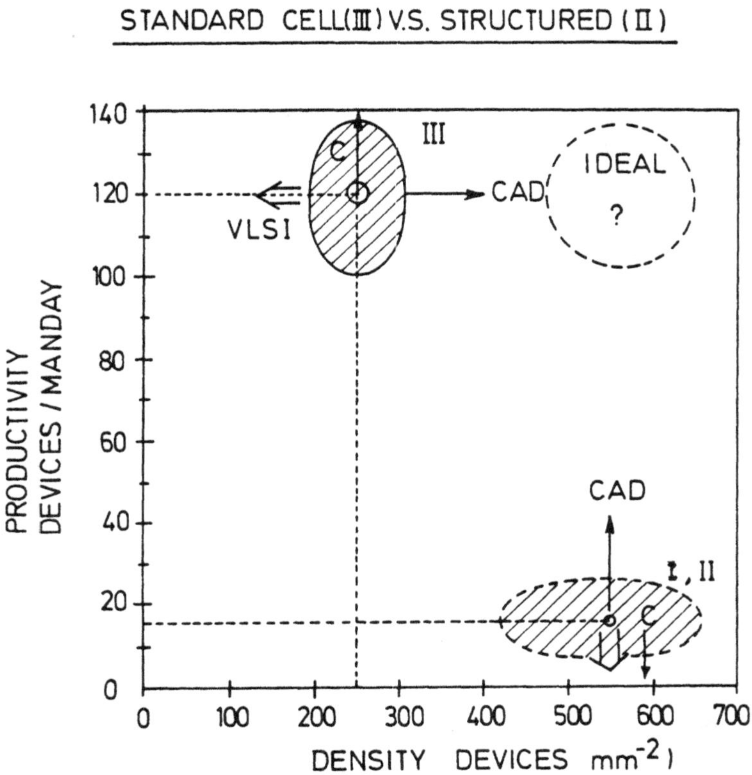
STANDARD CELL(III) V.S. STRUCTURED (II)
PRODUCTIVITY
DEVICES / MANDAY
140
120
100
80
60
40
20
0
0
100
200
300
400
500
600
700
DENSITY DEVICES mm-2)
III
C
VLSI
CAD
IDEAL
?
CAD
I, II
C

Fig. 4.

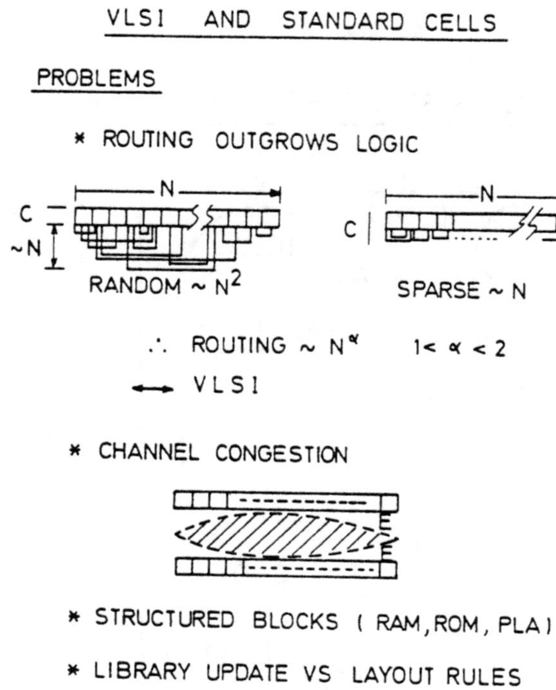

Fig. 5. Problems with standard cell layout for VLSI.

date from cell functions must be possible.
The following are actual tendencies to improve upon the above deficiencies without sacrificing throughput and correctness :

3.2.1. Hierarchical routing and placement.

Generally, as already discussed in 2., a digital system can be considered as an hierarchy of functional blocks with intensive internal interconnection patterns but sparse interblock communication e.g. ALU, RAM (connectivity is $\sim \log_2 N$ with N memory capacity) etc...
If we assign to such blocks estimates of rectangular geometrical structures then the layout problem can be reduced to a hierarchical placement and routing procedure of rectangular blocks of arbitrary size and fixed (groups of equivalent) interconnection positions.
If the constituting blocks at the lowest level are classical standard cell blocks then the design implementation can be fully bottom-up if the partitions are defined in the top-down design phase. If these blocks are limited to LSI-MSI complexity (i.e., 100...300 cells) then, since interblock routing is sparse, the layout density can be kept constant with complexity and automated VLSI layout should be feasable from one hierarchical structural

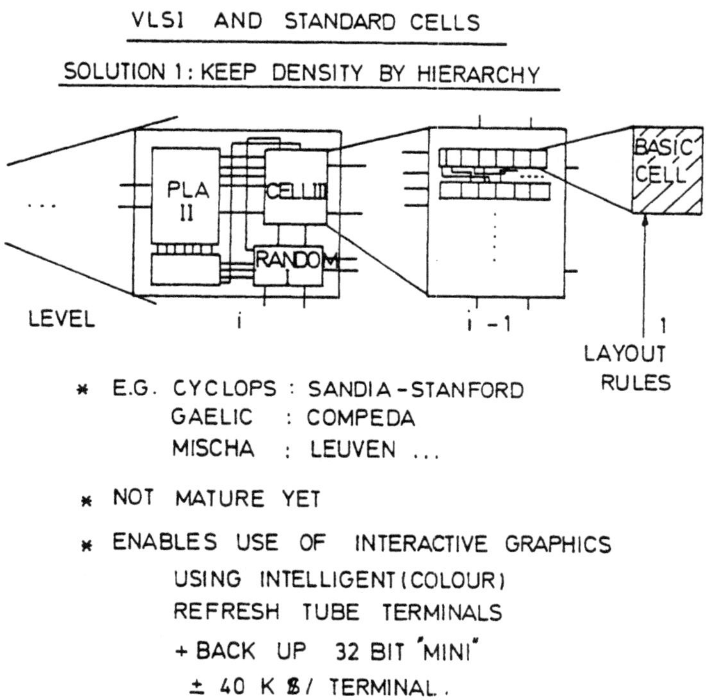

Fig. 6. Hierarchical standard cell design.

description language linked to a cell library.
However, since at the rectangular level no assumption is made about the block content, the blocks can also be of type I or II (Fig. 6.).
In such case, however, if these blocks are not predesigned and stored in a library, the system has to be made interactive to solve the global-local interaction problem discussed in 2.
This hierarchical-routing and placement problem, when coupled to a graphics editor is probably an absolute necessity to build the chip plan at the top-level since at that point a 2 m line on 1cm chip size requires a resolution of 1/5000 which is far above the limit of raster scan color tubes.

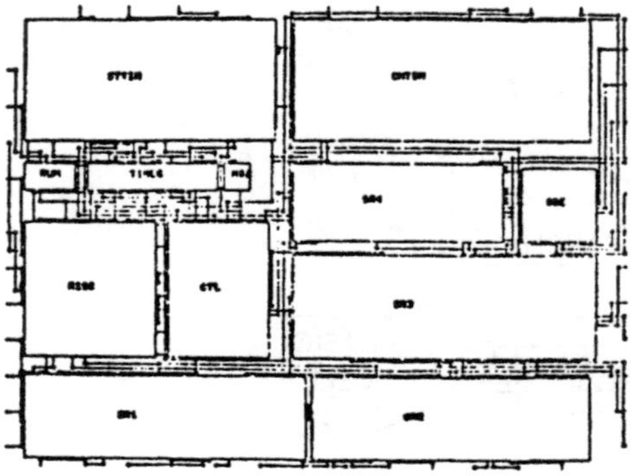

Fig. 7. Hierarchical layout CYCLOPS program example [27].

Since no details of interconnect lines are possible an algorithm with interconnect layout rules as input is an absolute necessity! Several algorithms have been proposed [27] [29]. Preas and Van Cleemput [27] propose a graph based method in which placement and routing are executed separately. The method builds up the substrate to minimize silicon-area and roughly routing space. Fig. 7. shows the layout of a circuit generated by this method. Savings up to 25% with respect to an all standard cell design are reported.
Heyns, Beke and Sansen propose a method based on a global partitioning of the substrate based on graphs and minimizing the routing wire length [30].
The above methods contrast with a method proposed by Loosemore [29] in which placement and routing are built-up in parallel but on a local basis. General problems with these methods still exist because optimal global trade-offs between placement and routing are much more complicated than for row placement and channel routing algorithms.
It is our belief that these methods, when becoming more mature, will become very useful for very quick turn-around custom VLSI design involving much random logic, PLA and memory and for chip planning [31] at the highest level of hierarchy [5].
The disadvantage still remains the lack of density with respect to manual structured design. Recently however a number of new methods similar to cell design have been proposed and can be called gate-matrix techniques.

3.2.2. Gate-matrix techniques.
Problem one of standard cell design is that routing is separated from the densily wired logic cells. When indeed interconnection is taking up most area one should then conclude that it is better practice to embed logic into the routing space and not the other way around. This principle is of course the same as employed to

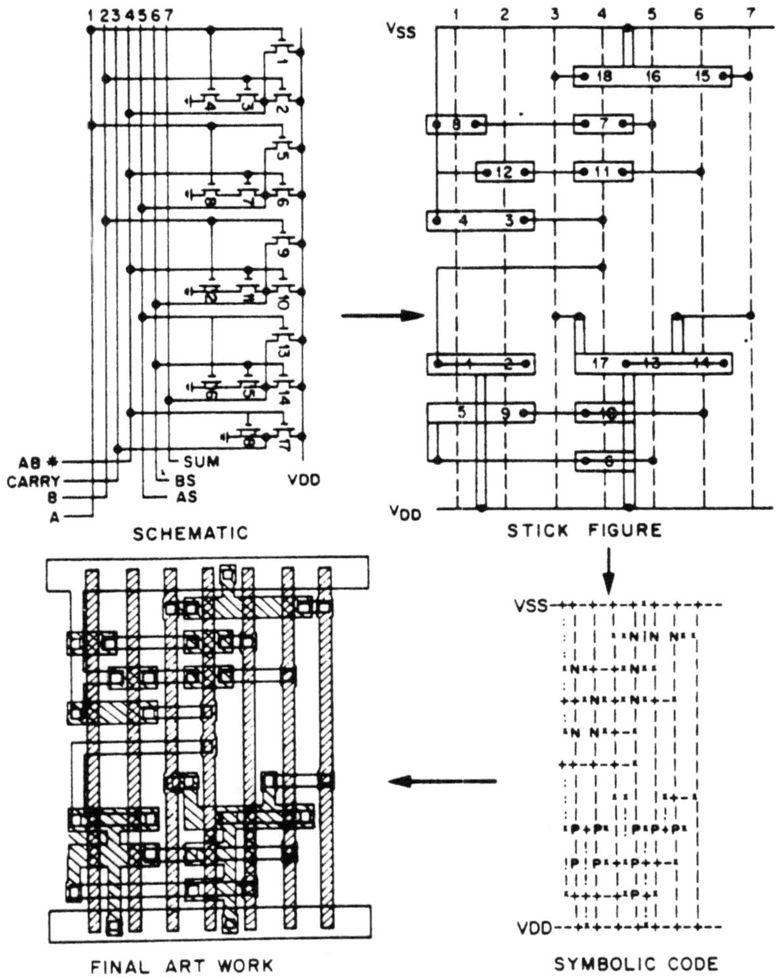

Fig. 8. The gate-matrix approach.

build e.g. RAM's and PLA's where the orthogonal interconnection matrix is "personalized" by single transistor cells.
In [32] Lopez and Law introduce the gate matrix to standard-cell like designs.
Fig. 8. illustrates this concept. The schematic illustrates the inefficiency of the routing channel concept since basically the orthogonal interconnects into the cell columns (here) are a waste of area since in polysilicon transistors are automatically generated at the orthogonal intersect of polysilicon and diffusion. The stick-figure shows simple rules to move the routing inside the cells which can be grouped between orthogonal power and ground power distribution lines.

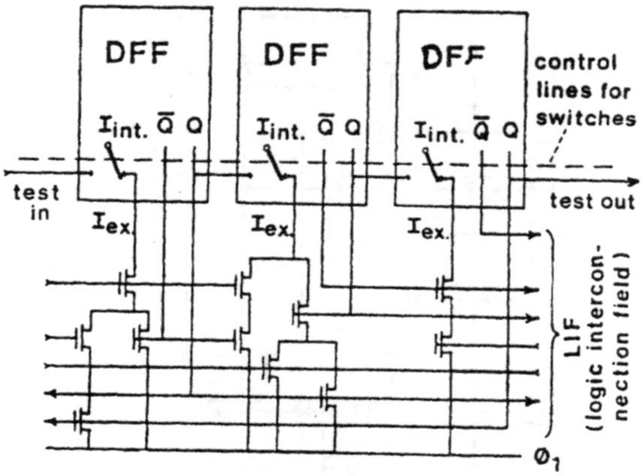

Fig. 9a.
4-phase n-MOS gate matrix with LSSD technique (schematic).

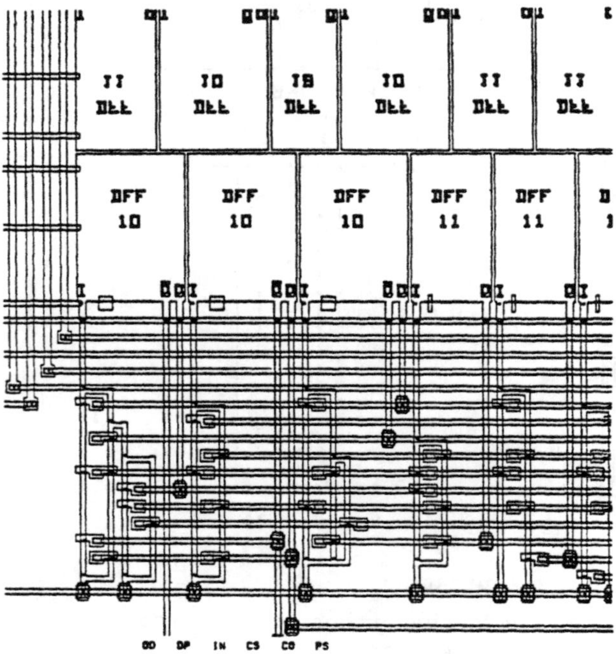

Fig. 9b.
Automatically generated layout from symbolic representation of Fig. 9a.

Although in the original proposal symbolic manual design techniques are used (as shown in Fig. 8.) it is conceivable that the technique can be automated based on existing row placement and channel routing algorithms.
Notice that this technique is most suited for bus oriented structures and less for pure random logic since in that case random communication to other columns takes up extra space. However we have seen that most VLSI systems will be bus oriented and this technique is certainly worth exploring.

Lopez and Law [32] report the layout with this technique of a 20.000 transistor circuit where a density of 600 transistors/mm^{-2} is obtained for 5µm features. From Fig. 4. one can see that this is as good as manual structured design style II. This is not amazing since basically the principles of exploiting interconnection structure is common to both methods and is an argument to the following rule :

take the randomness out of design

An other similar technique used in a semi-automated design system had been proposed earlier by Mulder et al. [30] and is illustrated in Fig. 9. The technique is exploiting the reatioless character of dynamic 4 phase MOS technology.
Exclusively synchronous logic is thus used. Each 4 phase level sensitive D-flip-flop is driven by a logic tree in the logic interconnection field (LIF : Fig. 9). This tree can be synthesised from the Boolean expression to be realized and due to its ratioless aspect and the orthogonal formation of transistors in poly-diffusion lines they can be laid out on a symbolic fixed grid (see also 3.2.4.). Layout is generated by computer with layout rules as input.

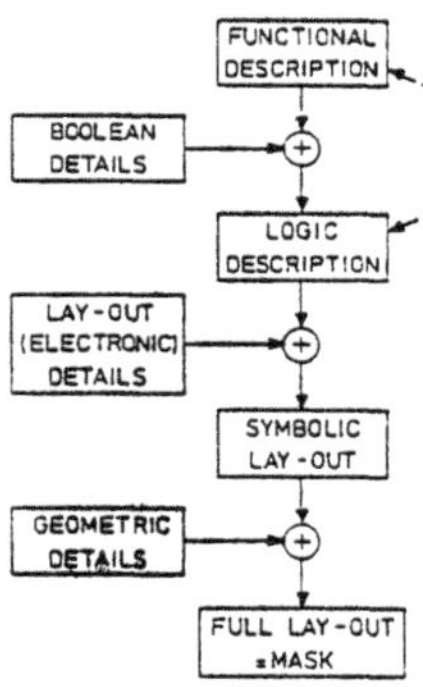

'Integrated" CAD system
or 4-phase n-MOS gate-
ıatrix layout.

The logic description is regenerated from the symbolic layout and compared to the logic description used for behavioral verification in a synchronous logic simulator (see 4.). After symbolic layout loading rules are verified and, if satisfying, masks are generated.
Densities obtained this way are of the order of 400 devices/mm^2 for 8µm rules which is again equivalent to hand-layout.
In this technique, for testing purposes, the D-Flip-Flops can be connected to a shift register mode such that, using the LSSD technique [9], a 100% testable (V)LSI chip can be designed in a very short time.
The method is well suited for serial type or pipelined logic as often occurs in arithmetic units for signal processing.
It is less suited for highly parallel (bus) oriented logic structures.

In summary one can say that the hierarchical technique is very general. The programs exist but the density is at best equal to the standard cell technique if the latter is used as a basis. The gate-matrix techniques show that for particular circuit techniques and logic architectures high density, layout can be quickly generated from logic equations. The gate matrix techniques are based on symbolic layout and therefore quickly amenable to changing layout rules (see 3.2.4.).

3.2.3. CAD techniques for structured design.

The space partitioning property of VLSI circuits leads to structured repetitive design using matrix-like structures. The basis is to put routing first and enter personalizing devices later. If the number of building blocks is restricted to e.g. the following :

- Control logic : PLA or standard cell
- Memory : RAM, ROM, PROM
- Registers : clocked pass transistor registers
- Bit-sliced ALU types;

then it is conceivable that, starting from a high level description language (HLL), a logic synthesis is made of these building blocks followed by a structural (hierarchical) synthesis which, finally, with layout rules as input is translated directly into layout on silicon.
Such a technique has been called "Silicon Compilation technique" by J.P. Gray [33]
Basically the creation of random logic layout by the standard cell technique can be considered as a form of Silicon compilation. The name results from a similarity with "structured programming" in software whereby a flexible HLL is translated into machine instruction by a computer.
Other techniques for random logic design make use of PLA techniques.
Clocked PLA's using multiphase clocks can easily be used to synthesis and design finite state machines testable by the LSSD technqiue [19] [34] such as shown in Fig. 10 in a simple example. Since PLA's are well structured AND-(N)OR configurations it is possible to synthesize a minimum (in a heristic sense) PLA [34] or a hierarchical set of clocked PLA's [35] from a textual HLL description of the state equations of the machine.
As illustrated by Fig. 11, a PLA structure can easily be generated from a limited set of basic cells when the I/O routing is fixed (6 cells are sufficient in n-MOS). Furthermore, due to the known physical and logic structure of PLA's it is possible to write dedicated automatic test pattern generation algorithms for PLA's as described in [36] [37]. These can be therefore be very efficient and complete and when combined with LSSD lead to the same testability as the 4 phase n-MOS gate matrix technique [20] mentioned above.
Fig. 10. illustrates how all the existing techniques could be

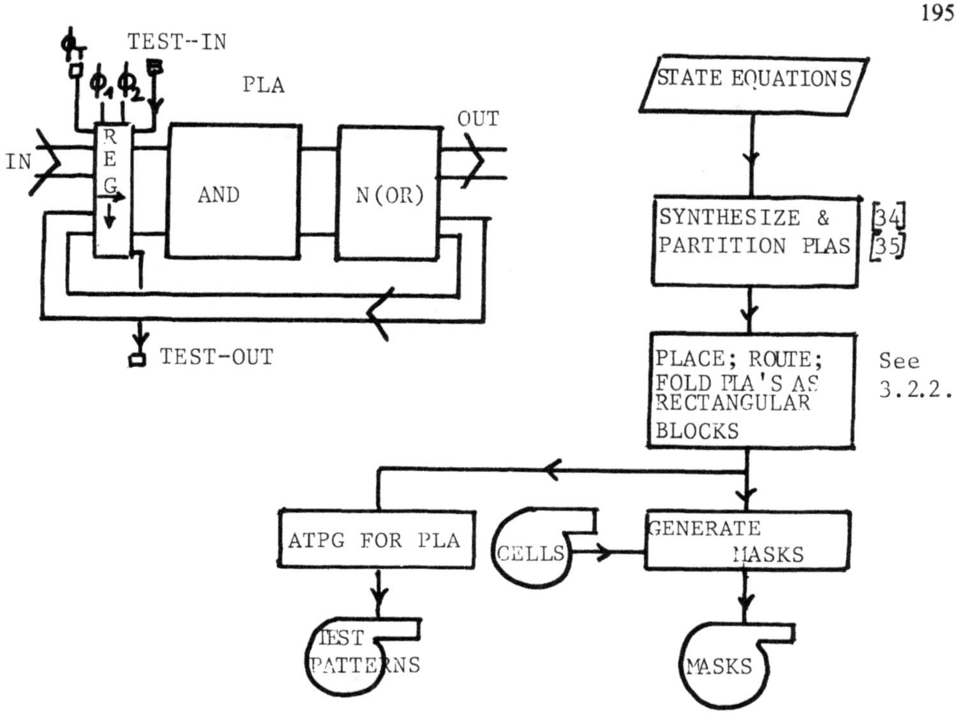

Fig.10. PLA based design system.

combined into a finite state PLA based VLSI compilation technique. The synthesis and partitioning technique has been described by Ayres [35] and a full design system based on the LSSD technique is being developed at K.U.Leuven.
Another approach by Johannsen is the so called Bristle Block Approach [38] which will be discussed in more detail elsewhere in this course.
The idea is illustrated in Fig. 12 and is obviously oriented toward computer structures. Such VLSI structures are built from bit-sliced parts such as ROM, RAM, ALU, REG etc... They form the CORE of the structure and are represented by rectangular structures built of "stretchable procedural cells" at the cross-points of horizontal buses and clock lines and vertical control lines. The pitch is adjusted by the "most compressed" cell in the corresponding direction. The position of vertical control lines is automatically generated.
The critical cell causing the pitch can then, as individual small cell, potentially be redesigned until the best compact layout is generated. When the core is completed, the control logic can be compiled e.g. by PLA synthesis discussed above.

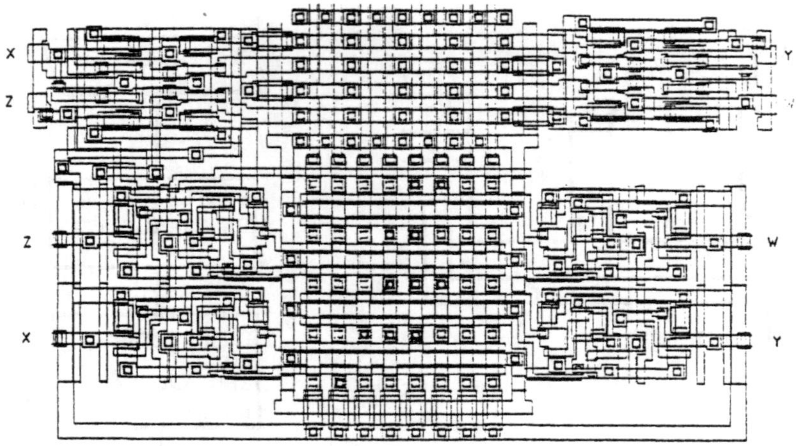

PLA modulo 10 counter.

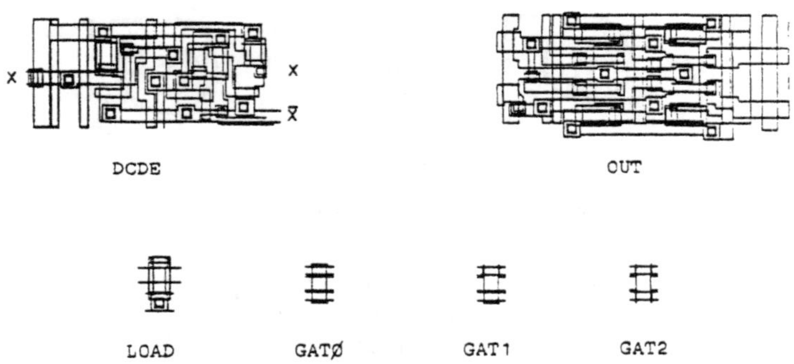

Basic cells for PLA synthesis.

Fig. 11.

This technique is a direct answer to the potential automation of manual structured design and possibly the way to dense and fast layout. It exploits the features of VLSI discussed in 1.
This concludes an overview of layout CAD techniques with the exception of ULA approaches which fall outside the scope of this paper. It should be noted that in the acutal state of the art only standard cell automated layout is used intensively. The other methods, with exception of the 4 phase gate-matrix still have a long way to go with hierarchical placement and routing making most progress towards designer friendly software.
No matter what method will be used, at some point the link between circuit-schematic, layout rules and rectangles has to be made. This relates to the problem of cell design.

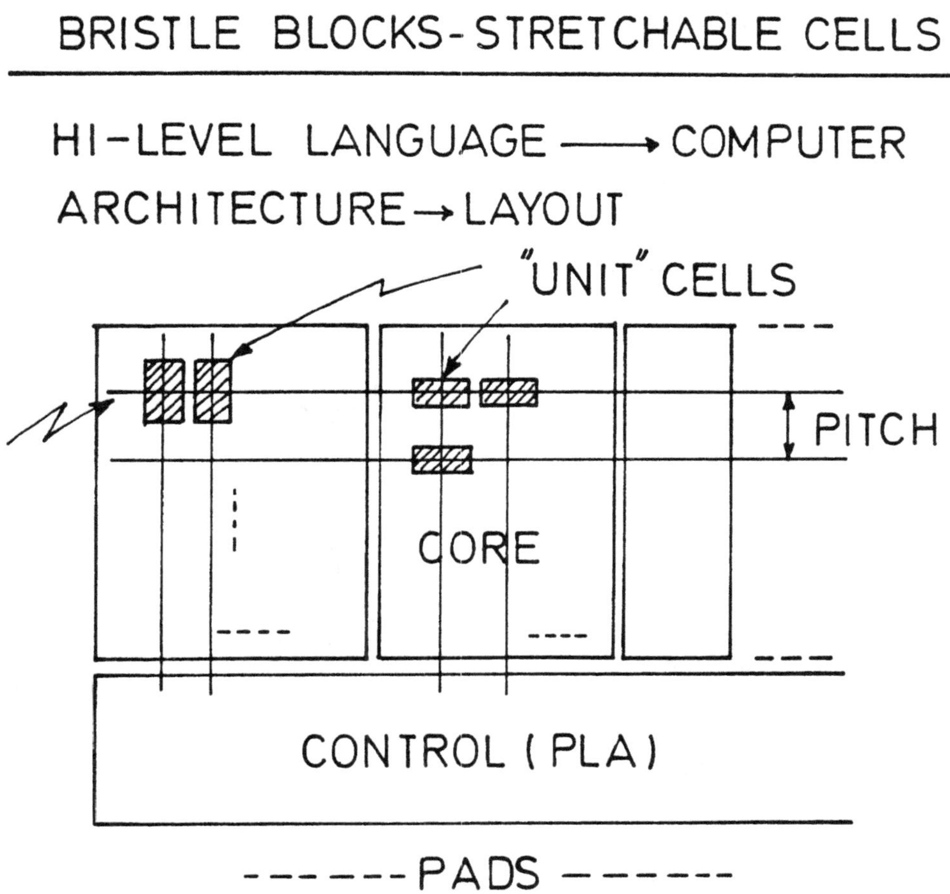

Fig. 12.

3.2.4. Designing the cells-symbolic layout.
At the lowest level of design, the circuit schematic has to be translated into rectangles on different mask levels respecting a number of geometrical layout rules.
This process is characterized by the fact that schematic information is functional whereas layout information is topological with a criterion of minimum space and (predescribed) interconnection pattern (pins).

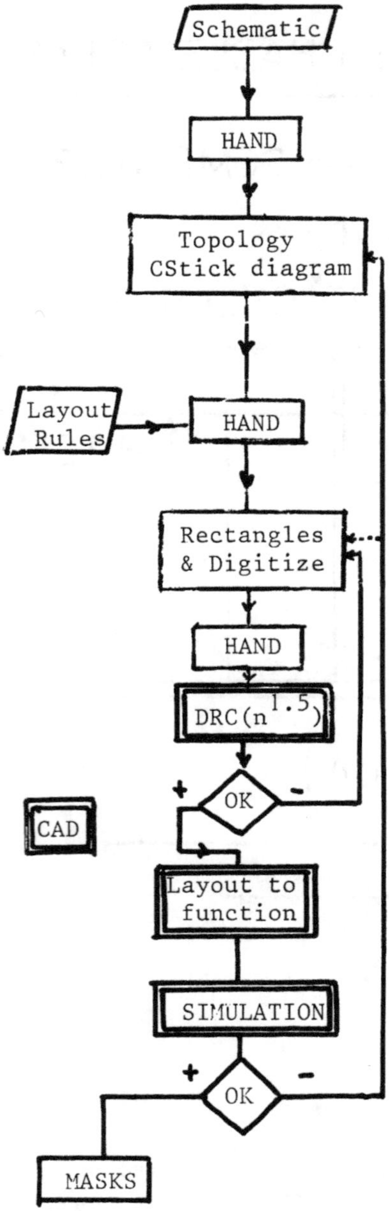

Fig. 13a.

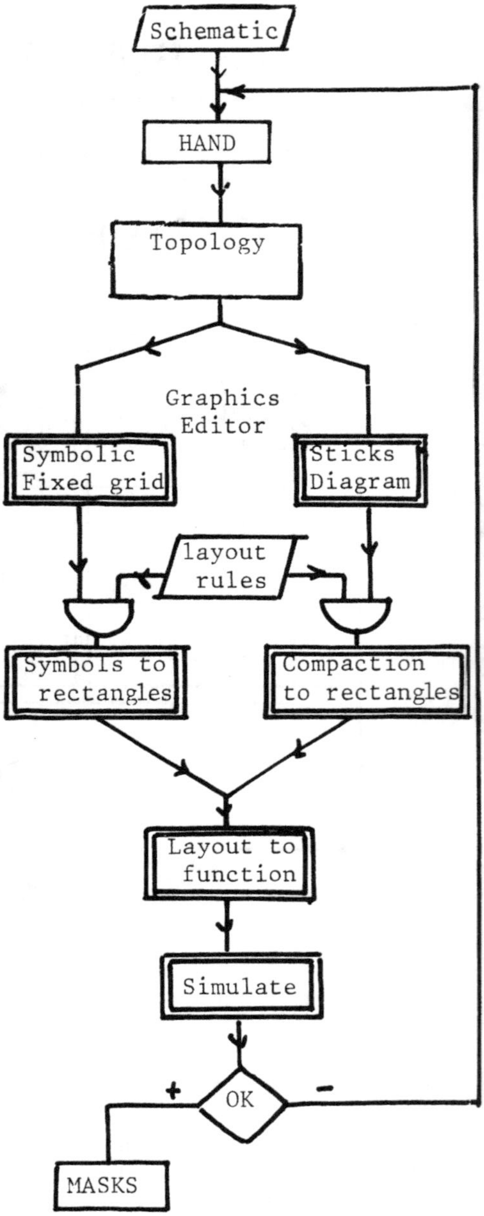

Fig. 13b.

Fig. 13a. illustrates the traditional and actual procedure in which three manual interventions are necessary i.e. three error prone steps.
Two checking steps are an absolute necessity

a) Design rule checking (DRC) software [17]
This is software based on set operations between different mask geometries (union, intersection, difference etc...) and on dimensional checks on these results. Theory and experience shows these algorithms to suffer from :

$$\text{CPU time} \sim n \log n \; \ldots \; n^2$$

such that DRC is excluded for full VLSI designs e.g. a DRC on 10^6 items on an IBM 370/158 requires ca. 8 hours CPU time. Therefore either hierarchical DRC is necessary or use has to be made of new computer architectures such as array processing on bit-maps for a "set operation computer structure".

b) Even if no design rules are violated no guarantee exists that layout maps into behavioral structure. Therefore Layout to function (LTF) software translators are necessary. A few experimental programs exist [15][16] to extract transistors and recognize logic functions from their interconnection pattern. The latter is a very difficult task if one is not restricting one self to simple elementary logic functions such as NAND-NOR-ANDORI-ORANDI.
Since these algorithms use the same set of operations as DRC, they suffer from the same cost and therefore are also limited to serial cells. Design verification can then be done by comparing interconnection patterns (not an easy task) or by simulation whereby a problem is which stimuli to use for a correctness check.
Fig. 13b. shows two evolving CAD techniques for cell design called symbolic layout.
Basically these methods create layout from a symbolic graphical topological representation of the circuit schematic either drawn on paper and then digitized [39] or drawn on an interactive graphics editor.
The basic principle is that detailed layout is produced by software with design rules as input. Two basic techniques exist as shown in Fig. 13b. and illustrated in Fig. 14. and 15.

a) The fixed grid system [39][40] (fig. 14). requires the layout plan be drawn on equally spaced grids. Here the grid spacings are determined on the basis of a set of dominant layout design-rules. The conversion program operates simply to replace each symbol in the user input with its actual geometric structure. Fig. 14. shows an example of a symbolic layout plan and the resulting layout from a fixed-grid symbolic layout system.

The fixed-grid symbolic layout has been used succesfully by a number of companies in the electronics industry. It has proven to be an effective method for generating reasonably compact layouts in a shirt time. Note, however, that the compactness of the resulting layout depends on two factors : First, the original

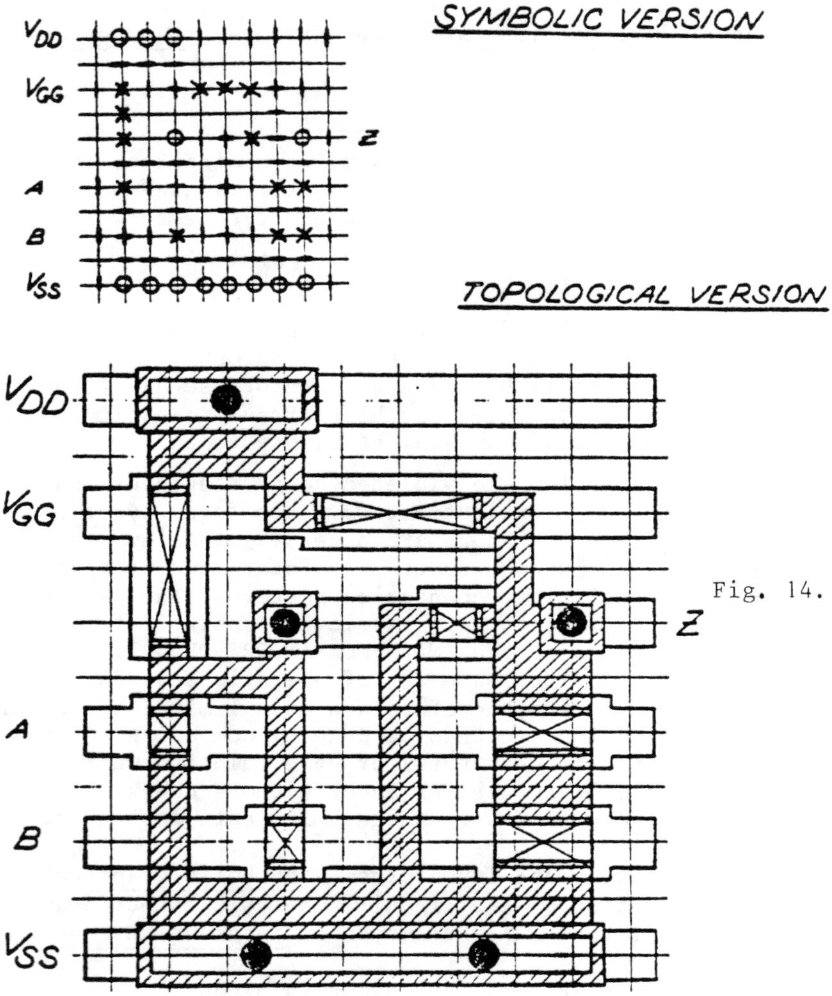

Fig. 14.

user input in the symbolic form has to be very compact as the program only performs symbol-to-geometry conversion. Second, since the design-rules used to determine the common grid spacing normally do not give rise to identical minimum pitch (the sum of the minimum line width and spacing) on all mask layers, layers with smaller minimum pitch are made sparser in order to match the larger minimum pitch of other layers. Alternatively, several grid spacings can be used for different mask layers to avoid this low-density problem at the expense of increased system complexity.

b) Sticks-compaction algorithms.
The second type of symbolic layout method avoids these difficulties by performing geometry modifications, such as compaction and expansion of elements, and limited topological changes, such as

introduction of jogs, on loose symbolic layout plans or other types of layout representations drawn by the user [41], [42], [43], [44], [45]. This type of method is termed the relative-grid method since the loose layout plan indicates only the relative placement and interconnection of element symbols with respect to other symbols.
Fig. 15. shows an example [46] of a cell design starting from a symbolic drawing on graphics terminal using symbols from a menu This so called stick diagram is then compacted into the layout shown from the layout rules. The results can be used directly for simulation and schematic verification. If sticks diagrams are stored in a data base, automatic standard cell update is conceivable.

The relative-grid method is more versatile because it lets the user ignore design-rule requirements almost completely. Thus the user can spend most of the design time wisely on developing good layout topologies based on his experience, cleverness, and common sense. Since the actual geometric layout of typical cells and simple building blocks can be generated with relative-grid symbolic layout systems (such as CABBAGE) in a matter of a few minutes on minicomputers, the user can receive almost immediate feedback on the topology he has devised.

The versatility of the relative-grid symbolic layout approach does require much more intelligence on the part of the computer program.

3.3. Conclusions for layout CAD tools.

The high cost of DRC at full chip level and the shear zero chance for design correctness for manual layout makes CAD indispensable in this step. Hierarchical design philosophy will be used and use will be made of a small number of cells (e.g. $<$100 transistors) probably designed by non-fixed grid symbolic layout with direct local functional checking capability.
Layout is a synthesis of these cells from a HLL language. The following techniques exist or emerge :

- Standard cell technique : reliably working up to 30000 transistors at expense of density and restricted to random-non-bus logic with embedded memory and PLA's.

- Hierarchical placement and routing : in principle possible for VLSI but still a long way to go in research. No particular style. Blocks can be any style.

- Gate matrix techniques : very dense for bus-oriented structures or clocked synchronous structures.
Linked strongly to given technology. Not very flexible.

- Bristle Block Approach : exploits VLSI computer architecture. Could improve throughput of manual structured design but still a lot of research needed.

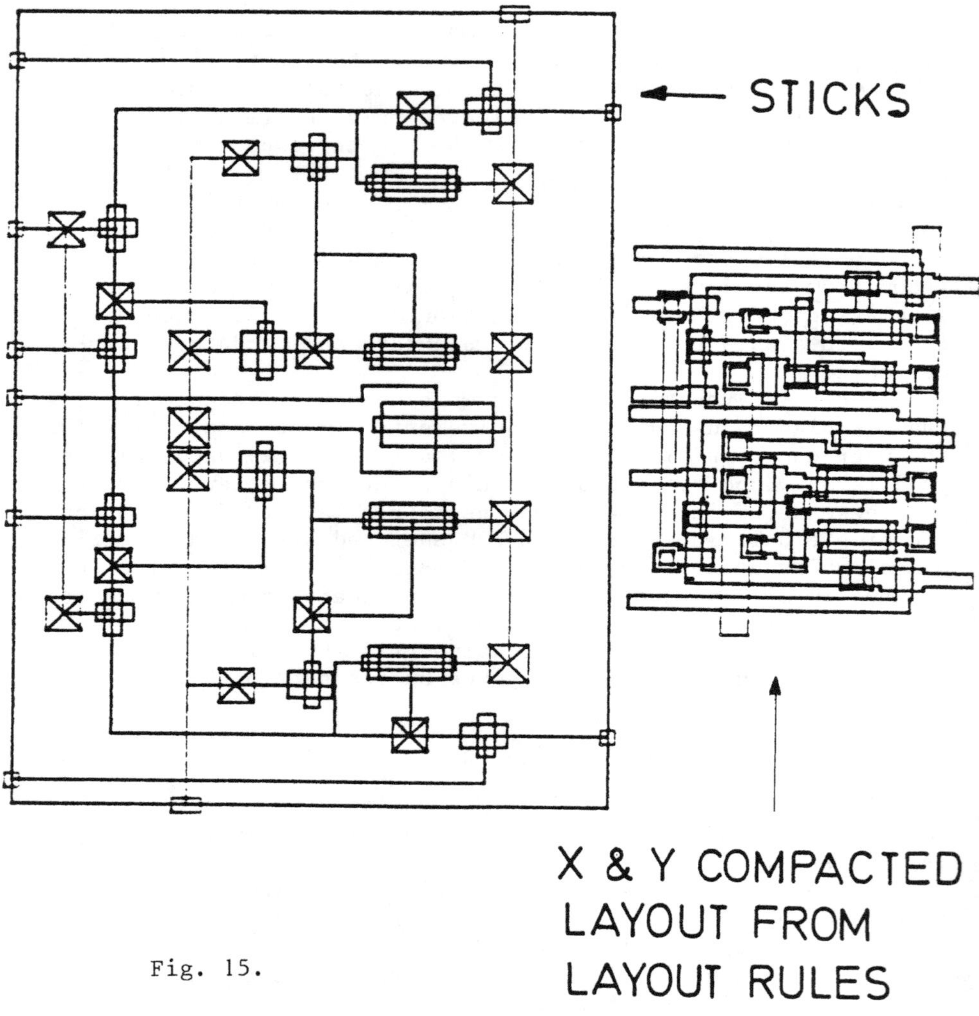

Fig. 15.

4. THE ROLE OF SIMULATION IN VLSI DESIGN.

4.1. Different forms of simulation and their origin.

Simulation can be defined as computer software for verification of the behavior in time and/or frequency domain of circuits and systems. Such a simulation always proceeds from a description in a behavioral or structural language (see table 1) using mathematical models operating on signal representations corresponding to the given level of abstraction.

Therefore many different forms of simulation exist as summarized in table 2.

With exception of circuit simulation, all tools find their origin in the area of PCB system design. Therefore most IC designers are

Level	Form	Model	Signal	Algorithms	Languages/program
System	System simulation	activities, processes	Numbers, bytes actions	Event-scheduling Sequential execution	SIMULA, GPSS, PMS ...
Register	Register transfer simulation	Register (1&2-D), Transfer logic, clocks	1,0	Synchronous event scheduling	CASSANDRE, RTS, DDL, ISP, AHPL ...
Gate	Functional & logic simulation	Gates, flip-flops, PLA ...	1,0,X,U,D,Z + delay	event-scheduling selective trace table look up	TEGAS, F-LOGIC, DIANA, SPLICE ...
Gate-transistor	Timing simulation	Transistor tables Gates	v(t)	submatrix & equation scheduling	MOTIS-C, DIANA, SPLICE ...
Transistor	Circuit simulation	R,L,C, Transistors	v(t), i(t) v(jω), i(jω)	matrix inversion Nonlinear iteration Num. integration	SPICE, MSINC, ASTAP, TRSIT, DIANA, SPLICE ...
Device	Device simulation	p-n-insulator structures	v(t), i(t), p(t), n(t)...	matrix inversion iterates; partial D-E solving	CADET, SITCAP, ...

Table 2.

only familiar with circuit simulation whereas now, as a result of the high complexity, higher forms of description and simulation are considered for IC design.
As will be shown below, this proces is rather problematic due to fundamental differences between PCB and VLSI design.
In what follows, first the limitations and role of circuit simulation will be covered in 4.2.
In 4.3. algorithms and role of logic simulation is considered and in 4.4. register-transfer and system simulation. In 4.5. the concept of timing simulation is introduced while in 4.6. the concept of mixed-mode simulation for MOSVLSI is considered together with some considerations on future CAD hardware.

4.2. Circuit simulation, role and limitations.

As circuit simulation is covered in another lecture we will only briefly expand on its role in VLSI design and concentrate on its limitations.
Since 1968 numerous circuit simulation programs have become available [1][2][3].
These programs are based on lumped algebraic-differential terminal models for devices which are now being extended down to the micron channel length level.
The algorithms allow for very accurate DC, temperature, time, frequency, sensitivity and noise analysis whereby the link to technology as well as geometry is very close. As discussed previously, the circuit description input for circuit simulation can be (and should be) automatically generated from layout files [5] [6]. From the circuit topology the set of coupled differential equations (if time domain) is generated and the generally accepted technique then is as follows :

1. implicit numerical integration techniques [46] with automatic time step control transform the network into an algebraic nonlinear network with a given time step h;
2. numerical iteration schemes such as secant and Newton-Raphson [46] transform the previous network into a linear algebraic one within an iteration step within a time step;
3. This network is described by a sparse matrix with a dimension n nearly equal to the node count of the original circuit. This has the following consequences :

a) No sacrifice in accuracy (except lumping distributed effects) is made. Therefore circuit simulation is an invaluable tool for the software breadboarding (the only one possible) of MOSLSI critical circuit blocks (cells) with :
- very thight coupling : e.g. analog circuits (MOS opamps), clock-generators, flip-flops, schmidt-triggers, sensing amplifiers in memory, dynamic RAM storage cells, ripple through transmission gate chains dynamic comparators etc...

- critical timing paths e.g. in PLA's, RAM decoders, output amplifiers, high speed dynamic MOS circuits, clockfeedthrough effects in dynamic logic etc...
- analog sampled circuits: e.g. switched capacitor filters, charge redistribution A/D, D/A converters, multiplexers, demultiplexers etc...

Examples of the first two applications will be given in another lecture. The area of analog sampled circuits is relatively new and only recently [47][48][49] circuit simulation programs became available to simulate sampled data-circuits both linear and nonlinear in time, frequency sensitivity and noise.
As an example Fig. 16, 17, 18 show respectively an elliptic switched capacitor filter, its input description for the DIANA [48] program and the simulated passband for three values of the opamp gain. The simulation techniques used in this new program have been used to simulate complete A/D converters and complex switched capacitor filters up to 8th order, including clock-feedthrough, stray parasitics, non-linearities, aliasing and sensitivity effects.
Also macromodels for digital filter simulation are implemented. The algorithms [47] are fully compatible with existing analysis techniques and therefore can handle this new type of analog LSI techniques.

b) High accuracy is paid for in terms of maximum circuit complexity that can be handled.
For a 1000 time point analysis and 4 iterations/time-point 4000 matrix <u>updates</u> (for nonlinear devices) and <u>inversions</u> are required of the <u>full</u> matrix of the circuit to be analysed whereby most of the time (ca. 80%) is spent in nonlinear element updates (computing FORTRAN functions).
Suppose we simulate a growing chain of N gates over a fixed time period T. Then the matrix grows proportional to N. However also the time for the rising and falling signal edges to ripple through the chain increases and due to time step control, a large number of intermediate time points within the period T have been calculated. Clearly this number is also proportional to N and therefore in this case :

$$CPU_{time} \sim N^2 \qquad (1)$$

For more "square" networks in average one can say :

$$CPU_{time} \sim N^{1.5} \qquad (2)$$

This leads to excessive simulation cost for complex circuits as illustrated in Fig. 19. where the approximate simulation cost in $ is shown versus the number of MOS transistors m per circuit. Data are extrapolated from experiments with TRSIT [13] and SPICE [1] on IBM 370/158. Full line is for "serial" networks, dotted line for "square" networks.
Data are for 1000 user requested time points.

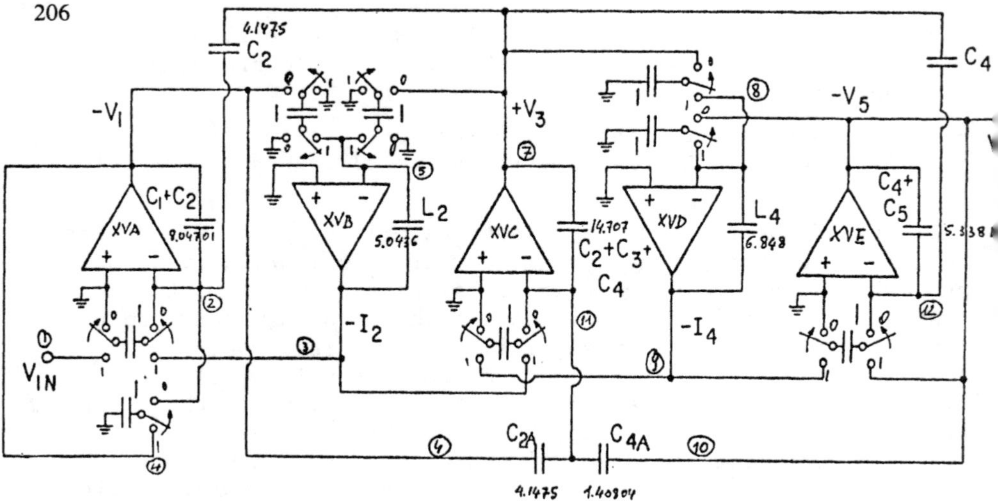

Fig. 16. Elliptic filter.

```
ESAT ELLIPTICAL FILTER - PASSBAND STUDY
.OPTIONS NODE OPTS
.SC IN=1 OUT=10 CL=CL TIME=3U FREQ
.TRAN NOUT=1 DELTA=1
; CLOCKING CIRCUITRY
INPUT CL V-T=0 0U 2 1U 2 5U 0 3U 0 10U CYC
; CIRCUIT
C12 4 2 8.4701
C234 7 11 14.707
CL2 5 3 5.0436
CL4 8 9 6.848
C45 10 12 5.3381
C4 12 7 1.40804
C2 7 2 4.1475
C2A 11 4 4.1475
C4A 11 10 1.40804
XVA 4 0 0 2 100
XVB 3 0 0 5 100
XVC 7 0 0 11 100
XVD 9 0 0 8 100
XVE 10 0 0 12 100
S31 0 4 2 CL 0 C=1 VT=1
S321 CA 1 0 CL 0 C=2 VT=1
S322 CA 3 2 CL 0 C=2 VT=1
S331 CC 0 4 CL 0 C=2 VT=1
S332 CC 5 0 CL 0 C=2 VT=1
S341 CD 0 7 CL 0 C=2 VT=1
S342 CD 5 0 CL 0 C=2 VT=1
S351 CH 9 0 CL 0 C=2 VT=1
S352 CH 3 11 CL 0 C=2 VT=1
S36 0 8 7 CL 0 C=1 VT=1
S37 0 8 10 CL 0 C=1 VT=1
S391 CG 9 0 CL 0 C=2 VT=1
S392 CG 10 12 CL 0 C=2 VT=1
.END
```

Fig. 17. DIANA input.

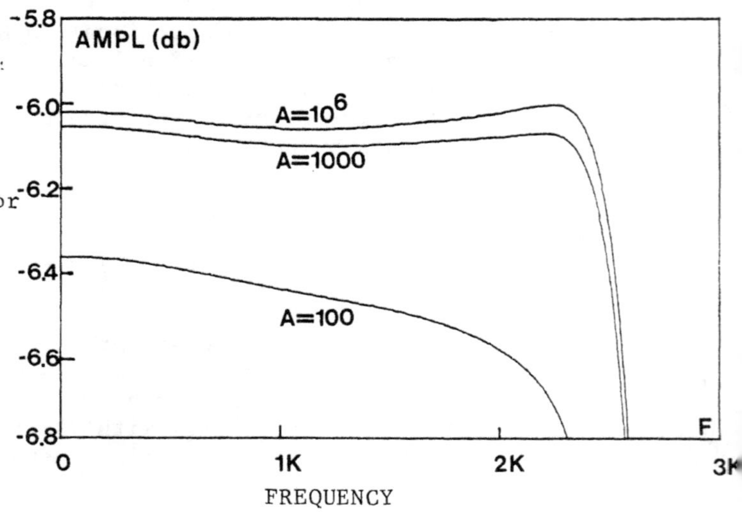

Fig. 18.
DIANA output for this filter.

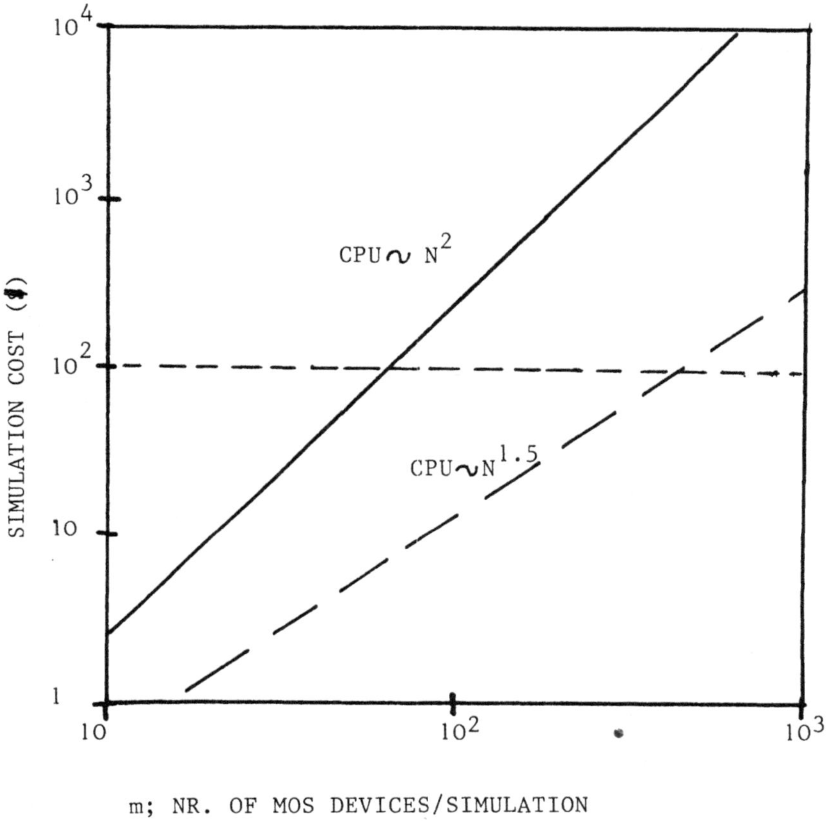

Fig. 19.

Clearly circuit simulation becomes uneconomical between 100... 1000 MOS transistors/network.
In reality the situation is even worse since usually the number of timepoints (complexity of stimuli) also increases with complexity !
It is thus seemingly impossible to simulate VLSI at the circuit level if no change in either algorithms or computer architecture is made.
It will be shown in 4.5. how the properties of space and time partitioning of VLSI can be exploited to get reasonably accurate circuit simulation up to 10.000 devices by using so called timing simulation techniques.

c) Clearly circuit simulation has to be used where accuracy is requested on small subcircuits and cells. Since a direct link to technology and geometry exists it is the excellent technique for cell behavioral verification and parameter extraction for higher levels of simulation.

4.3. Logic simulation and VLSI.

Since circuit simulation can not handle full VLSI complexity it is natural to look at logic simulation. This type of simulation is used intensively for PCB design using standard MSI packages. Logic simulation is simulation in the time domain at least at gate level rather than device level in circuit simulation. Models are no longer algebraic-differential but logic operators in the algebra of signal states followed by a more or less sophisticated delay description [50] [18].
Fig. 20. illustrates the concept of some widely used signal states. Boolean states {1,0} are seldom used. Most simulators use at least {1,0,X} where X is an undefined state e.g. for studying initialization problems or rise-fall transitions. Other simulators such as TEGAS [50] use {1,0,X,U,D} whereby U corresponds to a $0 \rightarrow 1$ and D to a $1 \rightarrow 0$ transition. Other simulators e.g. SALOGS [51] use a "tristate" state Z for buslines. Other simulators go as far as defining static and dynamic hazard states up to 16 states.
For an overview see [18].
Once the signal states defined, every logic operation is assigned a "thruth table" stored in memory. Fig. 20. gives a simple example for a NAND gate in {1,0,X}. Clearly these tables become very complex for higher order signal representations and high fan-in gates.
All modern simulators have a delay associated with a gate. These delays range from simple unit delay, differing rise and fall delays, min.-max. delays for U and D states and inertial delays which suppress transitions within the delay time of the gate and translate the transition into a spike state. The latter is important for detection of races, hazards and oscillations which may disturb logic circuitry.
More complex delay representations slow down the algorithms even as more complex signal representations and in most cases timing predictions are rather pessimistic [18]

The output of a logic simulator is usually a time table of state transitions as a response to input waveforms.

The algorithms used for logic simulation are much simpler than for circuit simulation and exploit the unilateral character of the models.
A very important aspect of logic simulation is its event-driven nature and the use of the principle of selective trace, as illustrated in Fig. 20b. An event is defined as a change of state of a signal path; e.g., $0 \rightarrow 1$ for i in Fig. 20b. As shown, this event is propagated to its fan-out, potentially creating new-future

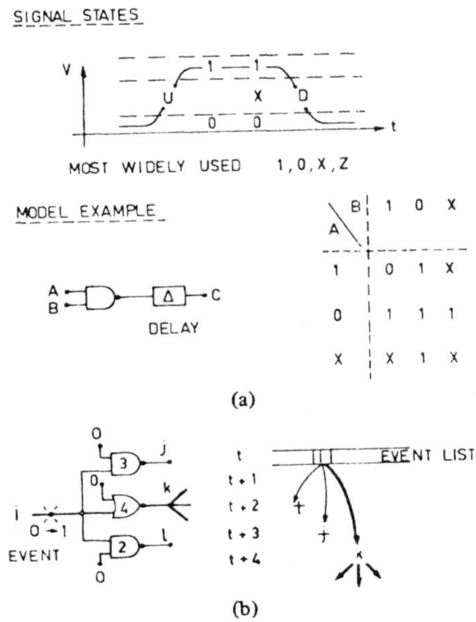

Fig.20. Some principles of logic simulation.

events if and only if propagation through the logic element can occur. At each time point the simulator stores the potential events in event lists, but at the time of execution unpropagated events (l,j) are cancelled and only active gates (k) are evaluated and their events are scheduled.

Following this scheme we see that only active (i.e., changing) circuit parts are evaluated. In most large digital systems the number of active gates may be as low as 1 percent. This selective trace property, also called latency, is an exploitation of the time sparsity of digital systems. Model simplicity and selective trace make LS of about $10^3 \ldots 10^4$ more efficient in execution and storage with respect to circuit simulation and as a result can cope with 10 to 20K gate complexity when using $\{1,X,0\}$ signal states. The price paid is a very weak link between performance, technology, and layout. This was no big problem for PCBMSI design as a result of I/O buffering of packages and the high degree of standardization allowing for the use of a library of models for most widely used MSI modules, as discussed in Section II. However, as part of the VLSI complexity is achieved as a result of "circuit cleverness", i.e., minimizing device count/function, no such standardization exists on-chip, especially when high-volume, high-performance products are envisioned. Thus, although logic simulation can in principle handle VLSI complexity, the following problems arise when applying it to VLSI :

1. difficulty of getting the modelparameters (delay) from layout & device data (parametrization problem).

2. no I/O buffering, therefore difficulty of taking capacitive load rules into account.

3. serious problems with bidirectional elements such as transmission gates in dynamic logic. Their representation is possible using three gates per transmission gate but then they are still unidirectoral.

4. in MOS gates full delay depends on the rise time of the input signal which is difficult to take into account.

5. levels and speed in dynamic logic circuit are dependent on threshold voltage, bulk effect and capacitive clock-feedthrough which cannot be taken into account.

6. Mixing with "analog" circuitry (e.g. sensing amplifiers in RAM) is not possible with existing simulators.
This does not mean that logic simulation cannot be used for VLSI design. In our opinion its use is restricted to :

a. Logic verification (unit or rising & falling edge delay) of functional blocks up to 5000 gates using {1,0,X,Z} before commiting to transistor level design.

b. In standard cell design where, in analogy with PCB design, cell characteristics and loading rules are known and logic simulation models exist in the library.

c. Test-pattern verification (fault simulation). It is well known [18] that, if ATPG is used, not all test patterns may be valid due to potential timing errors and/or redundanly in logic. Therefore all generated test patterns are checked by logic simulation. This involves often as many as $10^5 \ldots 10^6$ transitions for $10^3 \ldots 10^4$ faults ! For this job special simulation techniques such as parallel, deductive and concurrent simulation [52] have been proposed in order to either process as many faults as possible in parallel. Test pattern verification accounts for 90% of all logic simulation activities.
Clearly there is a gap of simulation between logic simulation and circuit simulation for complex circuits (100...5000 transistors). Such simulation should respond to the 6 problems mentioned above. This gap has recently been filled by so called timing simulation to be discussed in 4.5. Before discussing this, we will first consider register transfer level (RTL) simulation.

4.4. Register-transfer level simulation.

Register transfer level simulation is based on so called register transfer languages [5]. These languages are similar to high level computer languages such as PASCAL or ALGOL but instead of generating machine code, they generate, by compilation , ordered sets of Boolean expressions for synchronous finite state machines.

Such a language expresses the data transfer and transformation between registers synchronized by a set of synchronized multiphase clocks.
All RTL languages allow for multi-dimensional wire (buses) and register (memory) description (in contrast to logic simulators).
The following are examples of RTL primitives :

- Wires (signals) : A,B,C (0:15); "C=16 bit bus";
- Registers : 'REGISTER' INSTR (1:20), ROM (1:256,1:8);
  "8 bit adress, 8 bit word ROM";
- Clocks : 'MASTERCLOCK' C256,C64;
- Connections (Combinational logic) e.g. :

  S(1:4):=ALU(0:3)+IN(5:8); (3)

  All Boolean and arithmetic expressions are usually possible as well as sometimes comparisons, coding, decoding etc...
- Transfers : defined by <= at clock C
  describe Boolean operations on signals transferred to a predefined register e.g.

  <C> RADD <= S1.C + S2.C'

  or also :

  <C> 'IF' C 'THEN' RADD <= S1
  'ELSE' RADD <= S2

Dependent on the RTL, possibilities exist for rotation, shift, transposition (of wires), concatenation, vector operations etc...
Some programs are exclusively synchronous [54] , others allow for some asynchronism [53][55] although its use is very ambiguous since often results depend on the sequence of instructions.
Notice now the following properties of RTL's :

1. From the examples it is clear that RTL is <u>not</u> a structural, but a <u>behavioral</u> language in contrast e.g. to logic and circuit simulation language. Its use is <u>mainly</u> for <u>algorithmic description</u> of synchronous systems regardless of their hardware structural realization. Hardware synthesis of RTL is possible but very inefficient. Research goes on at Carnegie Mellon using ISP [28]. Therefore translation into hardware is <u>manual</u> and mutual verification of RTL <u>functional blocks (operators)</u> versus realization is a necessity (e.g. by <u>comparing RTL block stimuli & responses with Logic Simulation of the corresponding block</u>).
At Stanford University work is progressing to design a language for capturing both behavior and structure at the same time, in the ADLIB-SABLE system [3][7].

2. Simulation of RTL proceeds from the compiled ordered set of Boolean expressions using control structures to execute only active branches [54] . <u>Since all vector operations</u> up to the double precision <u>word length</u> of the host computer are <u>single ALU</u> opera-

tions, the speed of RTL simulation is ca. 5...50 higher than corresponding logic simulation and full multiprocessor systems have been succesfully simulated.

3. After RTL simulation of a system, the RTL description constitutes the high level behavioral reference for all further downward design.
If further developed to include structural information RTL will become an important tool in VLSI design at the system defining level.

4.5. Timing simulation-"logic" simulation for VLSI ?

RTL simulation is used to verify system behavior. The combinational logic blocks resulting from time partitioning in synchronous systems can then be realized at gate (PLA) level and verified by crude {0,1} , unit delay logic simulation. In 4.3. we discussed 6 problems to use logic simulation for timing verification of VLSI circuits and decided that a superefficient circuit simulator would be more appropriate. Such a simulation technique, based on the latency and decoupling property of digital MOSLSI has been proposed recently, first by Rabbat and Ryan [56] and later by Chawla and Gummel [57] in their MOTIS program.

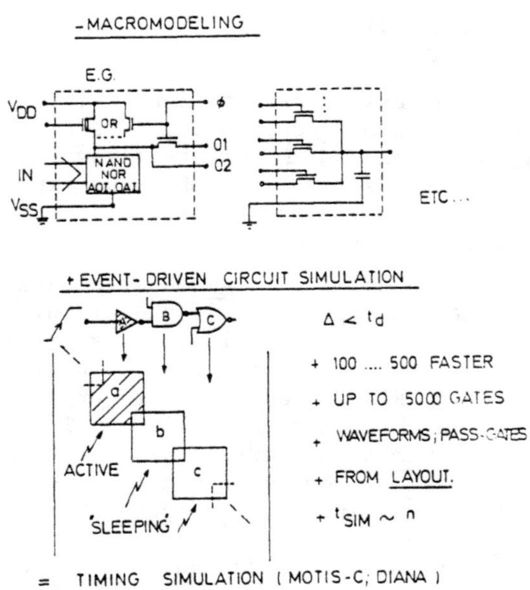

Fig. 21.
Timing simulation is based on table stored macromodels and event-driven circuit simulation.

Timing simulation is a relaxed form of circuit simulation based on four main principles (see Fig. 21) :

a) Device models are described by scalable tables or threshold functions [58] of voltages, currents and charges such that no FORTRAN nonlinear functions need to be computed in matrix updating which is 80% of computational activity (factor a=4).

f) If the time step Δ is smaller than the minimum gate delay t_d one can show [57] that, if at each time step only one secant iterate is used, the response still converges to the correct steady state while providing better than 20% waveform accuracy or timing withing a few percent (factor b $\approx$ 4).

c) Table form macromodels for the usual MOS and CMOS building blocks are used (e.g. NAND, NOR, ANDORI, ORANDI, transmission gate trees and stars, bootstrapped buffers etc...). These blocks are evaluated at the speed of one invertor (factor c $\approx$ 1...4).

d) Use is made of the event driven scheduling mechanism as used for logic simulation. As shown in Fig. 21, every gate circuit A,B,C... can be represented by a submatrix (a,b,c...). If now $\Delta < t_d$ one can decouple the submatrices and evaluate only the active submatrices until a sufficient input change makes it necessary to schedule another one and perhaps deschedule the previous one. This saves a factor d = 5...50 depending on circuit activity. Combining the above effects one obtaines a waveform timing simulator with the following properties :

1. 80...1000 times faster than circuit simulation depending on model complexity and network activity
2. layout and technology link possible
3. electrical loading and bidirectional transmission gates, clockfeedthrough etc... within 20% correct.
4. easy to interface with circuit- and logic simulation (see 4.5.)
5. economically feasable up to 5000 gates.

An advantage of the use of table-stored devices is that even fairly complex device models can be used. Tables are computed in a preprocessor program and this needs to be done only once. On the other hand, the use of the crude one-iterate models and equation decoupling imposes some limitations on the circuit types that can be simulated. As a result, much of the success depends on the willingness of the designer to understand these limitations. Therefore, most timing simulators are also optimized for specific circuit techniques, such as dynamic multiphase MOS, with triode load [57][58][9], depletion load, 4 phase MOS [58], and I^2L [59].

Fig. 22(a) and (b) illustrates a typical application and the performance of the timing simulator in DIANA [9] for a 2-phase MOS triode load logic. In this program, macromodels for clocked and unclocked NOR, NAND, ANDORINVERT, ORANDINVERT are provided. Transmission gates and logic MOS trees are modeled from device dimensions. Gate-source (-drain) capacitances are dependent on the

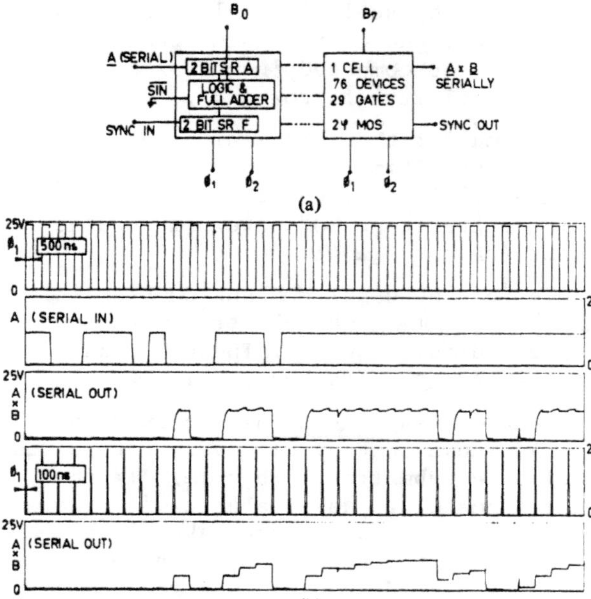

Fig. 22 **(a) Block diagram of an 8 bit serial-parallel multiplier simulated using DIANA-2φ. (b) Results of the simulation of a PMOS version of the serial-parallel multiplier for the multiplication shown. Voltages are plotted as positive. The upper two plots are for a normal clock width, the lower ones for a critical clock width, causing malfunctioning of the circuit.**

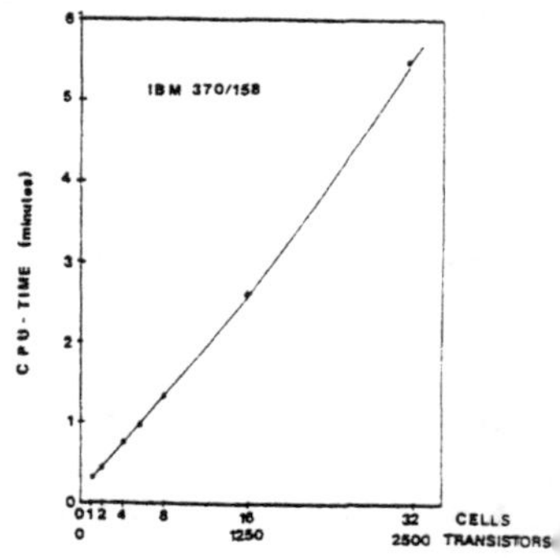

Fig. 23.

PERFORMANCE OF TIMING SIMULATION FOR MULTIPLIER EXAMPLE

	Timing SIM (DIANA-2φ)	CKT SIM (TRSIT)	Measurements
Predicted t_W (ns)	100	120	130
CPU/device/time point	0.05 ms	12.5 ms	
CPU/gate/time point	0.13 ms	31.5 ms	

Note : speed improvement is 250.

Table III.

transistor state. The latter is essential to predict clock feed-through effects with reasonable accuracy. Load capacitances are computed from layout.

Fig. 22(b) shows an output plot of the simulation with DIANA of an 8 bit serial-parallel multiplier in 2-phase PMOS, of which the block diagram is shown in Fig. 22(a). The circuit performs the multiplication of a coefficient B (8 bits) entered in parallel with a word A entered serially, one bit/clockcycle. The output consists of the truncated product delayed by 16 clockcycles. The circuit simulated contains 616 transistors, 235 gates, and 96 transmission gates.

Fig.22(b) shows a simulation of this circuit from layout data of the actual circuit. The plots in Fig.22(b) are, respectively, the serial input word, the 1 MHz clock with pulsewidth t_W=500 ns, and the resulting product output as simulated by DIANA. A total of 1700 time steps of 20 ns were used. Also shown is a simulation with a narrow clock pulsewidth t_W = 100 ns, which just causes wrong operation as shown on the corresponding response plot. Total CPU time for this simulation is 83 sec. on an IBM 370/158. As this minimum clock width depends critically on the timing of the circuit, it is a good measure of accuracy of the timing simulation.

Table III shows that agreement with measured data on the actual circuit and simulation is quite good. It also shows that in this case timing simulation is about 250 times faster than circuit simulation (if the latter could cope with the 616 transistor complexity).

The effects of charge redistribution and clock feedthrough on the signals are clearly visible. These effects cannot be obtained from logic simulation results.

Fig. 23 shows CPU time for the same example but now for different number of multiplier cells (same number of time steps). Notice that CPU time is nearly linear with complexity and 2500 transistor circuit simulation is still possible in a matter of minutes.

In our opinion timing simulation can, for VLSI, be a reliable substitute for logic simulation in timing verification since it is very closely related to device technology, layout and electrical interaction.

4.6. Mixed-Mode Simulation

In the previous paragraphs we have seen that RTL simulation, logic, timing and circuit simulation are all useful in VLSI design. In practice however a number of problems arise :

1. all disciplines exist as separate programs without common language, algorithms or data-structures.
2. output of simulation at a higher level is useful as input to lower levels for consistency checks. This is very difficult with existing programs.
3. often different levels of simulation are needed simultaneously e.g. for RAM circuits (decoder may be logic, sensing amplifier:circuit, output buffers:timing...), for mixed analog-digital circuits (e.g. A/D convertors), for critical paths in PLA's timing, for rest RTL etc...

Therefore there is a tendency to create CAD simulation programs which have a common data structure for different levels of simulation. They are fully event driven down to the subcircuit level and can be driven from a common hierarchical structural language in which multilevel models can be described.

Such technique has first been called "hybrid simulation" by De Man and Newton [60] but recently has been called mixed-mode simulation.

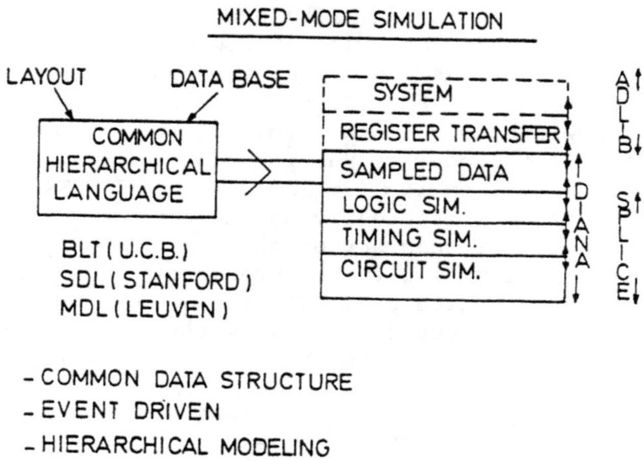

Fig. 24.

Fig. 24. shows the state of the art.
Actually two mixed-mode simulators are in the public domain i.e. SPLICE [61] of U.C. Berkeley, covering logic, timing and circuit simulation and DIANA [9] of K.U.Leuven covering the same plus sampled-data analog circuits. Extension towards RTL for DIANA is being studied.
At the higher level of systems design the ADLIB-SABLE [3,7] developed at Stanford University is a multilevel simulation system from concept to gate level based on PASCAL type language including behavioral and structural (interconnection) information.
As an example Fig. 25. shows the different levels in the DIANA [9] program, the primitives which are available, the signal representations and the types of analysis possible.
Fig. 26. shows a typical application of DIANA. Here an analog switched capacitor filter which is under digital control is simulated. The logic circuit is analyzed at the logic simulation level (level IV in Fig. 25) whereas the analog filter is simulated in the time domain by circuit analysis. One can clearly see how effects such as clock-feedthrough, sample-hold effects and even spike feedthrough from the digital to the analog circuit (spike A Fig. 26) is observed.
Simulation of the full circuit over 1000 time points of 1 μsec

DIANA PROGRAM

MODE	PRIMITIVES	VARIABLES	ANALYSIS
I) SAMPLED-DATA	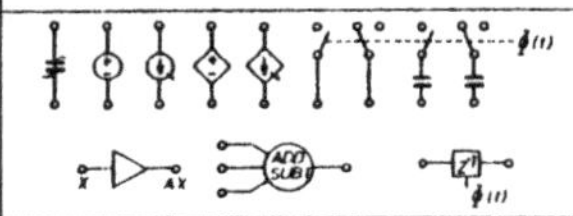	v, q, t, w	TIME (LIN. & NONLIN.) FREQUENCY SENSITIVITY NOISE (ROUNDOFF)
II) (SUB) CIRCUIT	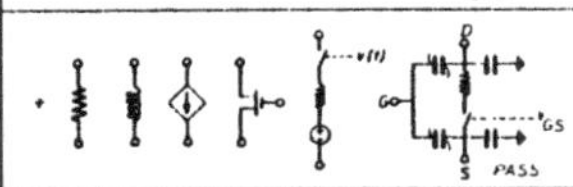	V, I, t	TIME, DC
III) TIMING	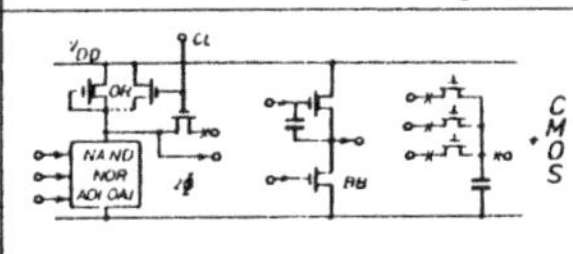	V, I, t	TIME
IV) LOGIC	(N)AND; (N)OR; (N)EXOR; AOI; OAI; J-K FF & MACRO'S	I, O, X, U, D, t	TIME

Fig. 25. Different levels of Mixed-mode Simulator DIANA.

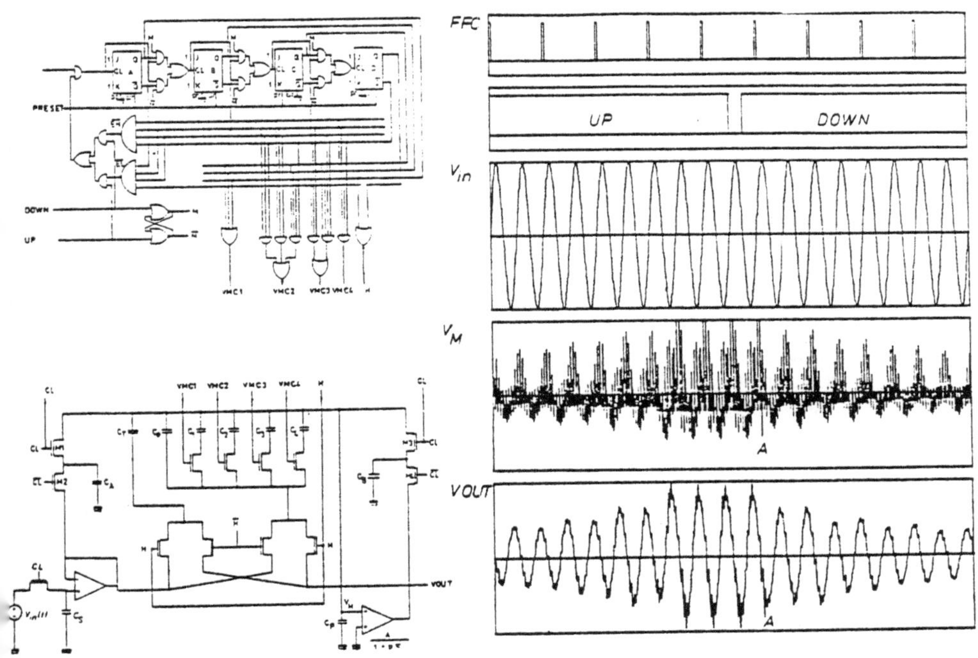

Fig. 26. Example of mised-mode simulation - logic circuit controling analog filter on the left. Response computed by DIANA at the right. Mixed analog and logic simulation.

takes 120 sec. IBM 370/158 time.
Of particular importance for the use of mixed-mode simulation for VLSI is the use of a "common hierarchical language" as shown in Fig. 24. This language should reflect the top-down VLSI design process and has to communicate with a "data-base" to store the design in progress.
Several such languages have been proposed and are under development such as B(erkeley) L(anguage) T(ranslator) [10], S(tructural) D(escription) L(anguage) [8] from Stanford and M(acro) D(escription) L(anguage) [9] from K.U.Leuven.
Fig. 27. shows a simple example of SDL whereas Fig. 28. shows the hierarchical principle of the MDL system being developed at K.U.Leuven. Fig. 28a. is an LSI circuit called CLGEN built of several blocks which are further subdivided etc... and interconnected.

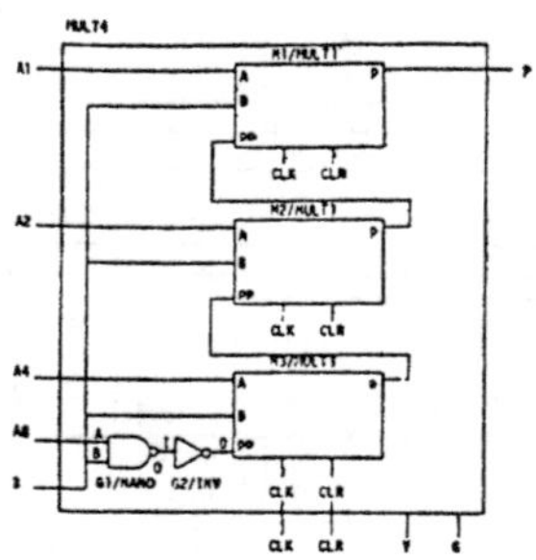

```
USER:'UMUC';

(* GENERAL CIRCUIT INFORMATION FOLLOWS *)
NAME:MULT4;
PURPOSE:ICDESIGN;
LEVEL:CHIP;

(* DESCRIPTION OF EXTERNAL CONNECTORS FOLLOW *)
EXT:EC:A1,A2,A4,A8,B,CLK,CLR,PRODUCT,V,GND;
INPUTS:EC.<A1,A2,A4,A8,B,CLK,CLR>;
OUTPUTS:EC.PRODUCT;

(* DECLARATION OF THE PRIMITIVE COMPONENT TYPES FOLLOWS *)
TYPES:MULT1,NAND,INV;

(* DECLARATION OF COMPONENTS AND THEIR TYPES FOLLOWS *)
MULT1:M<1:3>;
INV:G2;
NAND:G1;
END;

(* CONNECTIVITY BETWEEN COMPONENTS FOLLOWS IN TERMS OF NETS *)
NETSEGMENT;
N1=EC.A1,M<1>.A;
N2=EC.A2,M<2>.A;
N3=EC.A4,M<3>.A;
N4=EC.A8,G1.A;
N5=EC.B,G1.B,M<1:3>.B;
N6=M<1>.P,EC.PRODUCT;
N7=EC.C.K,M<1;3>.CLK;
N8=EC.CLR,M<1:3>.CLR;
N9=M<1>.PP,M<3>.P;
N10=M<2>.PP,M<3>.P;
N11=G1.O,G2.I;
N12=G2.O,M3.PP;
ENDNETS;
ENDC;
```

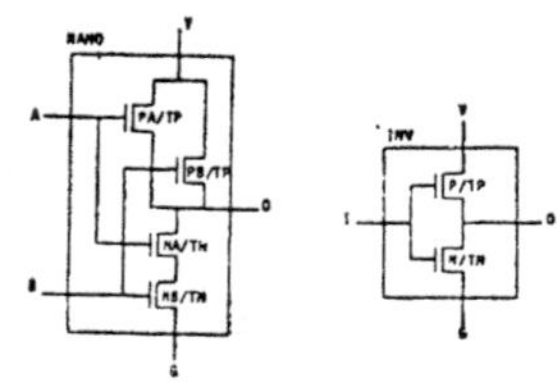

```
(* SDL NAND CIRCUIT DESCRIPTION *)
NAME:NAND;
PURPOSE:ICDESIGN;
LEVEL:TRANS;

(* EXTERNAL CONNECTOR INFORMATION *)
EXT:EC:A,B,O,V,GND;
INPUTS:EC.<A,B>;
OUTPUTS:EC.O;

(* COMPONENT TYPES AND COMPONENT DECLARATIONS FOLLOW *)
TYPES:TP,TN;
TN:NA,NB;
TP:PA,PB;

(* CONNECTIVITY OF CIRCUIT FOLLOWS *)
NETSEGMENT;
N1=EC.A,PA.G,NA.G;
N2=PA.D,PB.D,NA.D,EC.O;
N3=EC.GND,NB.S;
N4=EC.V,PA.S,PB.S;
N5=EC.B,PB.G,NB.G;
N6=NA.S,NB.D;
ENDNETS;
ENDC;
@
```

```
CURRENT STATUS
---------------------

  COMPILER/SYNTAX CHECKER OPERATIONAL ON
       DEC 20 SYSTEM
       IBM 370/168
       HP 3000
       HP 1000
  INTERFACES BUILT TO
       SPICE (CIRCUIT ANALYSIS)
       TESTAID (LOGIC SIMULATOR,FAULT
                TEST GENERATION)
  INTERFACES PLANNED:
       MOTIS (MOS TIMING SIMULATOR)
       CALMOS (IC LAYOUT)
       SICLOPS (IC LAYOUT)
```

Fig. 27.
SDL example.

The philosophy is that this hierarchical nature is stored in an hierarchical father-son type file management system (FMS) [62] similar to the editing part of a computer operating system.
Fig. 28b. shows the hierarchical nature of Fig. 28a. The names in lower case symbols are library types (if created previously).
Fig. 28c. shows the input of the circuit into the FMS system.
The designer may at first only create an RTL file of CLGEN to si-

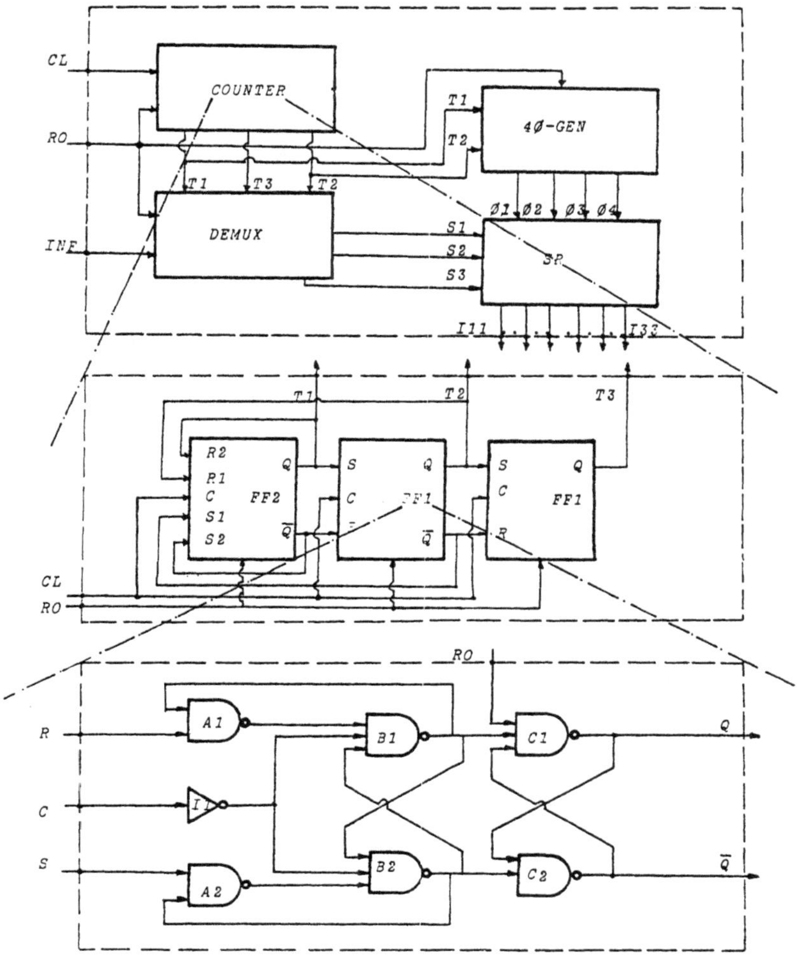

Fig. 28a. : LSI circuit CLGEN with hierarchical decomposition.

mulate it in an RTL simulator which in the future will also be in the mixed-mode simulator.
When approved he may create the subblocks one by one and create a gate description which in a mixed-mode can be imbedded block by block into the RTL description for verification of correctness etc... down to the creation of sticks diagramms and layout library descriptions.
When a full logic description is asked from e.g. CLGEN then expansion occurs down the hierarchical file structure until all logic

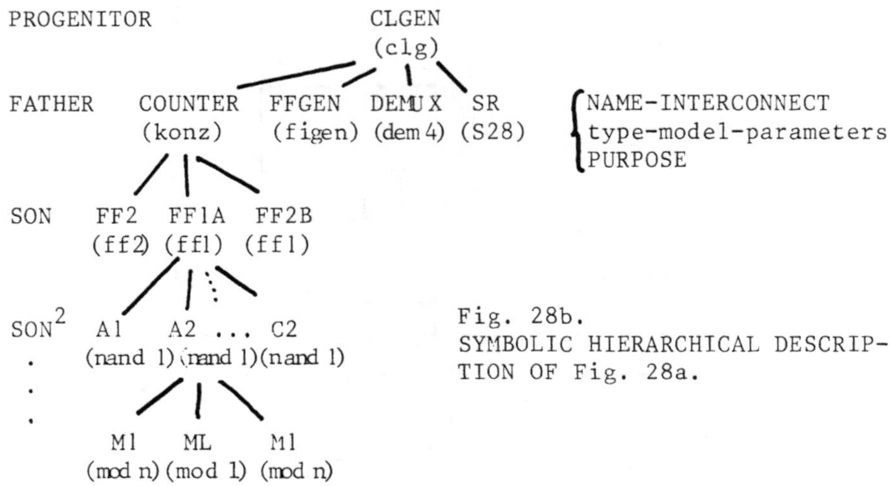

Fig. 28b.
SYMBOLIC HIERARCHICAL DESCRIPTION OF Fig. 28a.

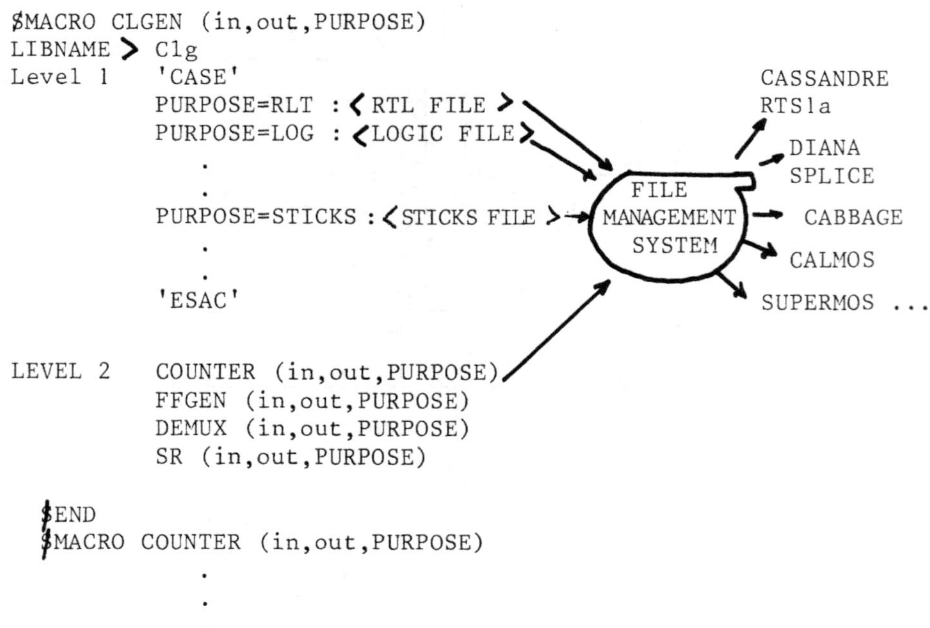

Fig. 28c.
ALPHANUMERIC DESCRIPTION TO FILE MANAGEMENT SYSTEM

files are found. Missing ones are reported etc... By the end of a design all relevant design information is stored in files and new types can be declared for further use by team members or future designers.
Notice that such a system is fully compatible with the principle of mixed-mode simulation and actually the only way to exploit its features.
Fig. 29. shows a flow shart of how all disciplines in CAD should be combined to provide a method for hierarchical design of VLSI circuits. It is easy to see how the above discussed CAD tools can fit into it.

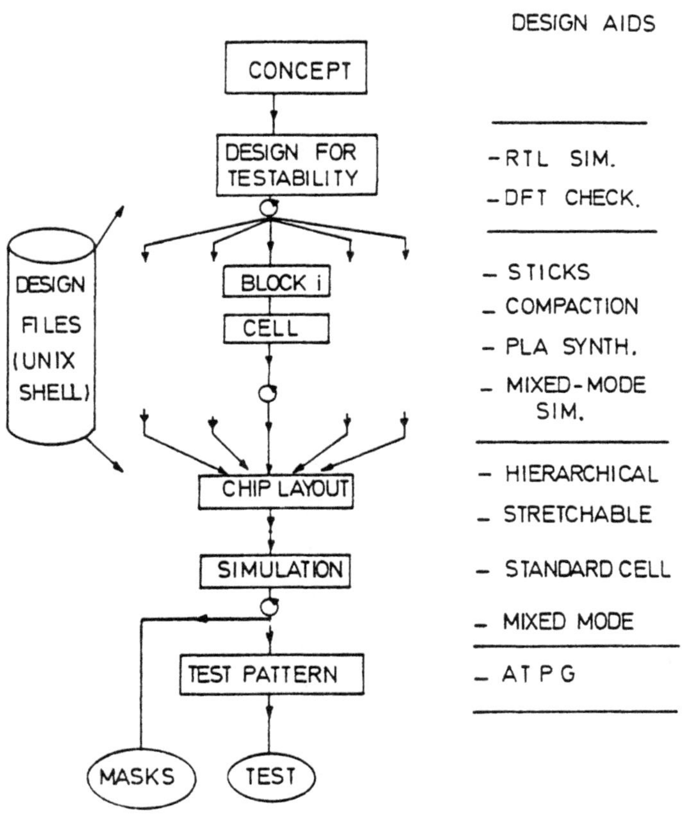

Fig. 29.

Finally it is interesting to notice that VLSI itself will perhaps have a strong "bootstrapping" effect on CAD tools to create the same VLSI. This will happen in two fields :

1. user interface : colored graphics;

2. dedicated hardware;

Today designers still use Mylar sheet drawing and digitizing. Graphics systems today are not much used because they have no local intelligence such that every three other stations a new computer is necessary. The advent of cheap multimicroprocessor systems, dedicated graphics operation VLSI chips and very cheap memory (bit-maps) will make high resolution colored graphics terminals cheap (4...40 K$)·and stand-alone i.e. accessible to every designer. These terminals can then communicate with a number crunching 32 bit megabyte local computer network for tasks such as simulations, routing, placement, testpattern generation, etc... Communication is possible by an intermediate graphics language such as e.g. CIF. The same effects (cheap memory) VLSI dedicated chips, multiprocessors) will also make it possible to create machines for the NP-complete and power-law type algorithms such as "set-operations" for DRC, device extraction etc..., ATPG, Test pattern verification, routing (Lee-type algorithms) and perhaps also timing simulation and matrix inversion.
By combining these hardware and software tools with hierarchical design methodology it will be possible to reduce designs costs back to a reasonable number. This is perhaps the only way to make custom VLSI feasable for the non-semiconductor processing industry (not neccessarily even electronic).

REFERENCES

[1]. G. Gordon, System Simulation, Prentice Hall, Inc., Englewood Cliffs, N.J., 1969.

[2]. O.J. Dahl, K. Nygaard, SIMULA - An ALGOL-base Simulation Language, Comm. ACM, Vol. 9, Sept.1966, pp. 671-678.

[3]. D.D. Hill, ADLIB-SABLE Users's guide, Computer Systems Lab. Techn. Report, Stanford University, Stanford 1979.

[4]. G.C. Bell, A. Newell, Computer Structures : Readings and Examples, McGraw Hill Book Co., Inc., N.Y. 1971.

[5]. M.A. Breuer, Digital System Design Automation, languages, simulation & data base, ch. 2-3, Pitman Publishing Ltd., London, 1975.

[6]. C.G. Bell, A. Newell, The PMS and ISP Descriptive Systems for Computer Structures, Proc. Spring Point Computer Conference, 1970, pp. 351-374.

[7]. D.D. Hill, W. Van Cleemput, SABLE : A tool for generating structured, multi-level simulations, Proc. 16th Design Automation Conference, june 1979, pp. 272-279.

[8]. W.M. Van Cleemput, An Hierarchical Language for the Structural Description of Digital Systems, Proc. 14th Design Automation Conference, New Orleans, June 1977, pp. 377-385.

[9]. G. Arnout, H. De Man, P. Reynaert, J. Rabaey, Users Manual for the DIANA program, ESAT Laboratories, Kath. Univ. Leuven, Belgium, April 1980.

[10]. A.R. Newton, J. Crawford, D.O. Pederson, A proposal for a Unified Input Syntax for CAD programs, Electronics Research Laboratory, U.C. Berkeley, June 1977.

[11]. L.W. Nagel, SPICE 2 : A Computer Program to Simulate Semiconductor Circuits, Electr. Research Laboratory, Memo No. ERL-M520, U.C. Berkeley, Calif.

[12]. - -, ASTAP General Information Manual, IBM Corp. Mechanicsburg, Penns. USA.

[13]. H. De Man, J. De Groof, Users Manual for TRSIT, ESAT Laboratories, Kath. Univ. Leuven, Belgium, 1978.

[14]. C. Mead, L. Conway, see ref. 21, pp. 115-127.

[15]. B.T. Preas, B.W. Lindsay, C. Gurjn, Automatic Circuit Analysis based on mask information, Proc. 13th Design Automation Conf., June 1976, pp. 309-317.

[16]. T. Akino, M. Shimode et al., Circuit Simulation and Timing Verification based on MOSLSI Mask information, Proc. 16th Design Autom. Conference, June 1979, pp. 88-94.

[17]. B.W. Lindsray, B.T. Preas, Design rule checking and analysis of IC mask designs, Proc. 13th Design Automation Conference, June 1978, pp. 301-308.

[18]. M.A. Breuer, Friedman, Diagnosis and Reliable Design of Digital Systems, Pitman Publishing Ltd. London, 1976.

[19]. E.B. Eichelberger, T.W. Williams, A Logic Design Structure for LSI Testability, Design Automation & Fault Tolerant Computing, Computer Science Press., 1978, pp. 165-178.

[20]. C. Mulder, C. Niessen, R. Wijhoven, Layout and Test of Synchronous LSI circuits, Digest of Technical Papers of IEEE ISSCC 1978, Philadelphia, pp. 148-249, Febr. 1978.

[21]. C. Mead, L. Conway, Introduction to VLSI Systems, Addison-Wesley Publ. Co., Mass. 1979.

[22]. G. Perskey, D.N. Deutsch, D. Schweikert, LTX-A Minicomputer Based System for Automatic LSI layout, Journ. of Design Automation and Fault Tolerant Computing, Vol.1, no. 3, May 1977, pp. 217-255.

[23]. A. Feller, T. Lombardi, S. Ozga, CAD VLSI Design Techniques and Microprocessor Application, Digest of Techn. papers of IEEE ISSCC Conference, 1978, Philadelphia, pp. 212-213, Febr. 1978.

[24]. H. Beke, W. Sansen, CALMOS - A portable software system for the automatic and interactive layout of MOSLSI, Proc. 16th Design Automation Conference, June 1979, pp. 102-108.

[25]. G.M. Klomp, CAD for LSI-production's interest is in its Economics, SIGDA Newsletter, Vol. 6, no. 3, Sept. 1976, pp. 11-15.

[26]. W. Lattin, ComputerAided Design : The problem of the 80's, Microprocessor Design, Proc. Symp. on CAD of digital electronic circuits & systems, EEC, Brussels, pp. 239-241, Nov. 1978.

[27]. B.T. Preas, W.M. Van Cleemput, Placement Algorithms for Arbitrary Shaped Blocks, Proceedings of 1979 ISCAS Conference, Tokyo, Japan, pp. 474-480.

[28]. M. Barbacci, D.P. Siewiorck, The CMU RT-CAD Systems : an Innovative Approach to Computer-Aided Design, Carnegie-Mellon University, Computer Science Review, Sept. 1975,pp.39-53.

[29]. K.J. Loosemore, IC Layout - The Automatic Approach, Proc. 5th ESSCIRC Conference, Southampton, Sept. 18-21, 1979.

[30]. U. Lauter, A Min-cut placement Algorithm for General Cell Assemblies Based on a Graph Representation, Proc. 14th Design Automation Conference, June 1979, pp. 1-10.

[31]. W.R. Meller, An Algorithm for Chip Planning, SSP Memo 2805, Comp. Science Dept., Calif. Inst. of Techn., May 1979.

[32]. A.D. Lopez, H.F. Law, A Dense Gate Matrix Layout Style of MOSLSI, Digest of Technical Papers, ISSCC, pp. 212-313, Febr. 1980.

[33]. J.P. Gray, Introduction to Silicon Compilation, Proc. 16th Design Autom.Conf., June 1979, pp.305-306.

[34]. L. Bienstman, H. De Man, An 8 channel 8 bit µP compatible NMOS converter with programmable ranges, Digest of Techn. Papers, 1980, ISSCC, Febr. 1980, pp. 16-17.

[35]. R. Ayres, Silicon Compilation - A Hierarchical Use of PLA's, Proc. 15th Design Automation Conf., June 1979, pp.314-326.

[36]. E.I. Muehldorf, T.W. Williams, Optimized stuck fault test pattern generation for PLA macros, Dig. Semicon. Test Symp., Cherry Hill, N.J., Oct. 25-27, 1977, pp. 88-101.

[37]. D.L. Ostapko, S.J. Hong, Fault Analysis and Test Generation for PLA's, IEEE Trans. Computers, Vol.C-28, no. 9, pp. 617-Q7, Sept. 1979.

[38]. D. Johannsen, Bristle Blocks : A Silicon Compiler, Proc. 16th Design Automation Conference, pp. 310-313.

[39]. O. Gibson, S. Nance, SLIC - Symbolic Layout of Integrated Circuits, Proc. 13th Design Automation Conference, June 1976, pp. 434-440.

[40]. J. Lusinchi, J. Redolfi, Grid Layout and Versatile Cells for Custom LSI Designers, Proc. ESSCIRC 78, pp. 160-162, Sept. 1978.

[41]. Y. Echo, A.J. Korenjack, D.E. Stockton, FLOSS : An approach to Automated Layout for High Volume Designers, Proc. 14th Design Automation Conference, June 1977, pp. 138-141.

[42]. A.E. Dunlop, SLIP - Symbolic Layout of Integrated Circuits with Compaction, Computer Aided Design, Vol. 10, no. 6, Nov. 1978, pp. 387-391.

[43]. J.D. Williams, Sticks - A Graphical Compiler for High Level LSI Design, AFIPS Conference Proceedings, Vol. 47, June 1978, pp. 289-295.

[44]. M.Y. Hseuh, D.O. Pederson, Computer Aided Layout of LSI Circuits Building Blocks, Proc. 1979, IEEE ISCAS Conf., pp.474-477.

[45]. A.D. Ivannikov, P.P. Sipchuk, Computer Aided Design of MOS Integrated Circuit Layout, Proc. IEEE - CADMECCS, Brighton U.K., July 1979, pp. 47-51.
[46]. L.O. Chua, P.M. Lin, Computer Aided Analysis of Electronic Circuits, 1975, Prentice Hall, N.J.
[47]. H. De Man, J. Rabaey, G. Arnout, J. Vandewalle, Practical Implementation of a General CAD technique for Switched Capacitor Circuits, IEEE Journ. of Solid State Circuits, Vol.SC-15, April 1980.
[48]. H. De Man, J. Rabaey, G. Arnout, J. Vandewalle, DIANA as a mixed mode simulator for MOSLSI sampled data circuits, Proc. ISCAS 1980, Houston, April 1980.
[49]. F. Brglez, Exact Nodal Analysis of Switched Capacitor Networks with arbitrary switching sequences and general inputs, Proc. of 1975 ISCAS Conference, Tokyo, July 1979, pp. 748-751.
[50]. S.A. Szygenda, E.W. Thomson, Digital Logic Simulation in a Time Based, Table Driven Environment - part 1, Computer, pp. 23-36, March 1975.
[51]. G.R. Case, The SALOGS digital logic simulator, Proc. 1978 ISCAS Conference, N.Y., pp. 5-10, May 1978.
[52]. See ref. 50, part 2, pp. 38-49.
[53]. G. Bogo, Mermet, Cassandre and the Computer Aided Logical Systems Design, Proc. IFIP Congress, Ljubljana, Yugoslavia, Aug. 1971, pp. 1056-1065.
[54]. H.J. Knobloch, RTS Ia : ein System formalen Beschreibung und Simulation komplexer Scholtwerke, Ph.D. Thesis, T.H. Darmstadt, D-17, Sept. 1978.
[55]. J.R. Duley, D.L. Dietmeyer, A Digital System Design Language, IEEE Trans. Comp., Vol. C-17, no. 9, pp. 850-862, Sept.1968.
[56]. N.B. Rabbat, W.D. Ryan, A computer modeling approach for LSI digital structures, IEEE Trans. Electron Devices, Vol. Ed-22, Aug. 1975.
[57]. B.R. Chawla, H. Gummel, MOTIS - An MOS timing simulator, IEEE Trans. Circuits Syst., Vol. CAS-22, Dec. 1975, pp. 901-909.
[58]. G. Arnout, H. De Man, The use of threshold functions and Boolean controlled network elements for macromodeling of LSI circuits, IEEE Journal Solid State Circuits, Vol. SC-13, June 1978, pp. 326-332.
[59]. G.R. Bogle, SIMPIL - A simulation program for injection logic, IEEE Proc. ISCAS 1978, pp.890-896, May 1978.
[60]. H. De Man, R. Newton, Hybrid simulation, IEEE Proc. ISCAS, pp. 249-252, July 1979.
[61]. D.O. Pederson, R. Newton, A simulation program with LSI emphasis, Proc. 1978 ISCAS, pp. 1-4, May 1978.
[62]. --, The UNIX Time-sharing system, The Bell System Technical Journal, Vol. 57, July-Aug. 1978.

RELIABILITY IMPROVEMENT IN MANUAL LAYOUT DESIGN

H. van der Gaag

Philips Research Laboratories,
Eindhoven, The Netherlands

ABSTRACT. Automated layout design is compared with manual layout design as regards important aspects like layout correctness, design time, chip area, circuit speed, adaptation to changing design rules and handling devices with various shapes. This is followed by a discussion about flexible layout design systems. Then the CAD-tool CADILL is introduced, an aid in manual layout design. It offers a full layout design freedom, performs special design rule checks and is able to reconstruct the logical network. This improves the reliability of manual layout design considerably. Special attention is paid to network reconstruction. Finally, we deal with the ways in which a layout can be structured, in which way the tool can be adapted to changing fabrication processes and interfaces with other CAD-tools have been implemented.

1. AUTOMATED VERSUS MANUAL LAYOUT DESIGN

A look at a layout of say several thousands of interconnected transistors needing 7 or more masks for fabrication and realizing that the complete layout obeys 20 or 30 technology design rules, makes it immediately clear that there is an urgent need for an adequate design automation. In the literature many papers concerning fully automated layout design of digital integrated circuits can be found and much work has already been done in the development of CAD-tools in this field [1] . In practice, however, a considerable quantity of circuit layouts are made manually, especially where circuits must be produced in large quantities. Obviously, the results of automated layout design are not in every aspect better than those of manual layout design. The quite different ways in which human beings and computer programs perform a layout designing task give different results.

Most of the automated layout design tools operate with standard rectangular cells of equal height which are placed in rows, after which the interconnections are made in the space between the rows. In manual layout design cells of arbitrary shape are used which are placed and interconnected simultaneously in such a way that all the elements fit together as well as possible. In both approaches there is often a choice from alternatives and numerous decisions have to be made.

Automated layout systems work systematically and each decision is recorded in such a way that all its relevant consequences for future decisions can be taken into account. In order to keep this manageable and to restrict program execution time, the number of alternatives is usually considerably restricted, thus simplifying the decisions to be made, which results in layouts with a rather simple structure.

Humans work intuitively rather than systematically and it is difficult for them to take the consequences of their decisions continuously into account, the more so because their bird's eye view tempts them to create rather complex layout structures.

The ways in which manual and automated design methods affect layout correctness, design time, chip area, circuit speed, design rule adaptation and handling of devices with various shapes will now be compared.

1.1 Layout correctness

Changing the layout, making new masks and processing new samples is always expensive and time-consuming. Therefore, the correctness of the layout is a very important factor. The systematic and accurate way in which relatively simple layout structures are created by automated design systems guarantees that the layout really represents the circuit required and that it fulfils all the design rules. The way in which human beings work makes a certain number of violations of the design rules almost inevitable and even a single missing or incorrect gate or connection results in an entirely wrong layout.

1.2 Design time

It is widely agreed that the use of automated design tools can save much time although a design time that is 50 times shorter than that for manual design, as is sometimes mentioned in the literature, is to be suspected. Occasionally reduction of product design time can be of vital importance, e.g. for competitive reasons. But apart from this it is the economic aspect of design time against which investments in CAD-tools and the cost of their use must be weighed.

1.3 Chip area

Chip area will be smaller with manual design because more complex layout structures can be created resulting in a higher circuit density. Because it is rather uncommon to design a circuit both manually and automatically, estimates given in literature must again be considered with some reserve, but it is not unrealistic to say that automated design may result in chip areas which can be twice as large as those obtained from manual design.

The importance of the chip area depends on at least two other factors, the quantities in which the circuit must be fabricated and the yield of the fabrication process. It will be clear that for high-volume production, the fabrication cost may be dominant. This is notably the field where layouts are designed manually.

The relation between yield and fabrication cost is perhaps not so clear. Therefore, suppose that a yield Y_0 is associated with an area A_0 and that the probability of a defect occuring in any part of a processed wafer is constant, then for an area A the yield $Y = Y_0^{A/A_0}$. For, an area A can be considered as A/A_0 areas A_0, each having the same probability Y_0 of having no defect, and to find the probability that all the A/A_0 areas have no defect the A/A_0 individual probabilities must be multiplied by one another. Because the number of chips per wafer decreases by a factor of A/A_0 and the number of perfect chips decreases by a factor of Y/Y_0, assuming that the fabrication cost per wafer does not depend on the chip area, then the fabrication cost per chip increases by a factor of $A/A_0 . Y_0^{1-A/A_0}$. This means that for $A/A_0 = 2$ and $Y_0 = 0.1$ the chip cost is 20 times higher ! So it is clear that especially for large chips with low yield a designer must take care to make the chip area not larger than is strictly necessary. Although in practice the occurrence of defects is not completely random and depends on defect density [2,3] , giving results not as worse as discussed here, the general conclusion holds that particularly in the case of a low yield an increase in chip area largely affects the production cost.

1.4 Circuit speed

The larger the chip the longer the interconnection lines will be and the more time the signals will take to travel through these lines. But what is certainly more important is that the interconnection lines are usually of different types, for instance metal and diffusion (or polysilicon) lines, the latter having a considerably higher resistance and thus introduce a longer delay time. In automated design there are usually fairly straightforward methods of choosing the requisite type of interconnection line, e.g. all the horizontal lines made of metal and all the vertical lines diffused. However, in manual layout design the number and length of diffused interconnection lines can be minimized.

1.5 Design rule adaptation

To be able to perform their layout task automated layout design systems must have all the relevant technology design rules at their disposal. Because these rules are frequently changed to follow the technological evolution, it is very important that layout design systems can easily and continuously be adapted to changing design rules. Whether a circuit already being fabricated must be redesigned is largely a cost problem.

Human beings can follow the change of design rules quite easily. But redesigning a manual layout may require as much effort as was needed for its first design. In practice, however, it will not often occur that this kind of redesigns are carried out, for, at the time the need for it arises the circuit itself probably becomes obsolete. And as long as a reduction in chip size is possible by multiplying all the mask dimensions by the same factor of less than unity, one can take advantage of technology evolution without redesigning the layout being necessary.

1.6 A device having various shapes

Automated layout design systems usually cannot cope with problems of a device whose shape or size depends on the remaining part of the layout. One example is an MOS-transistor for which the W/L-ratio defines the time it takes to switch the charges at the output. But the output capacitances depend on the gate areas of the transistors to be driven and the length of their connections. Proper adjustment of W/L needs an analysis of the circuit's electrical behaviour. Various useful transient analysis programs are available, but usually they can only handle small circuits of up to about 50 or 100 transistors [1]. So they are used in the design of cells and not in the design of complete layouts.

Another example is an I^2L-gate for which the shape of the gate and the location of its inputs and outputs can be varied within rather wide limits. This has nothing to do with switching times as in the case of MOS-transistors, but a human designer who utilizes this layout design freedom can compact the layout considerably. However, such a feature is very hard to realize in automated design systems. For example, the only freedom in composing an I^2L-gate incorporated in the automated layout design program CALDI [4] is that it can place an input, an output, a crossover or nothing at all at each of only 5 fixed locations. This does not look too bad but this restricted number of possibilities was one of the reasons, why a CALDI redesign of a keyboard-scanner circuit resulted in a chip area which was exactly twice that of the manually laid out circuit.

2. FLEXIBLE LAYOUT DESIGN

Some approaches to combine computer capabilities like accurate and fast processing of large amounts of data and human faculties like ingenuity, knowledge of the circuit function and a bird's eye view have been reported. In [5] a designer only gives a sketch of the layout, showing the relative positions of the cells and their interconnections, where a cell is usually a standard library element, but it may be any fixed collection of mask artwork. Then the program FLOSS transforms this sketch into an actual mask layout, packing the layout elements as closely as possible within the constraints of the technology design rules. A similar approach can be found in [6] where an interactive loop is formed with the computer aided design system STICKS doing all the tedious detailed work, while the human designer performs the necessary brain work.

Systems like FLOSS and STICKS try to minimize each spacing between adjacent components separately, but cannot identify the situations in which the overall packing density would be improved if a particular local spacing were not minimized. So it may be expected that the resulting layout contains unused areas or not optimally used areas. Then the usefullness of this kind of systems depends on the extent to which this waste of chip area is acceptable, which optimization possibilities the system offers and how much interaction with the human designer is needed. Moreover, these systems cannot choose the most suitable form of an element whose shape depends on the remaining part of the layout. The same holds for SLIP [7] and CABBAGE [8] , two other computer-aided layout design systems in this field. Till now not much has been published about experiences and comparative results.

It appears to be confusing to call layout design which requires only the relative positioning of the layout components "symbolic layout design" [1,7]. This term is already in use for manual layout design where components, like transistors, contacts, cross-overs and interconnections are represented by symbols [9], a kind of shorthand notation, but in which the design system does not change the locations of the components. Where relative positioning of elements suffices it is suggested to speak of "flexible layout design", implying the flexible possibility to compose the layout as well as to change it.

3. CADILL

When, owing to the circuit complexity a few hundred gates were obtained the weak points of manual layout design - layout correctness and design time - became annoying, but at that time, the beginning of the seventies, it was also clear that it still would take a long time before automated design systems would have been developed far enough to dispense with manual layout design. But with increasing circuit complexity the chance for an entirely correct first layout was decreasing and, moreover, redesigns were becoming more difficult. At that time CADILL, a Computer-Aided Design system for Integrated Logic Layout design was developed at the Philips Research Laboratories, Eindhoven. Its main goal was to improve on the reliability of manual layout design methods without constraining the use of the layout design rules.

Initially CADILL was intended to be used for the design of I^2L-circuits [10], but later extensions made it possible to use it as well for other bipolar circuit technologies like ISL [11], and after the introduction of black boxes it can in fact be used for the design of any circuit consisting of copies of a limited number of different cells and interconnections. Before CADILL became available manual layout designers at the Philips company were using CIRCUITMASK for describing layouts and for design rule verification [12]. Layout elements were described as rectangles or polygons in one or more masks and to ease coding layout elements could be grouped in patterns, which could be copied anywhere, mirrored and rotated if desired, allowing multi-level structuring of the layout. Various output processors were available for controlling displays, plotters and mask making machines. For logic design verification PHILSIM was usually used, a time dependent gate-level simulation program [12]. CIRCUITMASK and PHILSIM were rather powerful stand-alone programs, but had nothing in common, so there was not any guarantee that a CIRCUITMASK description and a PHILSIM description really represented one and the same circuit.

3.1 Layout design reliability improvement

One possibility to improve such a situation is to provide an automatic verification between the two descriptions. To take full advantage of the powerful operations on masks incorporated in CIRCUITMASK, preferably the final geometrical mask description must be the input for such a verification program. This description must allow circuit elements like gates and connections to be recognized. Recognizing implies knowledge of the relation between mask layout structures and circuit elements and this knowledge must be brought into the verifier beforehand. Because of the many possible relations it is clear why verifiers are mostly developed for particular applications and it is usually not easy to adapt them to new relations. A further drawback is that mask layout structures are very detailed three-dimensional structures and, for processing these structures to recognize circuit elements, computer runtime increases much sharper than linearly with the chip complexity.

Another possibility is to assure beforehand that both descriptions are consistent, thus excluding the need for verification at all. Automated layout design systems are just designed in this way and can make a mask layout description if the logical circuit description is given, but has the disadvantages mentioned above.

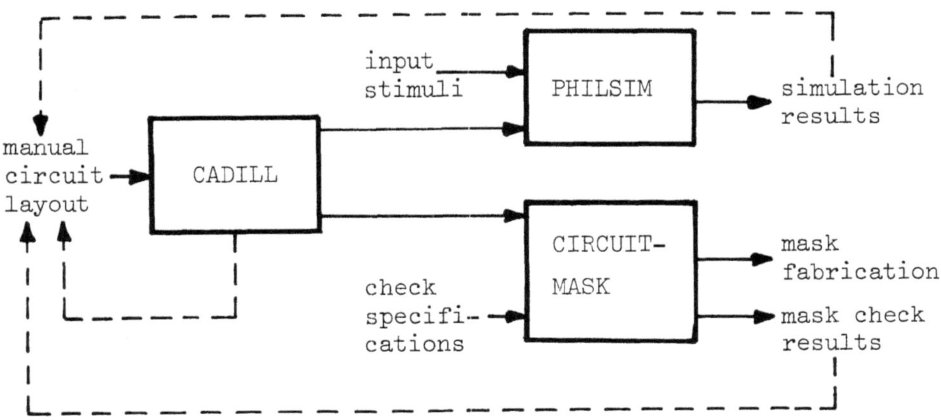

Fig. 1 The use of CADILL in conjunction with programs for circuit simulation and design rule checking.

A further possibility is to start with describing the manually designed layout such that it is quite easy to create both a mask layout description and a logical network description. Then, the latter can be used for verification with the original logical network description or for a functional simulation of the circuit behaviour, possibly a repetition of the simulation already done for design verification. This is the way in which CADILL can be used.(fig. 1). Time-consuming interpretation of all mask layout details is avoided, while a complete layout design freedom can be maintained.

A new layout description language had to be introduced and in the first instance this seems to be a disadvantage, but in practice it actually worked out as an advantage indeed. Why should a circuit designer, who is thinking in terms of gates, inputs, outputs, supply lines, connections, etc., have to know which mask details are exactly associated with these circuit elements ? In CADILL's descriptive language it was tried to approach this functional level of gates, connections, etc. as nearly perfect as possible, without cutting off any technological potentiality in composing layout structures. And designers did appreciate this.
Figure 2 illustrates how I^2L-gates perpendicular to an injector strip, a common structure in I^2L-layouts, are described in CADILL.

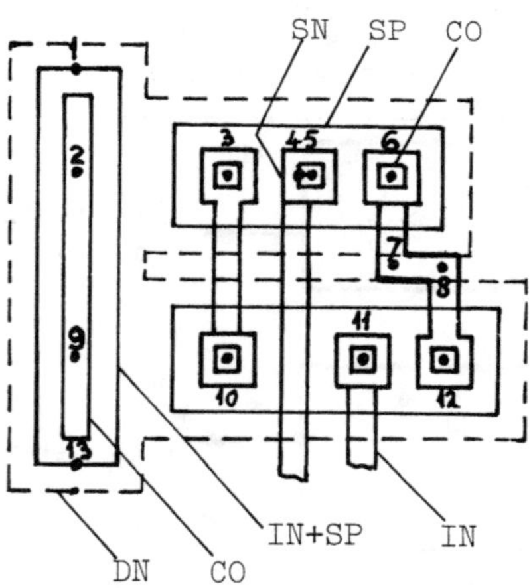


```
INJ     p1, p13
GATE    p2, p3(B),p5,p6
GATE    p9, p10, p11, p12(B)
CONN    IN  p3, p10
CONN    IN  p5, p4,....
CONN    IN  p6, p7, p8, p12
CONN    IN  p11,....
```

Fig. 2 I^2L-layout structure and its description for CADILL.

Except for the connections no mask at all appears in this description and only 13 points are relevant from which all the other points can be computed knowing the meaning of each layout element and knowing the technology design rules. Figure 3 illustrates that a description of the real mask layout requires about twice as much statements and almost twice as much layout points must be specified. The link between the functional description and the mask description is formed by patterns (BASE, COLL, etc.), each pattern containing all mask details of a layout element, and technology-dependent parameters (a, b, c, d, etc.).

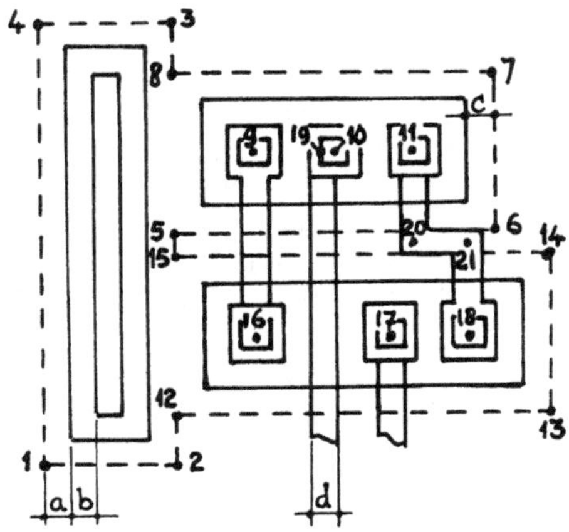


```
RECT  (DN, SP, -a, CO, -b, p1,
       p2, p3, p4)
RECT  (DN, SP, -c, p5, p6, p7, p8)
BASE  p9
COLL  p10
COLL  p11
RECT  (DN, SP, -c, p12, p13, p14,
       p15)
COLL  p16
COLL  p17
BASE  p18
CONN  (IN, d, p9, p16)
CONN  (IN, d, p19,....)
CONN  (IN, d, p11, p20, p21, p18)
CONN  (IN, d, p17,....)
```

Fig. 3 Same I^{2}L-layout structure with mask layout description. (BASE and COLL are calls of patterns containing all mask details of a base connection and a collector.)

3.2 Design time improvement

It will be clear that using CADILL's descriptive language can save much coding time and is less liable to errors compared with a description at the actual mask level containing all the details of the geometric elements. Evidently it is not even necessary to indicate such mask details in the layout.
Figure 4 shows a symbolic layout containing the same information as given in the mask layout.

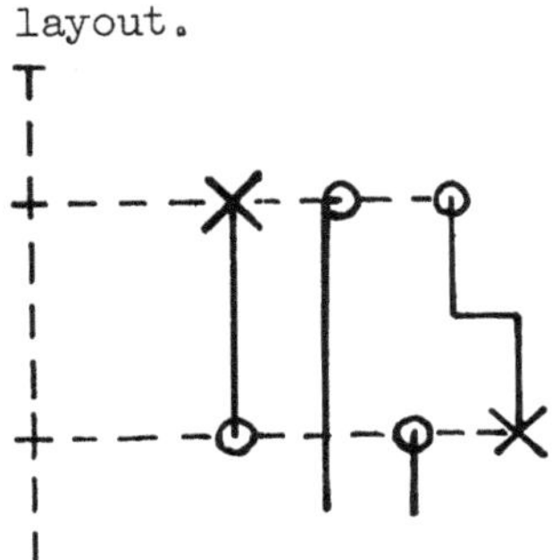

Fig. 4 Symbolic layout containing sufficient information for CADILL.

This again affects design time favourably while gaining knowledge of the mutual minimum and maximum distances between gates, gate inputs, gate outputs, connections and injectors in that symbolic form is not al all more difficult than learning all the design rules based on masks.

3.3 Special design rule checks

Because CADILL recognizes functional layout elements very easily it is also easy to check these elements against special design rules, checks which can hardly be performed at the mask level, like the maximum length of an I^2L-gate, the availability of an injector for each gate, non-connected inputs or outputs, etc. These checks, which otherwise must be performed visually, further improve layout correctness. They do not require much execution time, so they are done each time (a part of) a layout is created or updated, resulting in the automatic detection of violations against this kind of design rules in a very early stage of the layout design process.

3.4 Layout design freedom

Freedom in layout element shape cannot be obtained without imparting special features to the CAD-tool. Therefore, it will not be surprising that CADILL has built-in features to handle I^2L-gates of arbitrary shape, various kinds of injectors, vias and crossings, etc. Because one cannot foresee which features exactly have to be incorporated it is important to construct the tool such that future extensions can easily be made, especially if people in other design centres must be able to do that.

Polysilicon MOS-transistors are formed where a polysilicon path crosses a diffusion path and the layout designer is used to draw one continuous diffusion path to take advantage of the self-alignment feature of the polysilicon-gate process. For, during wafer processing a polysilicon area will prevent impurities to diffuse into the underlying silicon, thus forming the gate of the transistor. CADILL does not recognize the locations where a polysilicon path crosses a diffusion path, so all the MOS-transistors placed in a layout must explicitly be specified. However, CADILL automatically cuts a diffusion path into two pieces upon which an MOS-transistor is laid (fig. 5). So the self-alignment feature of this technology is maintained and the designer is freed from specifying all the actually disconnected pieces of a diffusion path, which otherwise would require the drawing and coding of a lot of additional details.
Unwanted MOS-transistors can be created wherever a polysilicon path unwittingly crosses a diffusion path. A check facility for

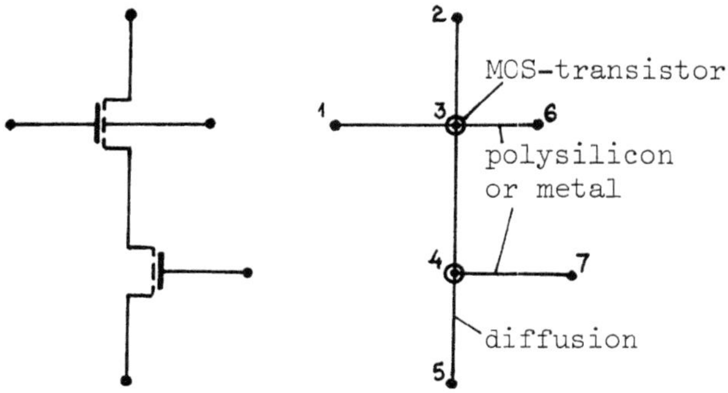


```
CONN  OD  p2, p5
CONN  PS  p1, p6
CONN  PS  p4, p7
MOS   (type)  p3, p4
```

circuit　　　　symb. layout　　　　description

Fig. 5 Polysilicon MOS-layout structure and its description for CADILL.

these situations has not been implemented in CADILL because the mask-operating and design rule checking facilities in CIRCUITMASK were sufficient for this purpose.

3.5 Mask layout generation

Of course the functional layout description must contain all the relevant information about the layout, from which all additionally necessary information for the final mask layout can be computed. Because CADILL allows the same multi-level structuring of the layout as CIRCUITMASK the generation of the final mask layout description is a fairly straightforward sequential process in which each functional statement can immediately be translated into one or more CIRCUITMASK layout statements.

One aspect will be discussed in more detail to illustrate how a designer can be freed from specifying tedious mask details by taking advantage of the mask operating facilities offered by CIRCUITMASK. In I^2L-circuits all the injectors are covered with metal to get low-resistance current supply tracks to the gates. But, if there is only one layer of metallization available, metal interconnection tracks cannot cross the injectors. Yet, sometimes it may be very advantageous to allow an interconnection track to cross an injector. In that case the designer can specify a hole in the injector and via this hole he can make the interconnection (fig. 6).

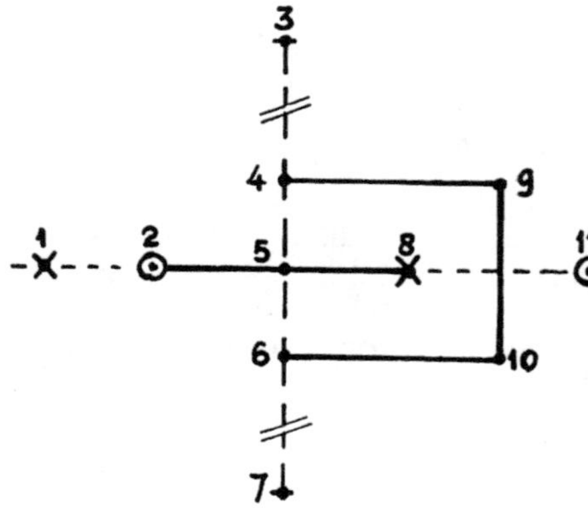


```
INJ   p3, p7
GATE  p5, p2, p1(B),....
GATE  p5, p8(B), p11,....
INJHOLE  p5
CONN  IN  p2, p8
CONN  IN  p4, p9, p10, p6
```

Fig. 6 Metal track crossing an injector.

During final mask layout generation all the injectors are initially specified in a dummy mask M1 instead of the final metallization mask IN. Each hole refers to a square in another dummy mask M2 with an edge-length equal to one standard connection path width plus two standard connection path separation widths. After all the circuit elements have been described a number of mask operating statements are added such that during CIRCUITMASK execution all areas common to M2 and M1 will be removed from the M1-areas and then the remaining M1-areas will be added to the IN-areas. (The same procedure is executed for the contact holes of the injectors.) Figure 7 shows the final mask layout.

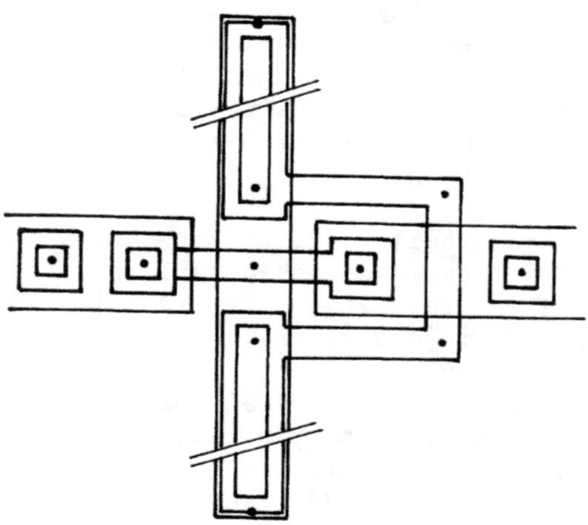

Fig. 7 Mask layout for metal track crossing an injector.

Because these mask operations are done afterwards it does not matter in which sequence connections, injectors and injector holes are specified. It is also not required to define the associated injector and holes at the same layout level or in the same pattern.

3.6 Logical network description

A logical network description shows all the logical elements of a digital circuit and the way in which these elements are interconnected. A logical element can be a simple logical function like AND, OR, NAND and NOR, or a more complex logical function like a flipflop, a register, a counter, etc. Each language suitable for describing digital circuits has a number of built-in primitives to describe certain basic logical functions and usually it offers a macro facility, which allows the user to construct new primitives by describing a more complex function as a network of already existing primitives. Whether this can be done in a multi-level way depends on the language. Such a macro facility is very convenient for the user, but in fact it is not an essential factor of a logical network language.

Each logical element has one or more inputs and one or more outputs. The way in which the values of the signals at the outputs depend on the current and possibly previous signal values at the inputs define the logical function of an element. Points where signals can enter or leave the circuit are the terminals of the network. They can be considered as logical elements with an output alone (input terminal) or an input alone (output terminal). Some languages require the explicit specification of the network terminals.

There are various possibilities to describe the interconnections of logical elements. One possibility is to give a unique name to each logical element and to number its inputs and outputs. Then a wiring list can specify which inputs and outputs are interconnected. Another possibility is to give a unique name to each connection and to describe each element by specifying its logical function and the names of the connections at the inputs and outputs. The latter possibility is more convenient for the user (fig.8).

Each language has its particular requirements as to the identification of the logical elements, the inputs and outputs of the logical elements, the inputs and outputs of the network and the connections, and the way in which the macro facility can be used. In order to postpone the effect of these peculiarities as long as possible CADILL first generates an intermediate network des-

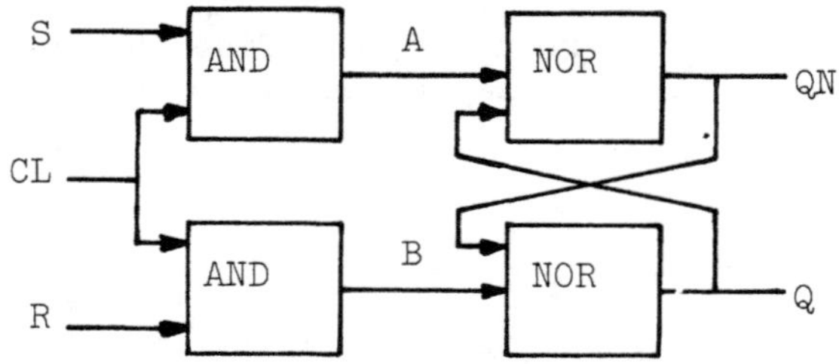


```
AND  IN(S,CL) OUT(A)
AND  IN(R,CL) OUT(B)
NOR  IN(A,Q)  OUT(QN)
NOR  IN(B,QN) OUT(Q)
```

Fig. 8 A simple digital circuit and its logical description.

cription, after which one of its network post-processors transforms this description into a logical network description for a particular language. Here, we will only deal with the creation of the intermediate network description, which is based on the second possibility discussed above.

The designer may give identification names to points of the layout to make the network description more recognizable. But to save memory capacity and time CADILL temporarily replaces these names by unique identification numbers and stores the real names in a separate file where they are kept until the network post-processing. All the other points of the layout that need an identification will automatically get temporary identification numbers which eventually will be replaced by systematically created names depending on the name-giving conventions of the particular network language.

A functional layout description as used in CADILL shows all the functional elements of a circuit and all its connection tracks. The functional elements have been chosen such that they can be easily translated into logical elements for the corresponding logical network description. But the inputs and outputs of functional elements and the connection tracks refer to layout coordinates and to masks. So the task is to decide from these explicitly or implicitly given location and mask references how the elements are logically interconnected.

The price which had to be paid for obtaining a full layout freedom is that not any ordering in the functional layout description could be prescribed and that different parts of the description can represent interfering parts of the layout. For example, it is possible that one logical interconnection physically consists of several connection tracks, each track described by a separate connection statement. This makes it necessary to collect all the logically relevant points, like the inputs and outputs of a gate, and to trace which of these points are associated with a particular connection. But in point of fact each relevant point and each connection corresponds with some layout area, so, strictly speaking, it would be necessary to store and compare the relevant areas instead of the relevant points. In case of circuits containing thousands of transistors this would require much memory capacity and much execution time. Therefore, to simplify the problem each input or output of an element is thought to be concentrated in one point and each connection track is thought to be reduced to its centre line.

3.7 Network reconstruction

In this section various aspects concerning the reconstruction of the logic network will be discussed in more detail.

For CADILL, recognition of all the logical elements is not difficult because they can be found in the functional layout description. Therefore, this description is scanned to collect and store all the relevant logical network points, like network inputs and outputs, gate inputs and outputs, vias, etc. This done, the functional layout description is scanned again to collect all the connection statements and for each connection the set of associated logical network points is retrieved and these points are coupled by an interconnection chain. Finally, the functional layout description is scanned for the third time to give each connection its unique identification number and to generate the logical description of each element. So the logical description is in the same sequence of logical elements as is the functional layout description, there being a convenient correspondence especially in case of fault locating.

The required memory capacity to store all the relevant logical network points depends on circuit complexity and in order not to limit the number of network elements by the available core memory a virtual memory concept has been introduced. All the data are stored in segments on a disc and only a limited number of segments is available in core. During program execution a swapping mechanism is responsible for the interchanging of segments between disc and core.

As mentioned earlier the layout may be structured by the use of multi-level patterns, but it was decided not to maintain this structure in the logical network description, although most logical description languages allow such a structuring by means of a multi-level macro facility. But a macro must be identified by a name and at least all its inputs and outputs must be specified. This information cannot simply be derived from the layout description, so the designer would have to supply it explicitly.

A certainly more serious problem would be to cope with situations where elements defined by a possibly mirrored or rotated copy of some pattern are overlapped by elements defined anywhere else without fully expanding the layout. Such situations could be forbidden, but that would restrict layout design freedom and a violation check would be necessary for which forbidden regions associated with each pattern must be available. Expanding the layout structure prevents these complications, does not oblige the designer to supply additional information, while complete layout freedom is maintained. Moreover, for a gate-level simulator this is immaterial because one of its first tasks will be to expand a structured network description completely.

Another point worth considering is which relations between the network points must be maintained to enable their retrieval during the various stages of the network reconstruction. At least four different relations can be distinguished :

1. a chronological relation indicating the sequence in which the network points are generated, namely as they appear in the functional layout description.
2. an element relation indicating all the network points belonging to the same element, for instance all the inputs and outputs of one logical gate.
3. a location relation, at least to retrieve a network point by its location, but it may also be useful to be able to retrieve neighbouring network points.
4. a connection relation indicating all the points belonging to the same connection.

Of these four types of relations the location relation is a rather complicated one because layouts are in fact three-dimensional structures. Since the functional layout description is based on location references one cannot completely dispense with a location relation. A logical network description is based on connections between elements, so a connection relation is important too.

Network points and their attributes, including their relations, can be stored in memory at successively higher addresses and this ordering can be used to represent one of the relations, in fact a two-directional relation. All other required relations must explicitly be added by means of tables or pointers. In virtual memory systems the use of tables of variable length is not very convenient and therefore pointers are used in CADILL to represent relations. But it will be clear that it would require much memory space to have all possible relations available and a choice had to be made, illustrated by the following discussion.

Initially all the networkpoints are successively stored in core memory as they are chronologically encountered in the expanded functional layout description and for each point a group of words is reserved to store all the relevant data. Here, such a group is called a bead. At the same time a location relation is constructed for which one bead is added for each x-value. These x-beads are linked with each other by forward and backward pointers. All the beads for the network points that have the same x-value are linked by forward y-pointers and the x-bead associated with such a group of beads gets a pointer to the bead with the lowest y-value (fig. 9).

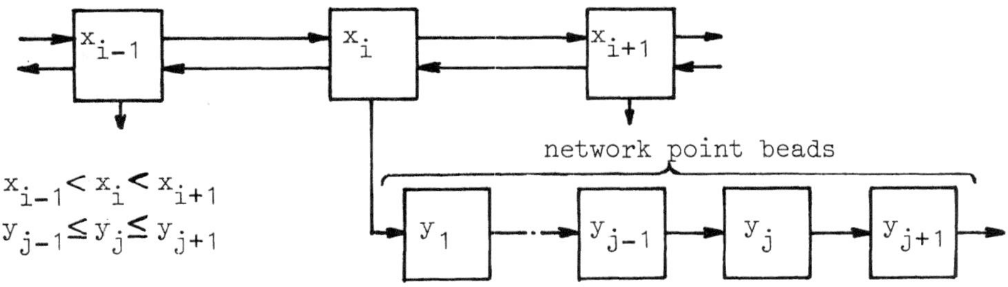

Fig. 9 Network point beads with initial location relation, stored in core in chronological sequence (not indicated).

When all the networkpoints have been collected or the core memory is completely filled all the beads, including the x-beads, are transferred from core to disc in the sequence of xy-values, where they are inserted between possibly earlier stored beads.
The chronological relation is lost but it is not necessarily needed, because in all further steps the network point beads can be retrieved by their location relation. For the final structure the forward y-pointers do not require memory space, moreover, storing y-beads with increasing y-value and the same x-value at subsequent addresses gives a two-directional y-relation (fig. 10).

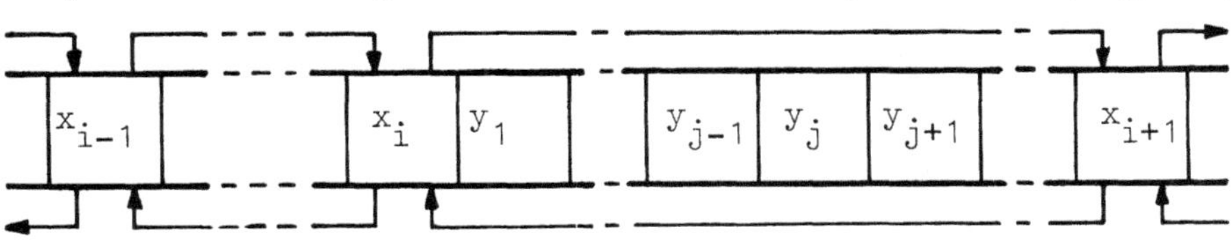

Fig. 10 Networkpoint beads with final location relation as stored in the virtual memory.

This very limited location relation scheme can easily be constructed and it saves memory space because it is not necessary to store x-values in the individual network point beads while it speeds up the retrieval of beads by means of their location.

If the string of beads is completed it is divided into fixed length segments and in a segment table, permanently residing in core, the highest xy-value and the location of the first x-bead is recorded for each segment. When a segment is swapped from disc to core its starting address is also stored in this segment table.

A pointer to a bead used for some relation may in principle be the address of that bead or of its first word in virtual memory. Then it must be possible to decide which segment contains the bead pointed to and with the help of the segment table the actual core address of the bead can be computed. To save space and time it is desirable that a pointer does not occupy more than one memory word and then the number of beads and hence the number of network points is restricted to the memory word capacity divided by the average number of words in a bead. If the beads are all of equal length it is also possible to number the beads and to use these bead numbers to point to them. Then the maximum number of beads is only restricted to the memory word capacity. Of course in that case converting a pointer into an actual core address will require an extra multiplication which takes additional time.

For each logical network point, space must be available to store the following data :

- its location in the layout, as has been discussed above the x-value is stored in a separate x-bead and the y-value in the network point bead.
- its type : input, output, connection point or switching point of a transistor.
- its mask(s) by which it can be connected to other network points
- a chain pointer to refer to another network point to which it is connected.
- its identification number.

The element relation between all the points belonging to one element is not strictly needed because these points can be retrieved by means of their location relation. Moreover, points belonging to one element are usually spaced a small distance apart, so it may be expected that they can be retrieved rather quickly.

In gate-level network descriptions, gates are in general of multiple-input one-output type, e.g. a NAND-gate, for which it is usually not allowed to interconnect several signal sources (gate outputs). But in layout practice signal sources are interconnected to save layout area or because it is inherent in the technology used, for instance in I^2L-layouts, where an I^2L-gate is a one-input multi-output device. The interconnection of signal sources represents an AND-function or an OR-function depending on the technology and the association of a high voltage with a logic 1 or a logic 0. The necessity to insert these additional logical functions is one of the reasons why the type of each logical network point is stored.

3.8. Layout structuring

A designer can structure the layout by defining and using patterns as well as black boxes. A pattern is any group of layout elements, connections and copies of other patterns which can be regarded as a unit. When it is necessary to go into the internal details of a pattern, as for example for reconstruction of the logical network, this pattern is completely expanded. So for a pattern it is not necessary to specify how it interferes with other elements in the layout.

For a black box it is just the other way round. Its internal details are not important but only its interference with other elements are. So for each input and output it must be specified where it is located in relation to the black box origin and in which masks it can be connected to other elements. Two or more points of a black box can also be directly interconnected by some internal connection. These points are indicated by xputs because they behave neither as inputs nor as outputs.

A black box can represent a via, a connection, a gate, a flipflop or any other logical function. It can be used for two reasons :

1. to represent some logical function which cannot be recognized as such from the layout elements, e.g. a memory bit, a shift bit or an input/output buffer,
2. to do away with the need for the program to take into account all layout details in order to reduce memory space and computer execution time.

It is the designer's responsibility that a black box really represents its supposed logical function and that it does not interfere with other elements except by means of its inputs, outputs and xputs. Forbidden areas in arbitrary masks can be associated with each black box to detect unwanted interference with other layout elements. However, such checks must be done with the help of CIRCUITMASK.

Because in the final mask layout a copied black box must be represented by a particular geometrical structure each black box can be related to a pattern containing that structure. To decrease plotter time and for better readability of a plotter drawing it can also be related to a symbolic pattern indicating for instance the outline of the geometrical structure, its type and the location of inputs, outputs and xputs. In general a black box will represent some logical function. Therefore, it can be related to a macro description defining that logical function or to a basic logical function of the simulator used, like a NAND-gate.

A black box can be simply defined by the designer, but it can also be developed as an ordinary part of a layout, including the descriptions of its geometrical mask structure and its logical function, and, after the geometrical and logical correctness have been checked, such a part can be defined as a black box and related to the descriptions developed. This has appeared to be a convenient way to structure a layout in one or more levels, thus reducing program run times considerably.

3.9 Adaptation to fabrication processes

As already has been discussed a number of patterns and parameters form the link between a functional layout description and the description of the corresponding final mask layout. The patterns contain layout descriptions of commonly used units like vias, transistors, a collector of an I^2L-gate, etc. The parameters are used to check the functional layout description against particular process-dependent design rules and in computations for transforming this description into the final mask description. The contents of these patterns and the values of the parameters possibly have to be changed when a fabrication process is changed. And because there is a rapid evolution in fabrication processes a flexible adaptation of a CAD-tool to these changes is highly desirable.

For CADILL an already existing general file system could be used in which two types of files are distinguished, i.e. common files and user files. One or more user files can be connected to one common file. This file system is used such that each designer has a user file of his own in which he puts all data concerning one layout. He can divide this user file into as many subfiles as he wants. All designers in one particular technology and fabrication process use the same common file and this file contains the common patterns and current values of the technology parameters.

It is easy to create a new common file or to change descriptions or parameter values and so adaptations to various fabrication processes or to changing design rules can easily be made. Of course, changing a common file is only allowed to the management at the design system level. If a user wants protection against changes, perhaps because he may expect to redesign a circuit in the future based on the current patterns and parameters, then he can copy the contents of the common file into his user file, since data will only be taken from the common file if it cannot be found in the user file.

3.10 Interfaces with other CAD-tools

The development of CAD-tools is expensive and time-consuming and therefore it may be expected that such tools have a reasonably long life, say 10 or more years. During that time it must be possible to make improvements and adaptations to the continuously changing field of designing integrated circuits. One possibility to keep development costs low is preventing duplications by taking advantage of already existing tools. This requires a good cooperation and flexible interfaces between them. Because a tool must be flexible in itself its internally used data structures are usually liable to changes and good documentation about these data structures is not always available or is not up to date. So it may be concluded that internal data structures are not appropriate for interfacing different tools.

In most cases a particular language is associated with a tool for describing input data and to control its behaviour and output. Such a language is in general very well defined and rather stable. And if it is changed indeed, one tries to do it in a compatible way in order to prevent all descriptions based on the previous language from becoming obsolete. Hence, such a language is an excellent tool for interfacing purposes, although it takes some extra time to convert an internal structure into a language-based description in one tool and to convert this in turn into an internal structure in the second tool. It is advantageous to have intermediate results available in a readable and familiar form to which additions can easily be made. And although automatically generated descriptions can be expected to be errorfree the syntax checking facility available in the second tool can be very useful during the development of the first one.

REFERENCES

1. H.J. de Man, "Computer-Aided Design for Integrated Circuits : Trying to Bridge the Gap", IEEE Journal of Solid-State Circuits, Vol. SC-14, No.3, pp.613-621, June 1979.
2. B.T. Murphy, "Cost-Size Optima of Monolithic Integrated Circuits", Proc. IEEE, Vol. 52, pp. 1537-1545, (1964).
3. G.E. Moore, "What Level of LSI is best for you ?", Electronics, pp. 126-130, Februari 16, 1970.
4. M. van der Woude, "CALDI : Computer-Aided Layout of I^2L", Digital Processes, Vol.2, No.2, pp. 143-156, (1976).
5. Y.E. Cho et al., "FLOSS : An Approach to Automated Layout for High-Volume Designs", Proc. 14th Design Automation Conference, New Orleans, pp. 138-141, June 1977.
6. J.D. Williams, "STICKS-A Graphical Compiler for High-Level LSI Design", AFIPS Conference Proceedings, Vol. 47, pp. 289-295, June 1978.
7. A.E. Dunlop, "SLIP : Symbolic Layout of Integrated Circuits with Compaction ", Computer-Aided Design, Vol. 10, No. 6, pp. 387-391, Nov. 1978.
8. M.Y. Hsueh and D.O. Pederson, "Computer-Aised Layout of LSI Circuit Building-Blocks", Proc. 1979 IEEE International Symposium on Circuits and Systems, pp. 474-477.
9. D. Gibson and S. Nance, "SLIC - Symbolic Layout of Integrated Circuits", Proc. 13th Design Automation Conference, San Francisco, pp. 434-440, June 1976.
10. K. Hart and A. Slob, "Integrated Injection Logic : A New Approach to LSI", IEEE Journal of Solid-State Circuits, Vol. SC-7, No. 5, pp. 346-351, Oct. 72.
11. J. Lohstroh, "ISL, A Fast and Dense Low-Power Logic, Made in a Standard Schottky Process", IEEE Journal of Solid-State Circuits, Vol. SC-14, No. 3, pp. 585-590, June 1979.
12. C. Niessen, "Computer-aided design of LSI circuits", Philips Tech. Rev., Vol. 37, No. 11/12, pp. 278-290, (1977).

DESIGN FOR TESTABILITY

G. Grassl

Siemens AG, Research Laboratories
D-8000 München 83, F. R. Germany

ABSTRACT

Most VLSI-circuits can only be tested economically, if they are designed for testability. A test concept must support testing during development, production and system maintenance with a minimum of cost. Basic principles for easy testable designs and some very important approaches are explained and their application illustrated.

1. INTRODUCTION

Only 10 years ago MSI-circuits consisting of less than 100 elementary logic functions could be completely tested by applying a very limited number of test vectors, which were easily generated manually because of the apparent function of the circuits. Therefore there was no need to think of testability on circuit level.

When the circuit complexity increased, longer test sequences could be managed by the use of faster Automatic Test Equipment (ATE). For the more difficult task of test generation and simulation CAD-programs were introduced. Some programs use the Boolean Differences /1/ to generate a function oriented test from the circuit equations. But this is often time- and memory consuming for larger networks. Other programs are based on the topological gate level description and use path sensitization and the D-algorithm /2, 3/.

But complete testing of LSI-circuits was often no longer possible. For some circuits economic test generation even turned

out to be almost impossible.
The cost of test generation is growing exponentially with increasing complexity of the circuits (see fig. 1.1). The efforts to reduce the cost of testing by improved CAD-tools for generation and fault simulation and more efficient test equipment can move the curve of costs only slightly towards larger circuits but new techniques are necessary to overcome the basic problem.

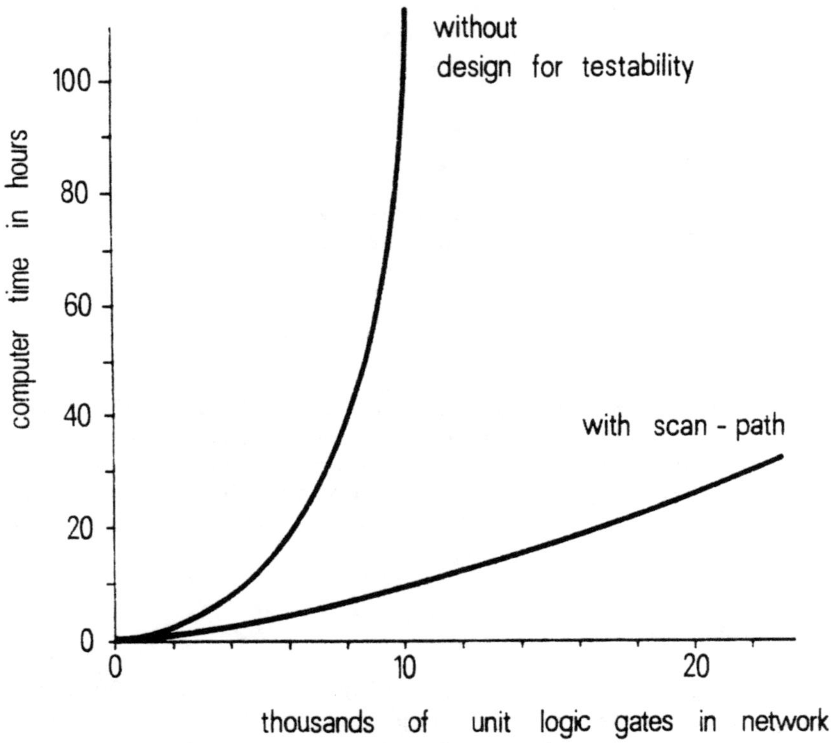

Fig. 1.1 Growth of the test generation cost /8/

In the past the main efforts of the designer were directed to fit a function into the smallest possible silicon real estate. For VLSI-circuits testability is now becoming a problem of major importance, which must be solved even if 10% or more additional circuits are needed.

2. THE REASONS FOR THE TEST PROBLEM

As shown in fig. 2.1 a typical logic circuit consists of combinatorial logic and internal memory elements. The primary inputs X_p and outputs Y_p are only a small part of all terminals of the combinatorial circuit. The response Y of this circuit is a function of the input vector, which is composed of the primary inputs and the internal states Z : $Y = f(X_p, Z)$.

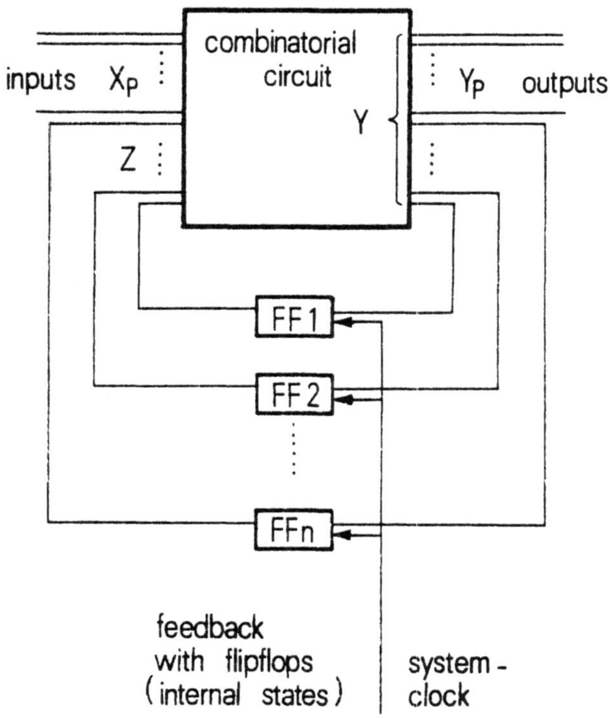

fig. 2.1 Synchronous sequential circuit

With the number of inputs and outputs the complexity of the combinatorial circuit is growing. That means

- the circuit contains more gates, which results in a linear growth of test generation time and
- the logical depth (the number of gates in a signal path between input and output) may increase, which causes an exponential growth of test generation time.

Even more critical is an increased sequential depth. To perform a certain test, first the internal memory elements must be set to a certain state by an initialization sequence. Then the checking sequence is performed and finally a read-out sequence is necessary to propagate the results to the primary outputs. These sequences may be longer or, in other words, the controllability and observability is worse.

Other test problems of high density VLSI-circuits are caused by dynamic effects such as hazards and spikes, critical races and crosstalk. The scaling principle affords also a reduction of the supply voltages, reducing the signal to noise ratio. These faults can't be described by the simple stuck-at fault model which is mostly used in computer aided test generation systems.

3. REQUIREMENTS OF DESIGN FOR TESTABILITY

3.1 REQUIREMENTS OF A TEST CONCEPT

An optimal and cost effective solution of the test problems of VLSI-circuits and systems can only be reached by consequent design for testability. This means to use design techniques such that

- the faults which are difficult to test are eliminated (e.g. to use synchronous logic to avoid dynamic faults)
- the test generation is simplified (e.g. by reduction of complexity of blocks).

Also a test strategy must be developed which considers the requirements of testing and test generation during the different test situations

- prototype debugging
- production test and
- maintenance test and system diagnosis.

For prototype debugging fault detection and localisation of all faults is desired. Also device characterization is necessary to learn the limits of operation. The faults may be process induced faults such as pinholes, opens or shorts, which are statistically distributed over the wafer. However the faults of main interest are the design faults like forgotten contact holes, critical or wrong ratios of transistors or too long delays on signal lines.
The tests are often carried out by the designer who has good knowledge of the circuit or some other qualified persons. They may use sophisticated automatic test equipment (ATE) and additional techniques such as microprobing and electron beam testing /4/. But these techniques are only for a better circuit analysis and can help to localize the faults.

The requirements of production tests are quite different. Most important are low cost of testing and test generation as well as good fault coverage. Therefore the tests should run on simple test equipment and/or short test time must be achieved. But only go/no-go information is needed because repairing of chips is usually not possible. This can only be done with circuits containing repetitive structures such as memories /19/. Some information on fault location can be useful for fault statistics, which is inportant for redesigns to improve yield.

During maintenance and system tests the VLSI-circuits are not as well accessible as during production tests. However, the go/no-go information about the chip or printed circuit board (PCB) is sufficient. In large systems the diagnosis is often performed by a service processor. Therefore important internal states must be made accessible to this processor. In some applications on-line fault detection capability is required to stop the system immediately if a fault occurs.

3.2 LIMITS FOR DESIGN FOR TESTABILITY

The freedom to fulfil these testability requirements is very limited. The main limitations are
- limited number of additional pins
- few additional circuits are possible
- minimum loss of performance.

The number of pins is an important factor of cost especially for low cost IC's. Additional pins also need more chip area and, if they are output pins more power dissipation. Larger packages require more area on the PCB.
The pins for testing can be reduced, if one or more inputs can be set to a second voltage level (e.g. 10 V) to control the test mode. Also input combinations which are not used in normal operation (e.g. read- and write signal active) can be used for testing. With one test input a set of additional test instructions may be activated to control the testlogic.

The overhead caused by design for testability must be as small as possible, because the yield is reduced rapidly with the increase of circuit complexity. For example the yield of a 8086 μ-processor with 33 mm² chip area would be reduced from 10% to 7% if the chip size is 20% larger (calculated with the yield modelling function /5/). Together with the reduction of systems per wafer this causes 70% higher manufacturing cost (fig. 3.1). Therefore the reduction of test cost is partly compensated by higher production cost.

Design for testability may also cause a reduction of performance, since additional gates and higher capacitive loads in critical paths may be necessary. Another reason may be the use of synchronous instead of asynchronous logic.

$$Z = \frac{A_W}{A_C} \left(1 + s^2 A_C \bar{D}\right)^{-\frac{1}{s^2}} .$$

A_W = wafer area

A_C = chip area

s = weighting function (0.5)

$\bar{D}$ = defect density (10 cm^{-2})

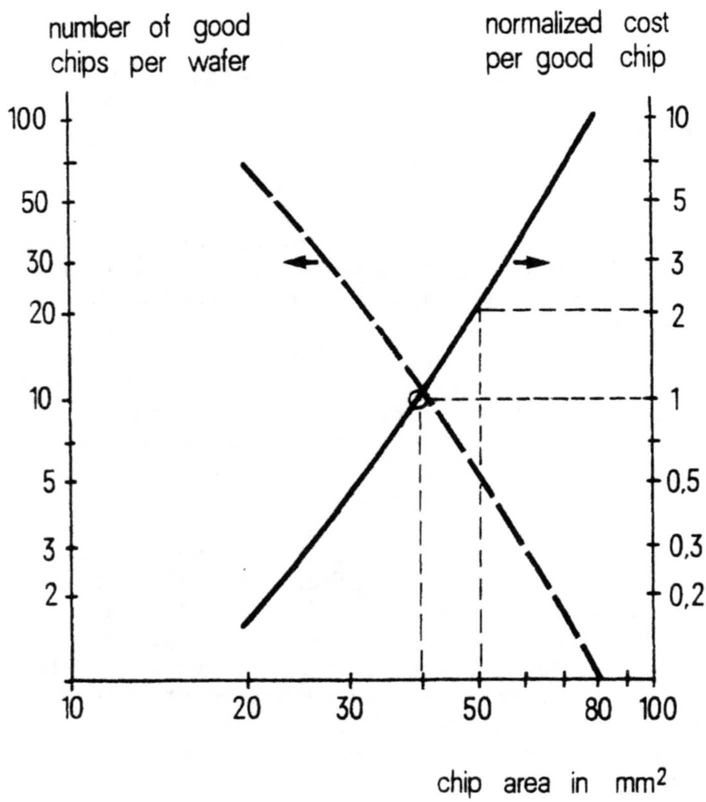

Fig. 3.1 Yield and production cost in relation to chip size

However the significance of these factors is quite different for low-cost mass-products such as general purpose microprocessors and circuits which are used in complex systems and mainfraim computers. Circuits of the first group are often not completely tested and testing costs are only relatively small. For the producer of those circuits only the rate of returns is interesting. Even if we assume test costs of 40% for future VLSI-circuits the most important benefit can be achieved with only 5 ... 10% overhead (see fig. 3.2).

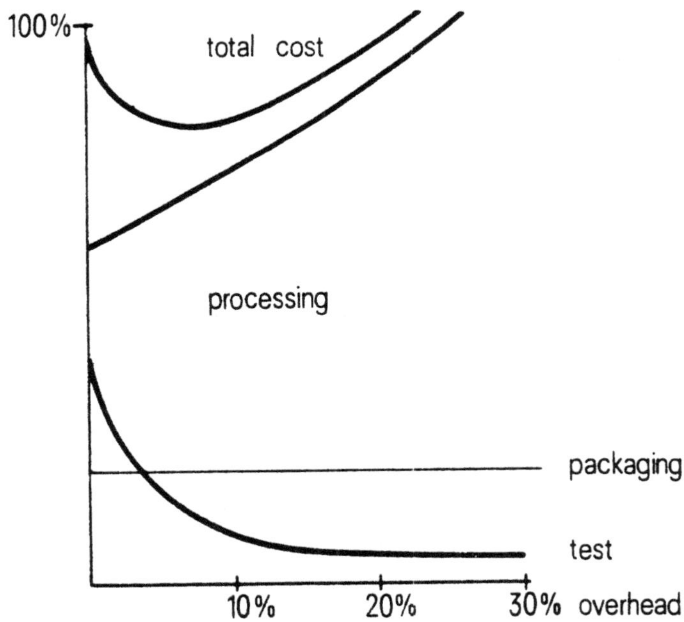

Fig. 3.2 Total chip cost as function of overhead for testing

On the other side circuit costs are only of minor importance for a whole system. Here an undetected fault which often behaves like an intermittend fault may cause considerable cost and delays. Therefore nearly 100% fault coverage is necessary even if the circuit is more expensive.
This may be a reason why up to now only manufacturers of mainfraim computers did considerable efforts of design for testability.

4. THE BASIC PRINCIPLES OF DESIGN FOR TESTABILITY

Many proposals to improve testability have been published during recent years. The basic principles of these techniques are the following:

- the improvement of controllability and observability
- partitioning of complex circuits
- the use of circuit techniques which provide easy test generation and avoid certain faults
- the implementation of self-test and self-checking circuits.

4.1 IMPROVEMENT OF CONTROLLABILITY AND OBSERVABILITY

Bad observability and controllability are mainly responsible for the exponential growth of the test generation cost and long test sequences. Also fault detection may be difficult in case of multiple faults because of masking effects.

A fundamental rule is to provide a general reset to all internal states for initialization. More advanced techniques use observation- and testpoints (observable and controllable points) /6, 7/ in conjunction with multiplexers to save pins. In systems with an internal bus, during the testmode some registers or single flipflops may be connected with the bus by multiplexers to set or read them.

A very efficient method is the scan-path, which was proposed for integrated circuits in /10/. In recent years variations became known as scan/set logic /8/ or as level sensitive scan design (LSSD) /11 - 14/.
The principle of the scan-path is shown in fig. 4.1. All storing elements except memory- and register-arrays are connected to a long shift register. In the test mode they can be set and read--out via this shift register. The sequential circuit is divided into a sequential part - the shift register - and a combinatorial part, which can be easily tested now. Only an additional serial data input and a clock control input are necessary. Detailed information about the scan-path will be given later.

4.2 PARTITIONING OF COMPLEX CIRCUITS

The scan-path provides a structure oriented partitioning on circuit level. VLSI-circuits are usually composed of several functional modules. Therefore partitioning of the chip into those modules, which can be tested separately is an adequate method to reduce the test problem to the level of testing several MSI circuits. Since there is an exponential relation between circuit complexity

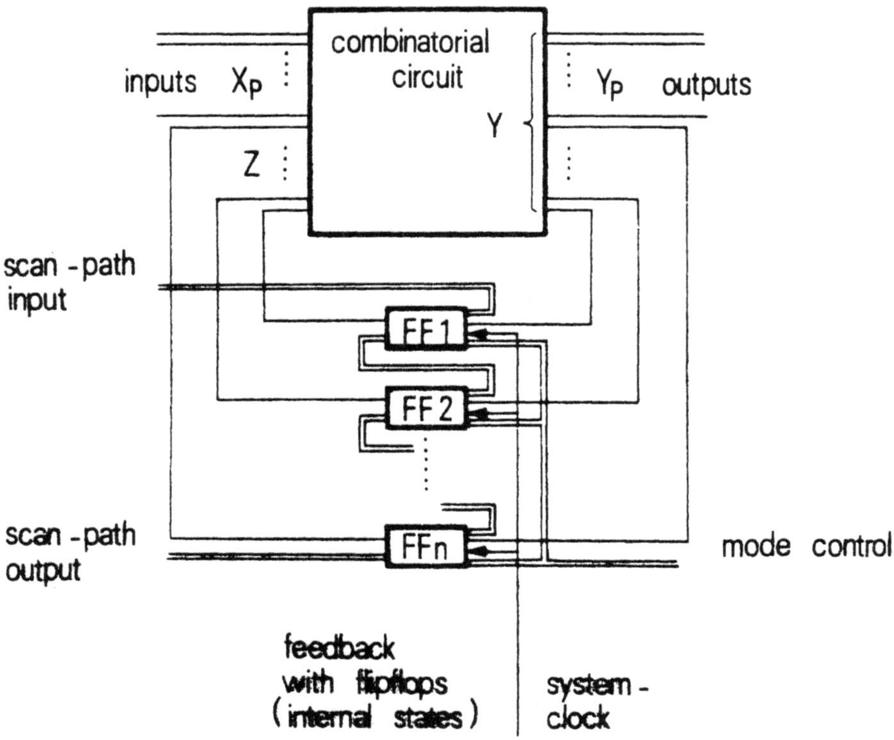

special flipflops :

- master - slave
- 1-3 clock - systems (1 system - clock, up to 2 scan - clocks or mode control signal)
- up to 4 additional pins

Fig. 4.1 Sequential circuit with scan-path

and computer time needed for test generation, the relation $\sum t_i \ll T$ is valid (t = test generation time for modules, T = test generation time for the complete circuit).

Circuit partitioning also meets the requirements of VLSI-design, which is usually composed of several parts made by different designers. If a function module is changed, only the test for this particular module has to be generated once more.
The method is especially convenient if the VLSI-chip contains memory- and register-arrays which cannot be integrated into a scan-path.
In many VLSI-circuits partitioning is already given by a bus--oriented structure. Therefore only few functional modules must be made accessable to the internal bus by additional multiplexers and reading circuits.

4.3 EASY TESTABLE CIRCUITS

The difficulty of test generation for certain types of faults can be avoided by a design which eliminates them. The common dynamic faults, which cannot be described by the simple static stuck-at fault-model, can be avoided by a synchronous logic where data signals and clocks are separated. A very advanced concept is the LSSD. As this system is based on level sensitive latches, hazards or spikes can't change the function because they are not latched. However the next clock signal must not appear before the data has become stable. In some circuits this will reduce the speed and therefore affect the system architecture. For example pipelining or parallel processing might be necessary in order to reach specifications.

Test generation can also be simplified if regular structures such as PLA's and ROM's are used instead of random logic /34/. Unnecessary redundancy within the circuit should be avoided whenever it is possible.

4.4 THE IMPLEMENTATION OF SELF-TEST AND SELF-CHECKING CAPABILITY

In larger computer systems self-tests are in common use. They are performed off-line by diagnostic programs. However if faults are not detected immediately, a great number of data might be destroyed Therefore redundant hardware and on-line tests are implemented in critical functions. If fault-security is necessary whole units are doubled or replicated /15 - 18/.

4.4.1 REDUNDANCY FOR SELF-CHECKING VLSI-CIRCUITS

Principally off-line and on-line test techniques can also be used for VLSI-circuits. However the application of double- or even triple-redundancy as it is necessary for error detection or correction in data processing circuits causes an enormous overhead. Therefore it can only be used in special applications on custom designed circuits. A more economic solution is to use three identical chips in parallel with an extra voting circuit.

In memories only 6 or 7 redundant bits for 16 or 32 bit word length respectively, are needed for single error correction and double error detection, if a hamming-code is used.
A circuit for generation of the checking bits and error detection and correction cause a signal delay of 50 - 75 ns /20/. In conjunction with the overhead of 25 or 40% this is usually not acceptable for on-chip memories.
Single error detection however needs only one parity bit. If a parity bit is added to each byte the overhead will be 12.5%.
The circuit for parity generation and checking consists only of a few EXOR-gates with relatively low signal delay.

4.4.2 BUILT-IN TESTS

Off-line self-tests can be useful for wafer tests as well as for diagnosis in the field. However they are not suitable for chip analysis because the results give only very few information about the fault location. The two major techniques are:

- self-test by a stored microprogram
- self-test with pseudorandom patterns.

In intelligent systems like microprocessors the proper operation of most of the circuit functions can be proved by running a suitable test sequence. If an error occurs, a conditial jump is carried out to a certain address which can give some references to the fault.

Bootstrapping is a very common strategy. First those operations are performed where only few functions are activated. If they work correctly they can be used to constitute a more powerful machine, which can be used to perform more complex operations. Since the program stops after the first detected fault, the other faults can't be detected.

Depending on the purpose and the required quality and complexity of the test, the program may be stored only in ROM or part of the test program can be loaded into a RAM for the test only. A self-test which is stored in ROM and checks 95% of the processor functions was published in /21/. The program needs 120 bytes and 1% additional chip area.

These results show that self-test of microprocessors is convenient not only for system diagnosis but also for wafer tests to select the chips for packaging. Sufficient propability that bad chips are detected can be achieved. The test equipment is very simple because only few pin electronic and no stored program or reference patterns are needed in the ATE.
However a different and more complete test must be performed before shipping of the parts.

It is clear, that self-tests by a stored program are restricted to microprocessor circuits. The limitation is not true for a built-in test using pseudorandom pattern and signature analysis. The test-pattern generator (TPG) and signature register (SR) are both easy to implement as cyclic shift registers by function conversion of existing registers. Some important characteristics of this method are:

- no cost-intensive test generation, only simulation to find out the correct signature,
- good fault coverage for small modules,
- partitioning of complex circuits is necessary to create modules with low fan-in and very low sequential depth,
- the test can be run with high internal speed,
- the test equipment to check the results is very simple,
- parallel testing on the chip is possible.

This method will be discussed in more detail later.

5. IMPORTANT METHODS OF DESIGN FOR TESTABILITY

After the overview of the basic principles of design for testability some examples of realization and experimental results are presented in the following. The techniques which will be discussed are:

- circuit partitioning and modular sensorialization
- the scan-path
- and the built-in test using pseudorandom patterns and signature analysis.

Of course it is often difficult to separate the different techques clearly because they are partly based on the same principles such as partitioning of a large circuit. There is also no common solution for every type of circuit because every technique has its disadvantages. Therefore a combination of different methods will often lead to the best results. At the end of this chapter I will discuss a circuit which is very suitable for such a combination.

5.1 CIRCUIT PARTITIONING AND MODULAR SENSORIALIZATION

Nearly all microprocessors and a great number of other VLSI-logic circuits consist of an ordered set of modules such as control unit, decoder, registers, ROM, PLA and data processing units. Often there is a system bus for internal data transmission between modules and connection to input- and output pins.

However most of the commonly used computer aided test generation programs don't care about the functional structure of the circuit. They use the gate-level description and path sensitization or other mathematical procedures.

An adequate test method that takes advantage of the circuit's functional structure is the modular sensorialization /22/. By this technique the circuit is not only tested for steady state faults but in its functionality. The circuit is first partitioned into its functional modules so that data can be applied to the input pins and results are propagated to the output by the use of microinstructions and controlsignals. Some of these modules may overlap.
Often there is good access to the inputs and outputs of modules by a system bus. However in many cases not a sufficient access will be possible for all modules, or isolation is necessary to avoid disturbing reactions between modules. Therefore additional logic such as separating gates, and multiplexers or reading circuits to provide additional bus connections during the test mode are necessary to improve testability.
With this additional logic the test may be generated for each module separately. Also the separation of the tests for data- and control-logic may help to simplify testing /26/.

Fig. 5.1 shows a microprogrammed processor control unit (PCU) with a two-step instruction decoding. The architecture, which is a small processor of its own, is difficult to test. The internal loops and only indirect access to several memories (mapping PROM, stack, μ-program-PROM and nanoprogram-ROM) are causing the problems. To improve testability, a direct input from the bus to the first multiplexer (MUX1) and an additional MUX3 at the entrance of the instruction PLA are introduced. The data flow at the sequencer output can be observed by an additional reading circuit. Other reading circuits are necessary to observe the output of the pipeline-register. In this way the mapping PROM can be checked directly.

The additional input to MUX1 allows the direct control of the microprogram-PROM and to set the registers of the sequencer. The functions of the sequencer may be controlled by MUX3 at the entrance of the instruction PLA. Therefore any test sequence can

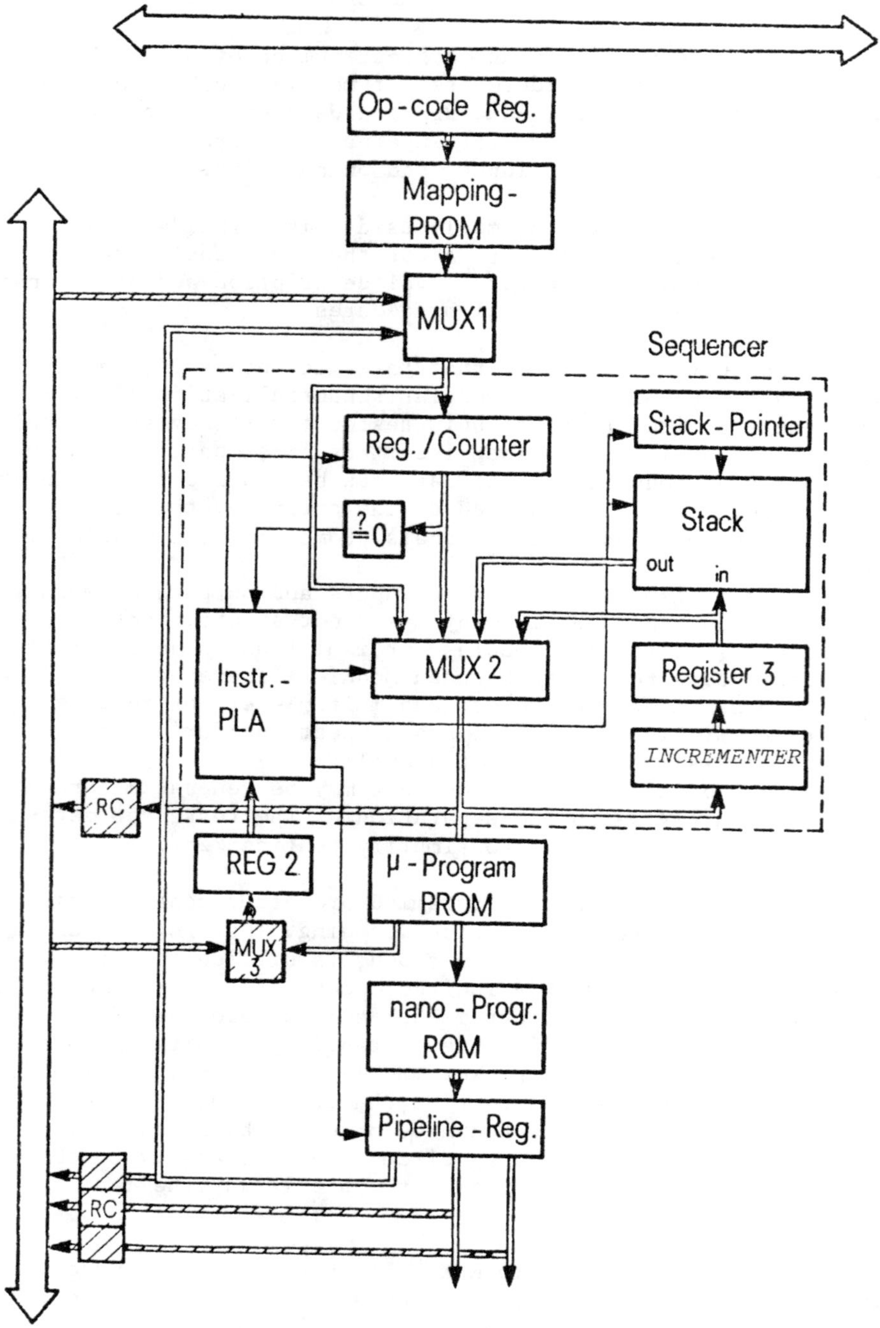

Fig. 5.1 prozessor control unit with additional multiplexers and reading circuits (RC) for improved testability

be performed independent from the microprogram. Such a test sequence will be much shorter and easier to generate, because no undesired loops must be performed.

Only few additional gates are necessary which need very few chip area. As they are not in critical signal paths during normal operation no loss of performance must be taken into account. The additional hardware may even be used to perform some special system functions such as saving a microprogram state after an interrupt.

However the method is very much dependent on a suitable architecture. It is also difficult to use it in conjunction with automatic test generation. Therefore it is not an adequate technique for general application.

5.2 THE SCAN-PATH

The main task of designing for testability is to increase controllability and observability of the internal states. A very efficient method for integrated circuits, which was investigated by several authors, is the use of a scan-path /8 - 11/.

5.2.1 THE PRINCIPLE OF A SCAN-PATH

The scan-path is an extension of the normal circuit function, allowing to connect all internal storing elements (except memory arrays) to a long shift register during the test mode.

Fig. 5.2a shows a normal circuit which is modified for the implementation of a scan-path by the introduction of some two--input multiplexers in fig. 5.2b. The flipflops are simple polarity hold latches. The purpose of these multiplexers is twofold:

1) to switch the data inputs of the flipflops so that they form a chain without combinatorial circuits between them,
2) to connect the two clocks C1 and C2 alternating to the latches.

Now the circuit can be devided into a simple sequential circuit (the shift register) and one or more purely combinatorial networks. In this example three additonal pins for shift register input and output and a control signal to select the mode of operation are necessary.

Any internal state may be set or read out by a simple shift operation. Therefore the scan-path can provide ideal controllability and observability.

The inputs and outputs of the flipflops are fictive outputs or inputs to the combinatorial circuit. This circuit can be obtained if the flipflops are removed from the original circuit

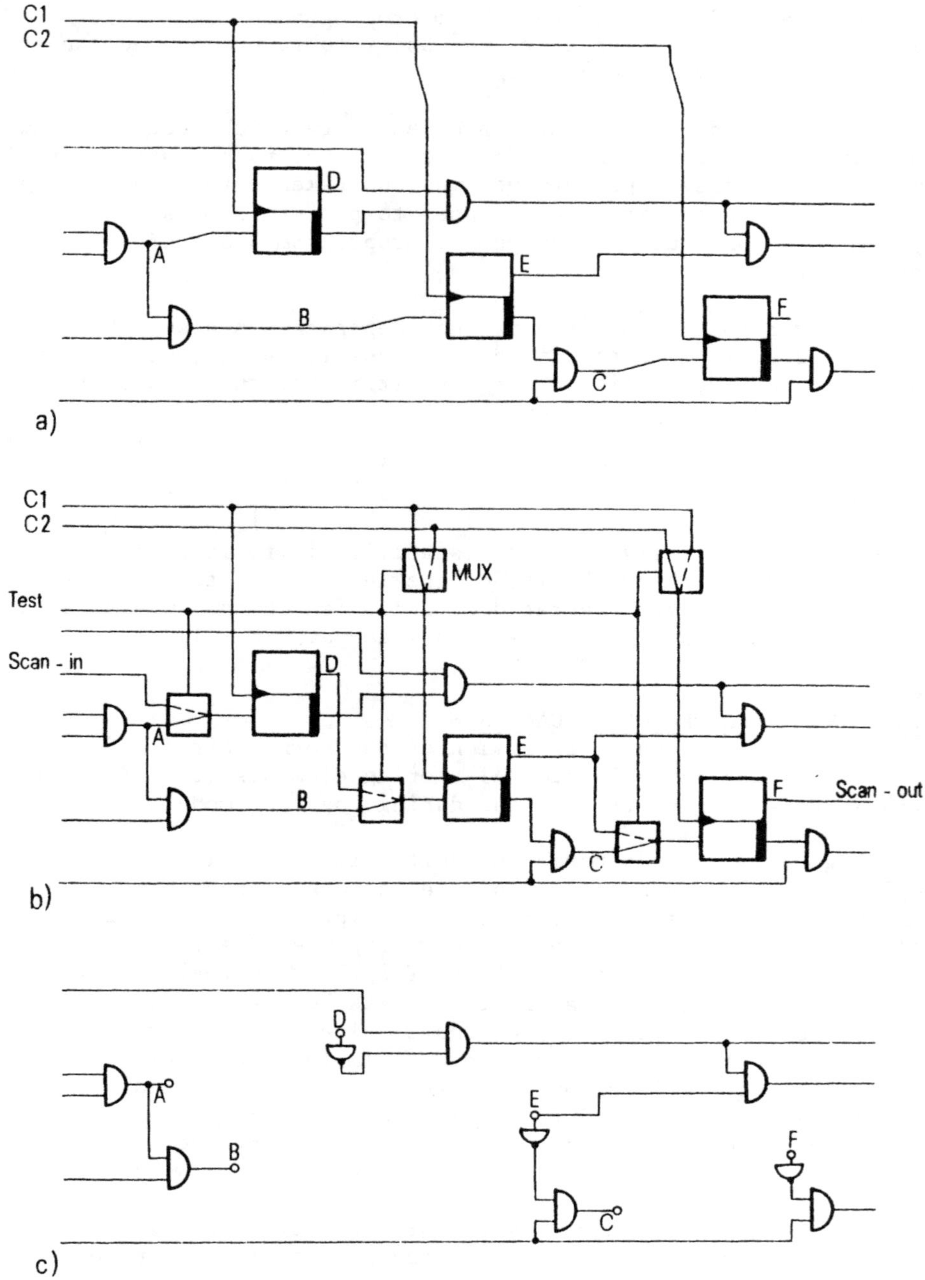

Fig. 5.2 a) sequential circuit without scan-path
b) circuit with scan-path
c) combinatorial circuit, scan-path removed

(fig. 5.2c). If the inverted output of a flipflop is used, an inverter must be inserted for correct modelling.

The test sequence for a circuit with scan-path is the following:

1) Test of the scan-path
   First the correct operation of all flipflops and their connections within the scan-path must be checked. Therefore the circuit is switched to the test mode. A characteristic sequence of bits is shifted through the scan-path applying alternating clocksignals C1 and C2. After a number of clock pulses equal to the number of flipflops the same sequence can be compared at the output.
2) Test of the combinatorial circuits
   The test of the combinatorial circuit is performed in three steps:
   - A test pattern is shifted into the flipflops during the test mode.
   - The circuit is switched to normal operation. One or more clock cycles are carried out while the adequate test vectors are applied to the primary inputs. The signal responses are checked at the primary outputs.
   - The internal states are shifted out after switching back to the test mode. At the same time the new sequence for the next test cycle is shifted in.

This sequence must be repeated until the whole combinatorial logic is tested. The number of test cycles is very much dependent of the fan-out and the logical depth of the combinatorial circuits. Usually these characteristics are almost independent from the total complexity of a VLSI-circuit. Therefore the number of test cycles will increase only slightly. The test time is mainly determined by the length of the shift operations which is equal to the number of flipflops.

For the remaining combinatorial network automatic test generation can be applied. Only relatively few computer time and a fault coverage of almost 100% can be reached.

The reduction of cost for test generation and the improvement of test coverage will be optimal for a <u>complete scan-path</u>. However speed requirements or subsequent introduction of the scan-path into an existing design may be reasons for the implementation of an <u>incomplete scan-path</u> which contains only part if the flipflops. Therefore the circuit parts remaining outside the scan-path are still sequential, and are subjected to the same test generation problems. Also the number of tests which are needed will increase with sequential depth of these parts. As there are long shift operations necessary to perform one test, the test time will increase very much. Therefore an incomplete scan-path can only be used if the sequential depth of the remaining circuits is very low, or if the not included flipflops are well controllable by the primary inputs /23/.

5.2.2 TYPES OF FLIPFLOPS

In principle level-sensitive as well as edge-triggered flipflops may be used. If level-sensitive latches are used, two successive ones must be driven by two non-overlapping clocks so that they act as master-slave flipflops. As shown in fig. 5.3 the same clock may be used for normal operation and the shift operation, if the data input of the flipflop can be selected by a multiplexer which is controlled by the test signal. If more than one clock is used on the chip all flipflops must be connected to the same clock for shifting.
In /11/ an arrangement with separate clocks for the shift operation is used.

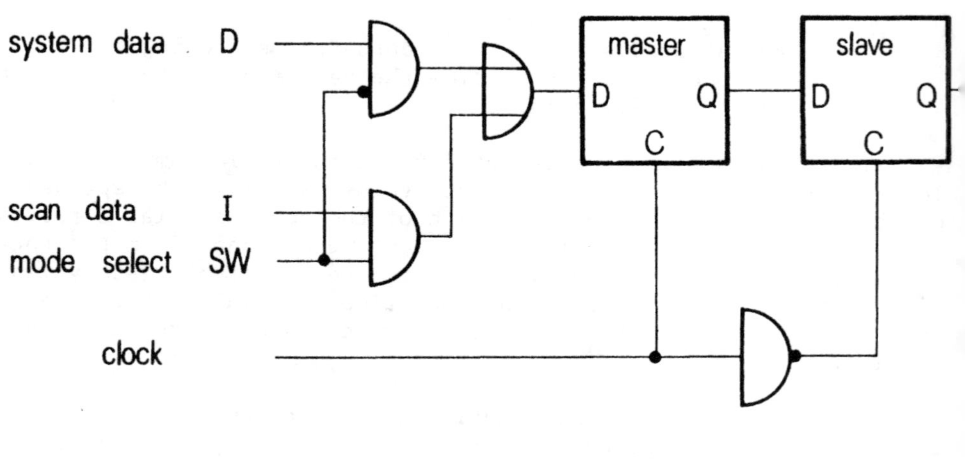

Fig. 5.3 Scan-path flipflop with common clock

In this case two realizations are possible:

- In the single-latch design (fig. 5.4a) only the master latch is in the data-path. Therefore the loss of performance is minimal. The state of the circuit can be stored in the second latch at any time. But there is a relatively large overhead.
- In the double-latch design (fig. 5.4b) both latches are used for the circuit function so that the overhead is a minimum.

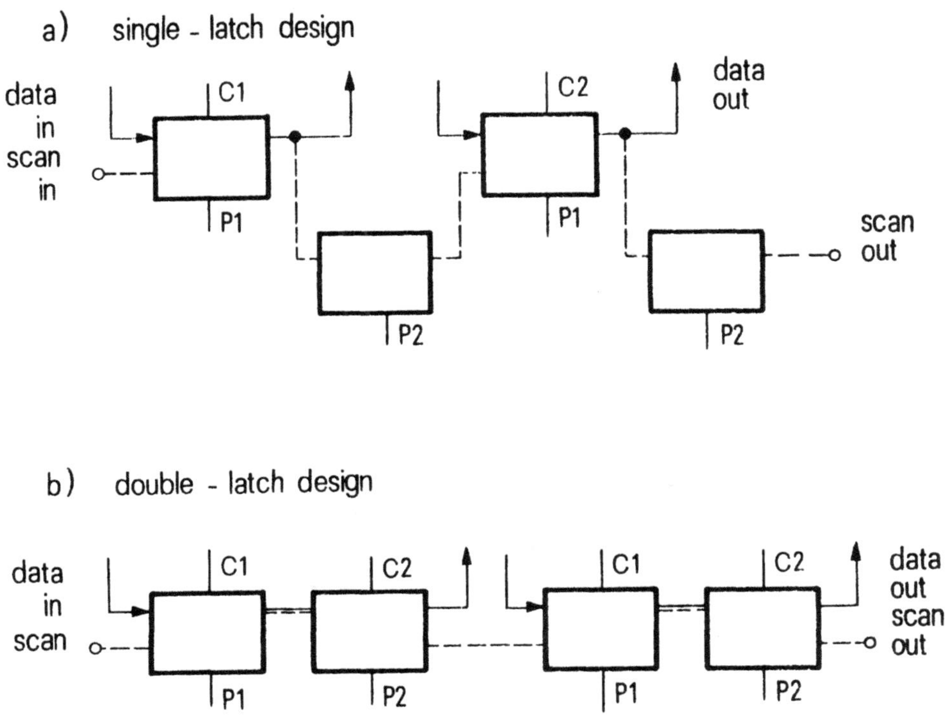

Fig. 5.4 Realizations of the scan-path

5.2.3 THE EFFECTS OF THE SCAN-PATH ON THE CIRCUIT

As it can be seen from the previous examples the overhead for a scan-path depends very much on the circuit structure. Other factors are the proportion of flipflops of the whole circuit and their distribution over the chip. The area of a shift register latch is about a factor 3 larger than a simple latch and a factor 1.5 larger than a master-slave-flipflop. Also some additional space for clocklines and connections between the flipflops must be taken into account. All together the overhead for a logic design with scan-path will be 4 to 20%.

A maximum of four additonal pins is needed. A reduction is possible if the shift register input and output are shared with existing primary inputs and outputs, respectively. The demultiplexer and multiplexer circuit can be switched by the control signal of the test mode. If the same clock is used for both modes only one additional pin for the control signal is needed.

The effect on the performance of the circuit is only of importance if master and slave flipflops are in the data-path. Otherwise a delay is only caused by the multiplexer circuit. This delay may even be avoided in designs where the multiplexer is integrated into the flipflops (fig. 5.5).

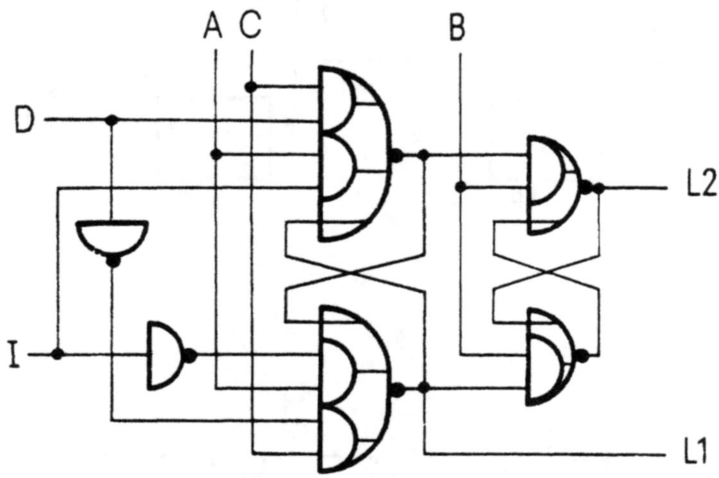

Fig. 5.5 Level-sensitive shift register latch
A, B scan-clocks, C system clock, D data input
I scan-path input, L2 scan-path output

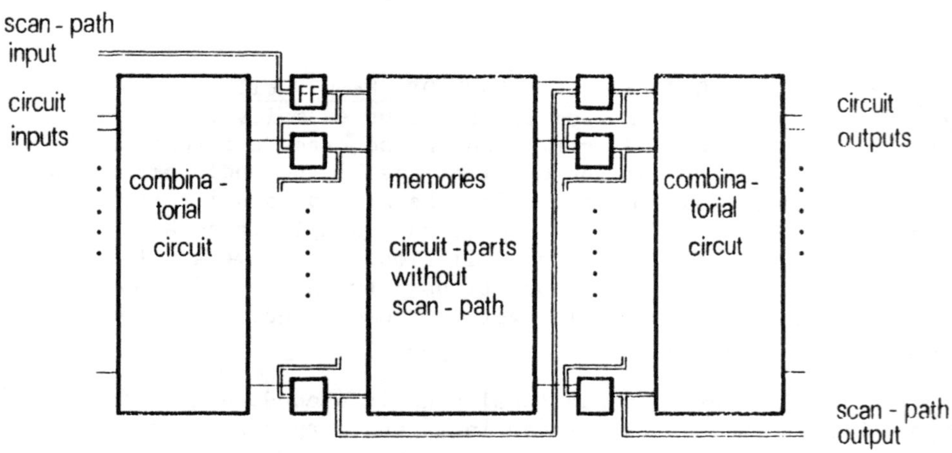

Fig. 5.6 Isolation of blocks by the scan-path

5.2.4 THE USE OF THE SCAN-PATH FOR DIFFERENT TYPES OF CIRCUITS

The scan-path is especially suitable for random logic with flipflops distributed unregularly over the whole chip. However a scan-path can't be used for memories. Also in circuits containing many registers and functional blocks which are easily accessible by a data bus (e.g. microprocessors), the scan-path might become very long. The shift operations will take most of the test time. These structures can be better tested by other

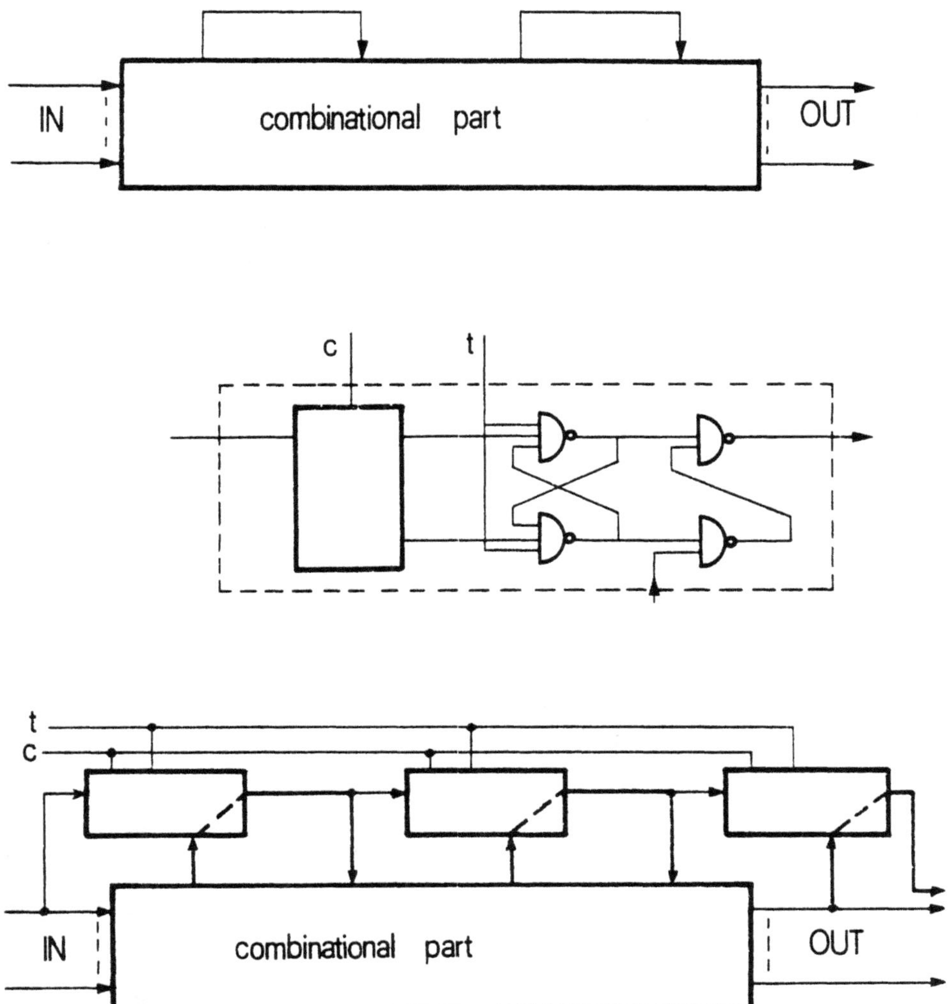

Fig. 5.7 Testing of asynchronous circuits /10/

methods such as functional partitioning. The isolation from the rest of the circuit can be done by the scan-path as shown in fig. 5.6.
Usually the scan-path is used in synchronous circuits. However also asynchronous feedback lines may be interrupted by additional gates which are controlled by shift register latches (fig. 5.7). This transforms the circuit to a synchronous one for testing. The state of the feedback signal can be set via the scan-path and clock signal t = 1. Otherwise the circuit functions in its asynchronous mode.

5.2.5 THE APPLICATION IN SYSTEMS

Although we are discussing design for testability on chip level the test problem must be seen under the aspectes of system testing and maintenance. For testing of assemblys of VLSI-circuits it is possible to connect the scan-paths of all chips in series. The scan-clocks or test control lines are connected to all circuits on the PCB. Theoretically all PCB's could be controlled by a service-processor via one very long system scan-path. However more efficient is the individual selection and loading of each PCB by the service-processor as it is shown in fig. 5.8. To perform the test of a PCB in such an arrangement only the unit under test (UUT) and those which are directly connected to it must be loaded.

5.2.6 REALIZATIONS OF THE SCAN-PATH

The scan-path is an effecitive method for DC testing to find out static stuck-at faults. In order to avoid most of the dynamic faults which are difficult to test, a complete design system called level sensitive scan design was presented in /11 - 14/. This system is based on a set of design rules and comprises automatic checking for violations of these rules as well as the generation of tests for static and delay faults.
LSSD includes two major features:

1) It is a level sensitive design.
   That means that the steady state response to any allowed input signal is independent of rise- and fall-times or delays on signal paths. Also the order in which several input signals are changing has no influence on the output.
2) All internal storage elements (except memories) can be connected to a shift register.

The first feature is achieved by a strictly synchronous design, which can be obtained if the following basic rules are observed:

- Only hazard-free level sensitive latches may be used.
- Successive latches must be controlled by different non-overlapping clocks.

- The clock-input of a latch is only stimulated by primary clock inputs which may be gated by internal states or primary inputs. AND-connections of clocks are not allowed.
- No clock signal may feed directly or through combinatorial logic the data input of a latch.

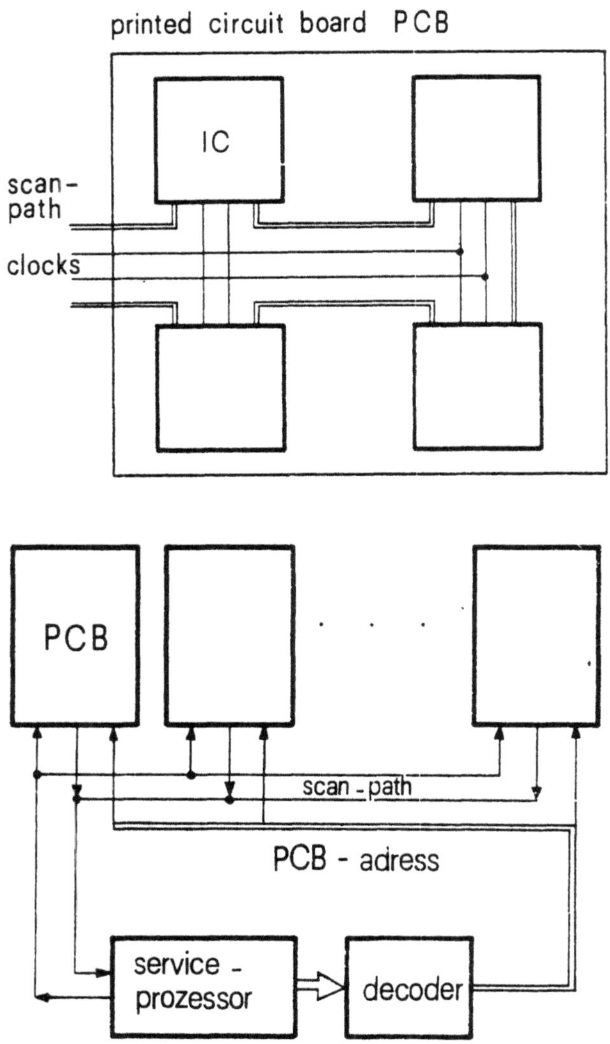

Fig. 5.8 Application of the scan-path for maintenance

To establish a scan-path the shift register-latches must be connected through a data path which contains no combinatorial logic, which may change the data. The latches must also be clocked directly from primary inputs.
The detailed rules are given in reference /12/.

Since no delay times or asynchronous feedback loops have to be observed, the circuit structure and its function are clearer if the design is made according to these rules, which will reduce the number of design faults and save time. The simulation of such a network can be done with a 3-valued logic-simulator without gate delays.
Although these design rules are limiting the designer's freedom to a certain extend, they are also a helpful guide for a good, that means clearly structured design.

Another realization of the scan-path method called scan/set was published in /24, 8/. In this approach the scan-path is not integrated into the circuits. The states of internal nodes may be stored into a shift register which is situated outside the circuit. The states can be read and shifted out without stopping the normal operation of the circuit. The internal flipflops may be set by a general set-signal which controls the data transfer from the shift register.

The scan/set design affords at least one additional line between each flipflop and the shift register-latch and two general read- and write-lines and some gates for the modified internal flipflops. The overhead for LSI/VLSI-circuits is estimated to about 10%.

5.2.7 EXPERIMENTAL RESULTS

Our investigations were made with an existing circuit built with an LSI-master-slice chip containing 36 cells. The circuit had 39 edge-triggered master-slave flipflops which could be connected to an incomplete scan-path. Another 4 RS-type flipflops were not included.

The most important data are given in table 5.1. Simulations were carried out with S-LASAR (logic automatic stimulus and response) on an SMC-computer which has an execution time of 0.2 to 0.4 μs per instruction.

Table 5.1

total number of nodes (S-LASAR-model)	1611
- combinatorial circuit	682
number of possible faults (according to S-LASAR fault model)	4947
total number of flipflops	43
- length of the scan-path	39
number of inputs	
- complete circuit	18
- inputs to the combinatorial circuit	56
number of outputs	
- complete circuit	28
- outputs of the combinatorial circuit	47

First a functional test was generated manually without taking care of the scan-path.
Then automatic test generation with S-LASAR was carried out for three different circuits:
1) The complete circuit without taking care of the scan-path.
2) The purely combinatorial circuit part (the connections to the RS-flipflops were replaced by primary inputs and outputs).
3) The sequential circuit parts which remains after removing the incomplete scan-path.

The results are shown in table 5.2.
The most important characteristics were:
- Automatic test generation for the complete circuit without scan-path was not possible because too many hazards and spikes occured. A manual correction and supplementation would have been ineffective.
- The computer time for automatic generation using the incomplete scan-path was reduced by 52% compared to manual generation.
- The number of patterns could be reduced by a factor of 2. But only 9% of the patterns were fault-simulated. The scan-path had to be loaded 11 times.
- The personal efforts can be reduced to 1/10.
- A complete scan-path could even reduce the number of patterns to 5%. The total CPU-time is reduced to 1/10, and the CPU-time for fault-simulation to 4%.

circuit	number of patterns	fault detection rate	total CPU-time (s)	fault simulation time (s)	remarks
1	1993	94%	2810	1267	manual generation, 1 month
1	1771	89%	7200		automatic generation, many races and spikes
2	97	95%	442	51	2 days manual work
3	992/186 *)	93%		656	iterative method

*) fault simulated patterns

Table 5.2

The fault detection rates of the different tests were nearly the same.
Form these results it can be concluded that
- the best results can be reached with a complete scan-path. Therefore it should be implemented whenever it is possible.
- The incomplete scan-path still reduces the test generation cost by an important factor if the sequential depth of the circuits outside the shift register is low.

However it must be noted, that these are specific results from the circuit which was investigated. From other circuits different figures may be obtained.
Other authors have reported savings of 50 - 80% for test generation with the scan-path.

5.3 BUILT-IN TEST WITH PSEUDORANDOM PATTERNS AND SIGNATURE ANALYSIS

The scan-path provides excellent controllability and observability of the circuit for an external tester. But much of the execution time is wasted for serial data transmission between the circuit and the test equipment. An alternative is to build the test equipment into the circuit.
A rather advanced test which uses the immanent intelligence of a circuit to execute a stored program was already mentioned. In this chapter I want to discuss in more detail a built-in test with integrated test pattern generator (TPG) and signature register (SR) on chip, which was proposed in /26, 27/.

The random pattern generator

Especially pseudorandom patterns are very suitable for on--chip pattern generation.
- They can be generated very easily by a linear feed-back shift register, which can be obtained from an existing register by function conversion (fig. 5.9) with a few gates.
- If the feed-back lines are well chosen, $2^n - 1$ different patterns are generated (n is the number of stages, which is usually equal to the inputs of the circuit under test). Only the pattern of all outputs zero or one must be performed by a reset after the test sequence. Therefore a complete test is possible if the circuit is partitioned into sections with not too many inputs (so that the number of combinations is not too large) and of low sequential depth. Usually good fault coverage is already achieved after a few hundred patterns.
- Most of the pattern dependent problems will be detected, because many different combinations are tested.

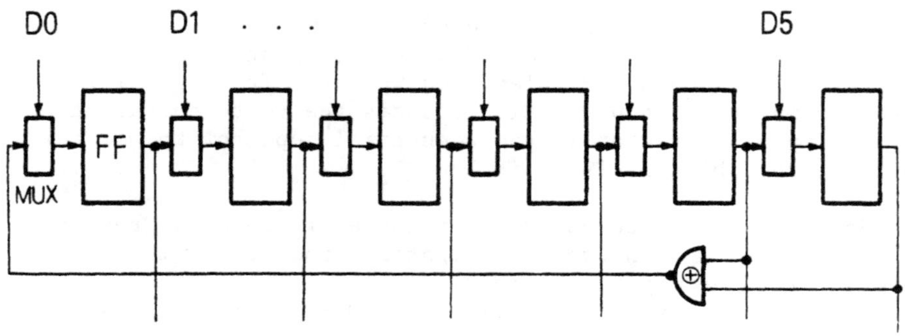

Fig. 5.9 Pseudorandom test pattern generator

Signature analysis (SA)

The voting of test responses can be done very efficiently by signature analysis. The principle was known from telecommunication techniques as cyclic redundancy check. In /25/ it was used with serial input for fault detection on selected points of microprocessor systems. A signature register with parallel input for in-circuit checking was proposed in /26/. Some important features which make this method attractive for test evaluation on chip are:

- SA is very easy to implement by linear feed-back shift registers (see fig. 5.10).
- The probability of fault recognition is very high and is given by the relation $p = 1 - 2^{-m}$ (m is the length of the SR). For $m = 8$ we obtain $p = 0.996$ and for $m = 16$, $p = 0.9998$. Single errors are always detected. The recognition of several errors depends on the selection of the feed-back lines. If they are well chosen, the recognition rate is nearly the same for all errors. However the quality of a test is much more dependent from the test coverage, which is usually only around 95%.
- It is possible to choose any desired signature by selecting a suitable starting word or adding a control word. For example the signature consisting only of zeros is very easy to check by a NOR-gate and can be obtained in ROM-tests by appropriately selecting the last word in the ROM.

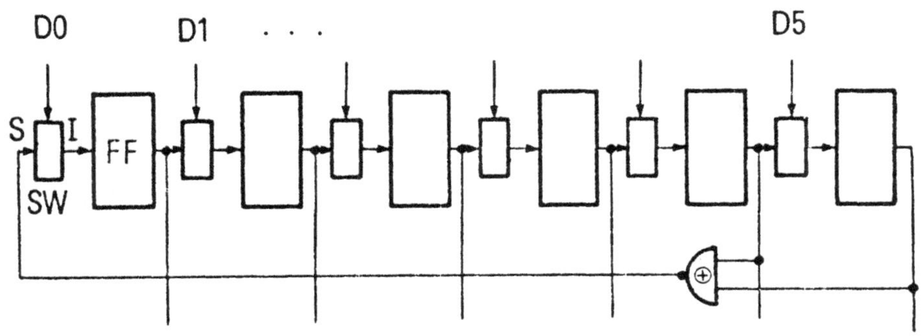

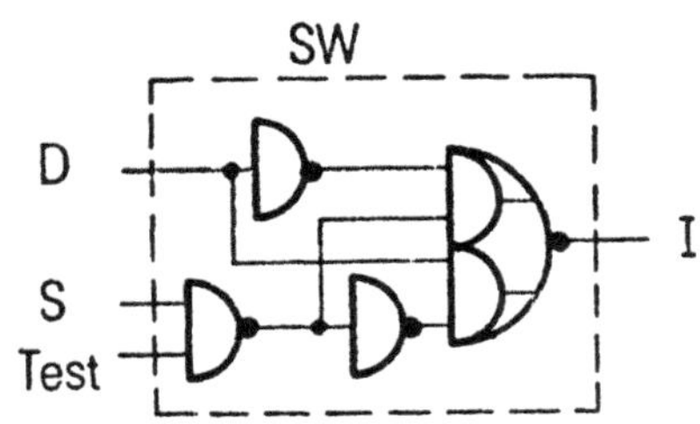

Test	I
0	D
1	D=S

Fig. 5.10 Signature register

Simulation

An advantage of the test with pseudorandom patterns is that no computer-time consuming test generation is necessary. However for circuit parts with a large number of inputs it is impossible to use the whole sequence for the test. Therefore a fault-simulation must be carried out to verify if a sufficient test coverage is reached with the applied patterns. If all $2^n - 1$ patterns are used, only a simulation to determine the correct signature is needed. This is often the cheaper way even if more patterns must be simulated because fault-simulation for complex circuits is much more time consuming.

Quality of random pattern tests

Experiments with several different circuits have shown, that usually a test coverage of approximately 95% can be reached already after 100 test patterns /28/. (With the same number of clock cycles only 2 tests with deterministic patterns which have to be shifted into a scan-path, containing 50 flipflops can be carried

out). The circuits were only combinatorial (up to 1000 logic gates). Even circuits of low sequential depth may be completely tested by random patterns without an important increase of the test length as was shown in /27/.
Experiments with PLA's however showed that random patterns are not efficient for combinatorial circuits with high fan-in. The reason for that are faults which are only detected by specific input vectors. If there is for example a NOR-gate with 20 inputs, a stuck-at-0 fault at the output can only be detected, if all inputs are low. The probability that such a combination occurs during a sequence of 1000 patterns is only $0.95 \cdot 10^{-3}$. If a PLA contains many of those terms, the test coverage will increase only very slowly.
For sequential networks the sequence of test patterns is of importance. Their length increases with the sequential depth. Therefore the probability that a suitable sequence is generated by the pseudorandom pattern generator within the first few 100 or 1000 patterns is reduced rapidly.
For these reasons the circuit must be partitioned into blocks of limited fan-in and low sequential depth. This is also favorable for fault localization, which is limited to the module level.

Test strategies

Since there is only an overhead of a few gates for functional conversion to form a TPG or a SR, many of them can be used in the circuit. The test can be decentralized and the different modules are tested in parallel. Each module has its own signature register and TPG. The initial patterns are selected so that the correct signature is zero and can be easily checked by a NOR-gate. This go/no-go signal from each module is transmitted to a status-register from where it can be read by a simple test equipment or be transmitted to a service-processor. The random pattern generators may be individual ones for each module or one provides patterns for several modules. In particular a centralized generator is efficient in systems, where several modules are connected to an internal bus. The test length is then determined by the desired test coverage of the most complex module.
However the direction of signal flow is always the same from the TPG to the SR. Therefore bidirectional elements are tested only in one direction. Also some interconnections between modules cannot be tested at the same time. For complete test a second run is necessary to test the remaining elements. For this test some SRs of the first run are now converted to TPG and vice versa. This type of built-in test is also very suitable for testing the first modules if a self-test with bootstrapping is applied. In this case different modules are checked one after another and the data may be checked by few SRs, which are connected to important points such as the internal bus.

The built-in logic block observer (BILBO)

From the previous discussion we learned that circuits are needed, which allow functional conversion of registers into pseudorandom pattern generators or signature registers. Such a circuit which needs only a few additional gates is the BILBO (see fig. 5.11). This circuit can even be used as scan-path. The mode of operation is selected by the control lines B1 for parallel data input and B2 for the shift operation. By the combination of B1 and B2 the circuit may work in the following modes:

B1	B2	flipflop input	mode of operation
0	0	1	reset
1	0	$\overline{D}$	normal mode
0	1	$\overline{I}$	scan-path or TPG
1	1	D = I	signature register or TPG (if all data inputs are constant)

The feed-back lines for the TPG and the SR are the same. Of course master-slave flipflops must be used for the registers. The signal delay for the data is only one gate transition time.

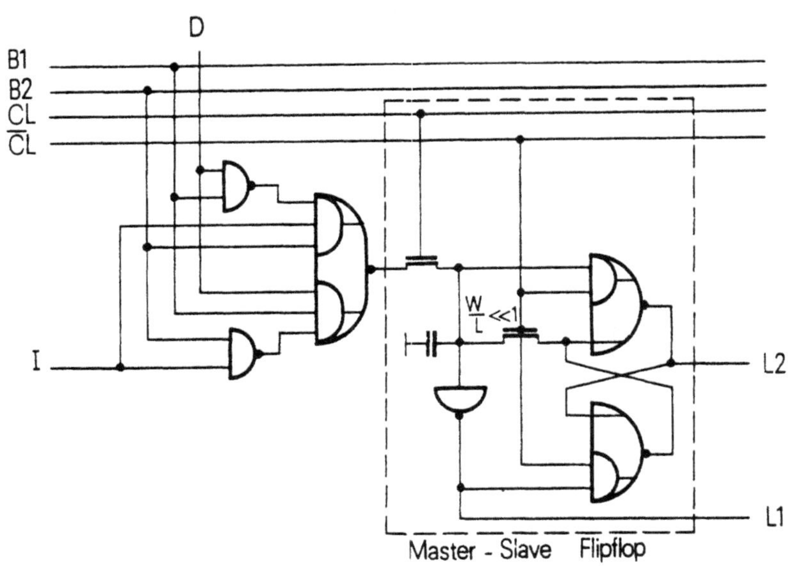

Fig. 5.11 Built-in logic block observer (BILBO)-flipflop

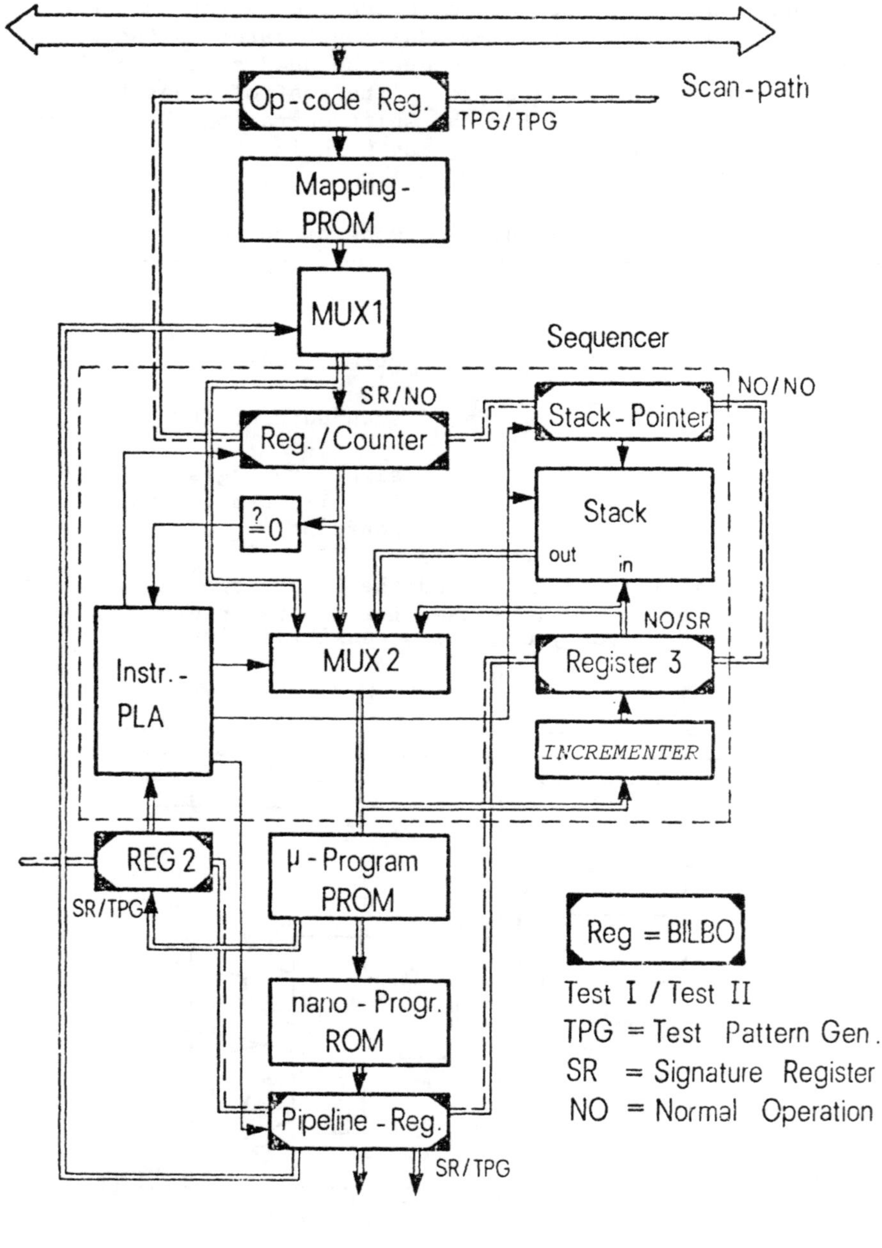

Fig. 5.12 Application of the BILBO in a microprogrammed PCU

The application of built-in test and the BILBO

The practical application of the built-in test with signature analysis and the BILBO may be demonstrated at the microprogrammed controller of a microprocessor (see fig. 5.12). This circuit can be partitioned into three different blocks for testing:

- the mapping-PROM
- the micro- and nano-program memories and
- the sequencer.

The read only memories are easy testable structures. The only condition for a complete test is that all words are read out. Therefore the test sequence, which may be generated by a counter or a complete sequence of pseudorandom patterns, must consist of all addresses including the combinations: all bits zero and all bits one. The sequencer however contains a stack-register with the stack-pointer. It is a sequential circuit of considerable depth. All registers except the stack are layed out as BILBOs.

In the first run, the mapping-PROM and the micro- and nano--program memories are tested at the same time. The OP-code-register is converted to a TPG for the mapping-PROM and the signature is formed in the register/counter. The zero-detector at the output of this register is used for the go/no-go decision at the end of the test. The loop consisting of the decrementer, register 3 and multiplexer can be used as address-generator for the micro- and nano-program memories. The results are condensed to a signature in the pipeline register which is connected with register 2. During this test run the data of the mapping-PROM and of register 3 must be switched through the multiplexers 1 and 2, respectively.

During the test of the sequencer it must be checked that all data paths, all registers and all control lines operate correctly and that all instructions are carried out. Register 2 is used as TPG providing a pseudorandom instruction sequence. Data sources are the mapping-PROM which is driven by the OP-code register and the pipeline register, both running as random TPGs. The signature is formed in register 3 which is also data source for the stack--registers.

The problem of such a circuit is that the stack-pointer must be incremented and decremented by the pseudorandom instruction sequence several times so that various data are written into all stack-registers and read out. For reading the multiplexer must also switch through the stack data. Simulations with an instruction set of the sequencer Am 2910 showed that these conditions are given for a maximum depth of four. However during the sequence of 63 ($= 2^6 - 1$) instructions, some stack-registers are read only once. To get a good fault coverage the instruction sequence must be repeated several times with various data.

These tests only indicate if there are faults in the three blocks. For exact localization of faults the registers are switched to the scan mode. As to be seen from fig. 5.12 the stack-registers are not included in the scan-path. But each of them can be selected by the stack-pointer which is part of the scan-path.
The different modes of operations are controlled by the lines B1 and B2 of the BILBOs. A third line is needed to select the first or second test run.
This example shows how rather complex circuits can be made self-testing with built-in TPGs and SRs only by functional conversion of existing registers. However it is necessary to study the circuit carefully if it contains sequential parts and if they are to be tested completely by this method. Of course other parts of the microprocessor such as the ALU may be tested in parallel in the same way.

6. CONCLUSION

Future VLSI-circuits will only be testable if they are designed for testability. However this implicates also additional logic gates, limitation of the designer's freedom and possibly a loss of performance. The requirements for testing are also quite different and depending on the application of the circuit and the production volume. Therefore there is no general solution of the testing problem which offers optimal efficiency.
Two very important methods which can be implemented without large overhead are

- the scan-path providing very good controllability and observability. It is also an adequate method for small volume productions.
- Built-in tests using stored programs or pseudorandom patterns and signature analysis.

For all test methods the complexity of the large VLSI chip is reduced by some kind of circuit partitioning.

ACKNOWLEDGEMENT

The author likes to thank E. Trischler, who carried out the investigations on the circuit with incomplete scan-path.

REFERENCES

/1/ M.A. Breuer, A.D. Friedmann, Diagnosis and Reliable Design of Digital Systems, Computer Science Press, Inc. 1976.

/2/ J.P. Roth, Diagnosis of Automata Failures: A Calculus and a Method, IBM J. of Res. and Develop., July 1966.

/3/ W.G. Bouricius et al., Algorithms for Detection of Faults in Logic Circuits, IEEE Trans. Comp., vol. C-20, Nov. 1971.

/4/ E. Wolfgang et al., Electron-Beam Testing of VLSI-Circuits, IEEE J. of Solid-State Circuits, vol. SC-14, no. 2, April 1979.

/5/ H. Murrmann, D. Kranzer, Yield Modeling of Integrated Circuits, Siemens Forsch.- u. Entw.-Ber. Bd. 9, 1980, Nr. 1.

/6/ J.P. Hayes, A.D. Friedmann, Test Point Placement to Simplify Fault Detection, IEEE Trans. Comp., vol. C-23, July 1974.

/7/ S.B. Akers, Partitioning for Testability, Fault Tolerant Computing Symposium 1976.

/8/ J.H. Stewart, Application of Scan/Set for Error Detection and Diagnosis, Proc. 1978 Semiconductor Test Conf.

/9/ A. Yamada, N. Wakatsuki, Automatic System Level Test Generation and Fault Localisation for large Digital Systems, Digest of Papers 15th Annual Design Automation Conf. 1978.

/10/ M.J.Y. Williams, J.B. Angell, Enhancing Testability of Large-Scale Integrated Circuits via Test Points and Additional Logic, IEEE Trans. Comp., vol. C-22, no. 1, Jan. 1973.

/11/ E.B. Eichelberger, T.W. Williams, A Logic Design Structure for LSI Testability, Proc. 14th Annual Design Automation Conf., June 1977.

/12/ H.C. Godoy et al., Automatic Checking of Logic Design Structures for Compliance with Testability Ground Rules, 14th DA Conf., June 1977.

/13/ P.S. Bottorff et al., Test Generation for Large Logic Networks, 14th DA Conf., June 1977.

/14/ E.P. Hsieh et al., Delay Test Generation, 14th DA Conf., June 1977.

/15/ J.B. Clary, R.A. Sacane, Self-Testing Computers, Computer, Oct. 1979.

/16/ R.G. Bennetts, Fault Tolerance and Digital Systems, Microprocessors and Microsystems, vol. 3, no. 8, Oct. 1979.

/17/ J.A. White, R.E. Pope, Microprocessor Flight Control, System Architectural Concepts, AIAA 2nd Computers in Aerospace Conf., Oct. 1979.

/18/ J.J. Stiffler, A Fault-Tolerant Communications Processor, AIAA 2nd Computers in Aerospace Conf., Oct. 1979.

/19/ R.P. Cenker et al., A Fault-Tolerant 64K Dynamic RAM, Digest of Papers Int. Solid-State Circuits Conf., 1979.

/20/ R. Koppel, RAM Reliability in Large Memory Systems - improving MTBF with ECC, Computer Design, March 1979.

/21/ J. Boney, Let Your Next Microcomputer Check Itself and Cut Down Your Testing Overhead, Electronic Design, vol. 18, Sept. 1979.

/22/ R. McCaskill, Test Approaches for Four Bit Microprocessor Slices, Digest of Papers 1976 Symposium on Semiconductor Memory Testing.

/23/ E. Trischler, An Incomplete Realisation of the Scan-Path, to be published in Digest of Papers, 1980 Semiconductor Test Conf.

/24/ J.H. Stewart, Future Testing of Large LSI-Circuit Cards, Digest of Papers 1979 Semiconductor Test Symposium.

/25/ G. Gordon, H. Nadig, Hexadecimal Signatures Identify Trouble-spots in Microprocessor Systems, Electronics, vol. 50, no. 5, March 1977.

/26/ B. Könemann et al., Signaturregister für selbsttestende ICs, NTG Fachberichte, Bd. 68, NTG-Fachtagung: Höchstintegrierte Schaltungen, 1979.

/27/ G. Zwiehoff et al., Experimente mit einem Simulationsmodell für selbsttestende ICs, NTG Fachberichte, Bd, 68, NTG-Fachtagung: Höchstintegrierte Schaltungen, 1979.

/28/ T.W. Williams, Random Patterns within a Structured Sequential Logic Design, Digest of Papers, 1977 Semiconductor Test Symp.

/29/ G. Crichton, Testing Microprocessors, Digest of Papers, 1978, 14th European Solid-State Circuits Conf.

/30/ J. Mucha, Testfreundliche VLSI-Schaltungen - Entwurfs- und Prüfprinzipien, NTZ, Bd. 32, Nr. 7, 1979.

/31/ E.B. Eichelberger, E. Lindbloom, A Heuristic Test-Pattern Generator for Programmable Logic Arrays, IBM J. Res. Develop., vol. 24, no. 1, Jan. 1980.

/32/ T.J. Frechette, F. Tanner, Support Processor Analyses Errors Caught by Latches, Electronics, Nov. 8, 1979.

/33/ W.C. Carter et al., Cost-Effectiveness of Self-Checking Computer Design, Dig. 1977 Int. Conference Fault-Tolerant Computing, IEEE Publ. no. 77CH1223-7C.

/34/ D.L. Ostapko, S.J. Hong, Fault Analysis and Test Generation for Programmable Logic Arrays (PLA), Digest of the 8th Annual Int. Conf. on Fault-Tolerant Computing, June 1978.

VLSI AND PETRI NETS

F.H.J. Feldbrugge

Philips Research Laboratories
WB 349, Eindhoven, The Netherlands

1. INTRODUCTION

Integrated circuit technology is developing very rapidly. Starting with only a few components on a chip, to-day's chip-makers are able to reach component densities which are orders of magnitudes higher. The naming evolved accordingly : LSI, VLSI and what next? In these course notes the main topic is VLSI-design and the question may rise as to why it is contrasted to LSI-design. It is more than just another order-of-magnitude step towards higher densities, higher speeds and/or larger chip areas. It also implies an increase of complexity of those systems which are potentially implemented on a chip or a family of chips. As a consequence, integrated circuit design is confronted with the already longer existing general problem of how to cope with complexity in system analysis and design.

Among the mathematical tools which are available today for that purpose (although still in a developing stage), Net Theory is getting a growing interest. What started in the early sixties by its founder Carl Adam Petri as an investigation of inter-automaton communication [6] has become a rather general system theory treating basic system properties such as the partial ordering of conditions and events in space and time, the notions of concurrency and causal dependency, etc.

Net theory will not be treated here in extenso, but we will focus upon one special Net theoretic formalism, namely Place-Transition Nets, more widely known as Petri Nets. We will first give a semi-formal description of Petri Nets. Then the practical relevance of Petri Nets will be illustrated by means of some examples followed by some concluding remarks.

2. PETRI NETS.

2.1. Net definition.

A <u>net</u> is a triple N=(S,T;F) where S and T are two non-empty disjunct sets and F is a relation between the elements of these two sets, called the <u>flow-relation</u> of N, such that all members of S and T belong to a member of F, i.e. there are no isolated members of S and T.

More formally:
(1) $S \cap T = \emptyset$
(2) $S \cup T \neq \emptyset$
(3) $F \subseteq (S \times T) \cup (T \times S)$
(4) $\mathrm{domain}(F) \cup \mathrm{codomain}(F) = S \cup T$.

Usually the S-elements of a net are graphically represented by circles (○), the T-elements by squares (□) and the members of F by arcs (○→□,□→○).

Example : N=(S,T;F) with S={s1,s2,s3,s4}
T={t1,t2,t3}
F={(t1,s1),(t1,s2),(s3,t1),(s2,t2),(t2,s3),(s4,t2),(s3,t3),(s4,t3),(t3,s4)}

is a net which is graphically represented as :

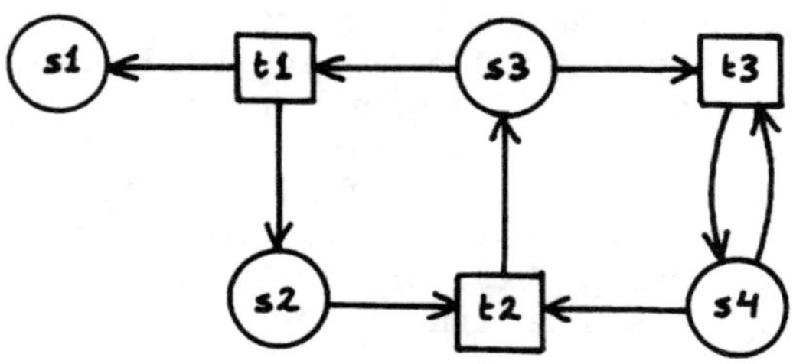

<u>Fig. 1.</u> Example of a net

A <u>pure net</u> is a net where $\forall x,y \in S \cup T : (x,y) \in F \Rightarrow (y,x) \notin F$ or to translate this into graphical terms : the net does not contain subnets of the form ○⇄□.

2.2. Petri Net definition.

A <u>Petri Net</u> consists of a net with some specific properties. The S-elements are called <u>places</u>, which may contain zero or more markers, called <u>tokens</u>, indicating the number of validations of each place. The T-elements are called <u>transitions</u>, which may <u>fire</u> according to some firing rule, thereby changing the token distribution over the places of the net. Such a token distribution is called a marking of the net. Formally :

A Petri Net is a 6-tuple (S,T;F,k,w,m0) where
(1) S is the set of places,
(2) T is the set of transitions,
(3) F is the flow relation between places and transitions,
(4) k is the token capacity function,
(5) w is the token width (or : multiplicity) function,
(6) m0 is the initial marking,
such that
(1) (S,T;F) is a net,
(2) $k:S \longrightarrow \mathbb{N} \cup \{\omega\}$
(a positive, possibly infinite, token capacity is associated with each place),
(3) $w:F \longrightarrow \mathbb{N}$ with $\forall (s,t) \in F: w(s,t) \leq k(s)$ and $\forall (t,s) \in F: w(t,s) \leq k(s)$
(a positive token width is associated with each arc which respects the place capacities),
(4) $m0:S \longrightarrow \mathbb{N} \cup \{\omega\}$ with $\forall s \in S: m0(s) \leq k(s)$
(initially each place contains a number of tokens which respects the place capacities).

A marking is a function $m:S \longrightarrow \mathbb{N} \cup \{\omega\}$ with $\forall s \in S: m(s) \leq k(s)$.

A place s is called an input place (output place) of a transition t iff $(s,t) \in F$ $((t,s) \in F)$.

A transition t is said to have concession at marking m iff all input places of t carry enough tokens and all output places of t have enough space :
$\forall s \in S: [(s,t) \in F \Longrightarrow m(s) \geq w(s,t)] \wedge [(t,s) \in F \Longrightarrow m(s) \leq k(s)-w(t,s)]$.

A transition t is said to have contact at marking m iff all input places of t carry enough tokens to grant concession to t but there are output places of t which do not have enough space:
$[\forall s \in S: (s,t) \in F \Longrightarrow m(s) \geq w(s,t)] \wedge [\exists s \in S: (t,s) \in F \wedge m(s) > k(s)-w(t,s)]$.

The firing of a transition t which has concession at marking m1 leads the system to a marking m2 such that
(1) the number of tokens of all input places of t is decreased by the corresponding token widths:
$\forall s \in S: (s,t) \in F \Rightarrow m1(s)=m2(s)+w(s,t)$
(2) the number of tokens of all output places of t is increased by the corresponding token widths:
$\forall s \in S: (t,s) \in F \Rightarrow m2(s)=m1(s)+w(t,s)$
(3) the number of tokens of all other places is not affected:
$\forall s \in S: (s,t) \notin F \wedge (t,s) \notin F \Longrightarrow m1(s)=m2(s)$
Firing transitions and moving tokens accordingly is often referred to as playing the token game.

Example:
Black dots within a place denote the number of tokens contained by that place. Numbers associated with places denote their capacities. No number stands for capacity ω. Numbers associated with arcs denote their token widths. No number stands for token width 1.

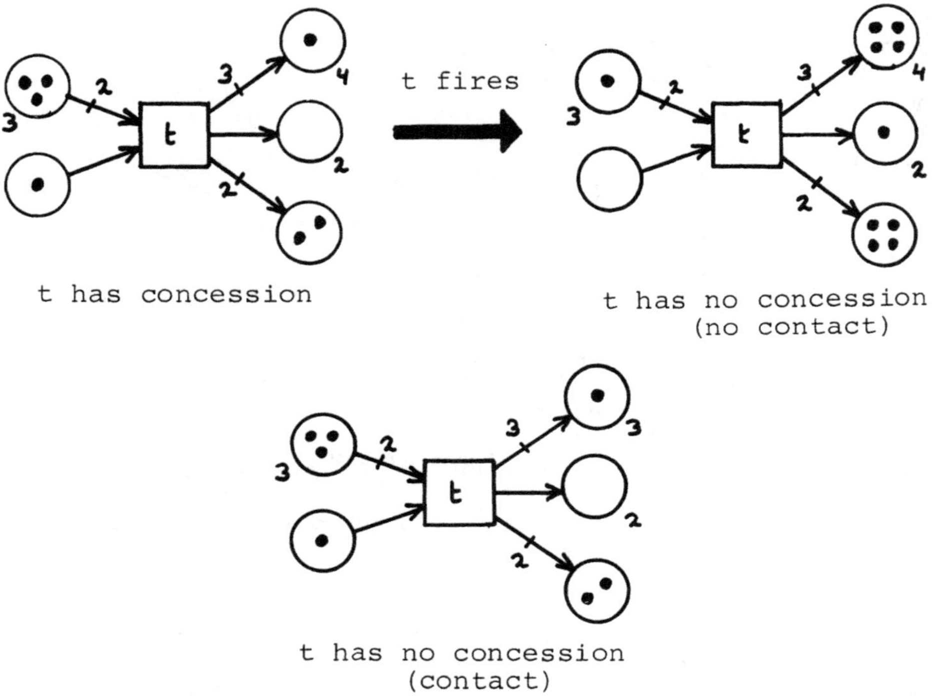

Fig. 2. Concession, contact and firing.

A firing sequence which has concession at marking m is an ordered sequence of transitions $\langle t1,t2,..,tn \rangle$ such that ti ($1 \leq i \leq n$) has concession after the consecutive firing of t1,...,ti-1 from m.

A marking m' is said to be reachable from a marking m iff there is a firing sequence which has concession at m and results in m'.

$[m\rangle$ denotes the set of all markings which are reachable from marking m.

The markings which are reachable from the initial marking can in turn be partially ordered in a net ($[m0\rangle$,T';F'), called the marking graph, where

(1) $T'=\{t \mid t \in T \wedge \exists m \in [m0\rangle$: t has concession at m$\}$,
    i.e. T' is the set of transitions which have concession at a marking reachable from m0,

(2) $F' \subseteq [m0\rangle \times T' \cup T' \times [m0\rangle$
such that $(m,t) \in F'$ iff t has concession at m
and $(t,m) \in F'$ iff the firing of t which has concession at a marking reachable from m0 leads to m.

Note that each transition of the marking graph of a Petri Net has exactly one input arc and one output arc.

A Petri Net is called <u>k-safe</u> w.r.t. a marking set M iff
(1) $\forall s \in S$: $k(s) \leq k$ and ($k \in \mathbb{N}$)
(2) $\forall m \in M, \forall t \in T$: t has no contact in m
i.e. all places have at most capacity k and no transition has contact in any marking of M.

A Petri Net is called <u>bounded</u> w.r.t. a marking set M iff
$\exists k \in \mathbb{N}: \forall m \in M, \forall s \in S: m(s) \leq k$
i.e. all places contain only a finite number of tokens under M.

A Petri Net with initial marking m0 is called <u>live</u> iff
$\forall m \in [m0\rangle, \forall t \in T: \exists m' \in [m\rangle$: t has concession at m'
i.e. for all transitions there is always a marking reachable such that they have concession.

2.3. Algebraic net representation.

Pure nets are fully represented by their <u>incidence matrix</u> $C: S \times T \longrightarrow \mathbb{Z}$ with $|S|$ rows and $|T|$ columns, the elements of which have the following values :

$$C(s,t) = \begin{cases} -w(s,t) & \text{iff } (s,t) \in F \\ +w(t,s) & \text{iff } (t,s) \in F \\ 0 & \text{otherwise.} \end{cases}$$

Let a marking m be denoted as a vector over the places, then the effect of firing a transition t which has concession at m can be expressed by adding the t-column of C to m:

$$m' = m + C(*,t)$$

Hence a marking m' being reachable from a marking m implies that there is a vector x such that

$$m' = m + C.x$$

An <u>S-invariant</u> is an integer vector over the places $\sigma: S \longrightarrow \mathbb{Z}$ with $C^T.\sigma = 0$. A <u>T-invariant</u> is an integer vector over the transitions $\tau: T \longrightarrow \mathbb{Z}$ with $C.\tau = 0$.

The algebraic net representation is a suitable one for machines. The requirement that the net should be pure is not really a restriction because there are two purification transformations rendering nets which are equivalent to the original ones in operational sense. The transformations are given here in graphical form:

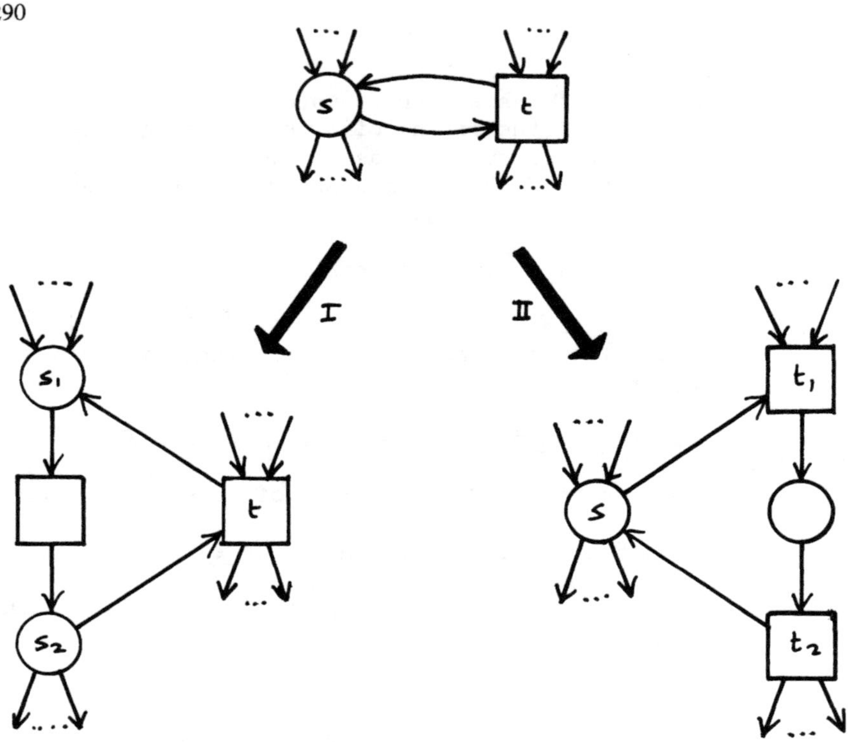

Fig. 3. Net purification transformations.

3. PRACTICAL USE OF PETRI NETS.

It is impossible to give a complete overview of the practical relevance of Petri Nets. In this chapter some application areas will be indicated and illustrated by means of examples. The reader, once he is familiar with the basic concepts of Petri Nets, will possibly discover more application areas by himself.

3.1. System description.

Sometimes (Petri) Nets are used for describing systems only. In this case it serves as a communication help, a language, which forces people to be more exact than natural language or block diagrams and similar notations. In this respect nets may even serve as a system specification language. Examples:

E1. On one of the first pages of a book a net may give a description of how to proceed through the book, given the interest and prior knowledge of the reader.

E2. The coherence and interdependencies of pieces of knowledge may be expressed in terms of nets, e.g. a system of axioms and theorems.

Elements: ○ = axiom or theorem
□ = deduction
ⓐ→[b] = a is used by b
[a]→ⓑ = b follows from a

{A1,A2,A3} is the set of axioms, {T1,T2,T3,T4,T5} the set of theorems and {p1,p2,p3,p3',p4,p5} the set of proofs which are related as follows:

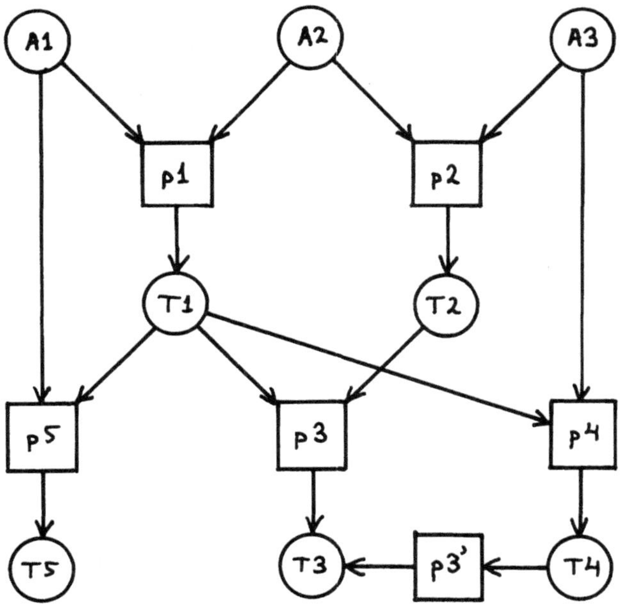

Fig. 4. Axiom-theorem system described by a net.

E3. A controller CR may make use of the services of a clock CK in order to take care of "time-outs", one at a time. The two processes communicate via one-bit mailboxes (e.g. flip-flops). There are four message types, each with its own mailbox, so the messages are identifiable by their location: "set-clock" (S), "acknowledgement-of-receipt" (A), "time-out" (T) and "cancel" (C).

The protocol in words is as follows:

CR starts with an S-message (the time value is given simultaneously via another channel) which is acknowledged by CK. If the operation under control is ready in time, a C-message is given, otherwise the T-message which came in is accepted. Since the operation may be finished concurrently with the going-off of CK, the C-message will enforce a T-message which does not have the effect of a true time-out alarm, and conversely the T-message will enforce a C-message which is considered an acknowledgement of the T-message rather than a cancel order. If in this case the transfer of a message is correctly expressed as

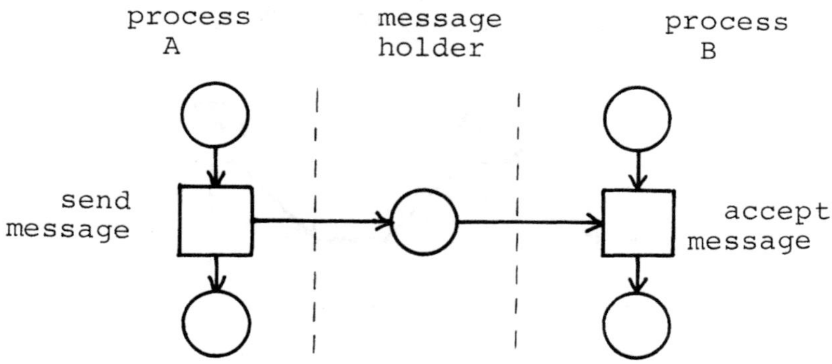

Fig. 5. Net representation of message transfer.

(e.g. if one process is able to set a flip-flop which is then reset by the other) then CR is represented by the net of fig. 6 and CK by the net of fig. 7.
The complete system w.r.t. this communication scheme is given in fig. 8.

3.2. System analysis; simulation, correctness proving.

Once a system has been described in net form, simulations can be performed by playing the token game. This can also be done by a machine.
Example:

E4. Let a marking be denoted as a sequence of identifiers p^i which indicate that place p contains i tokens. If i=0 the identifier is omitted. If i=1 then i is omitted. Now a simulation of the system represented by the net of E3 is given in fig. 9.

Simulation is only of limited value with respect to the question whether or not the system functions properly, unless it is tested exhaustively. Net descriptions allow for formal proofs of various system properties such as boundedness, absence of total or partial deadlocks, (non-)reachability of certain states, etc.

Two examples will be given here. The first one will show a capacity analysis, the second a total-deadlock analysis. These ana-

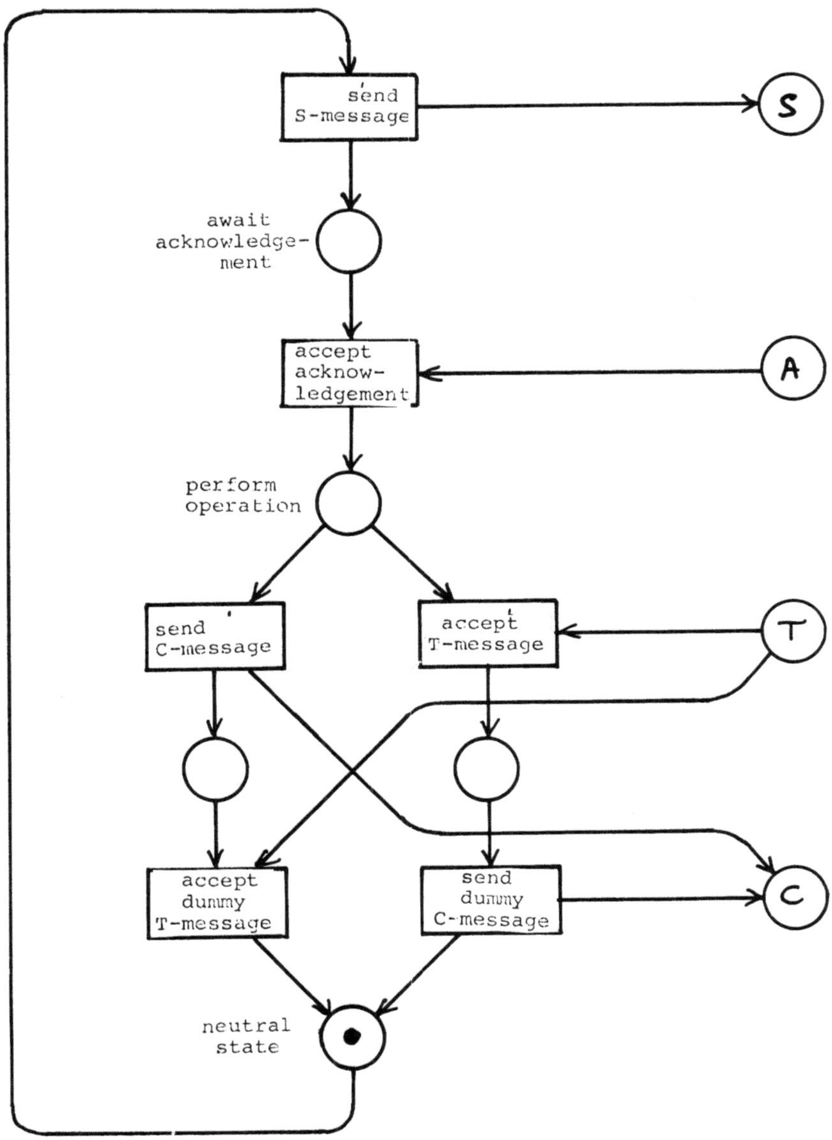

Fig. 6. Net representation of CR.

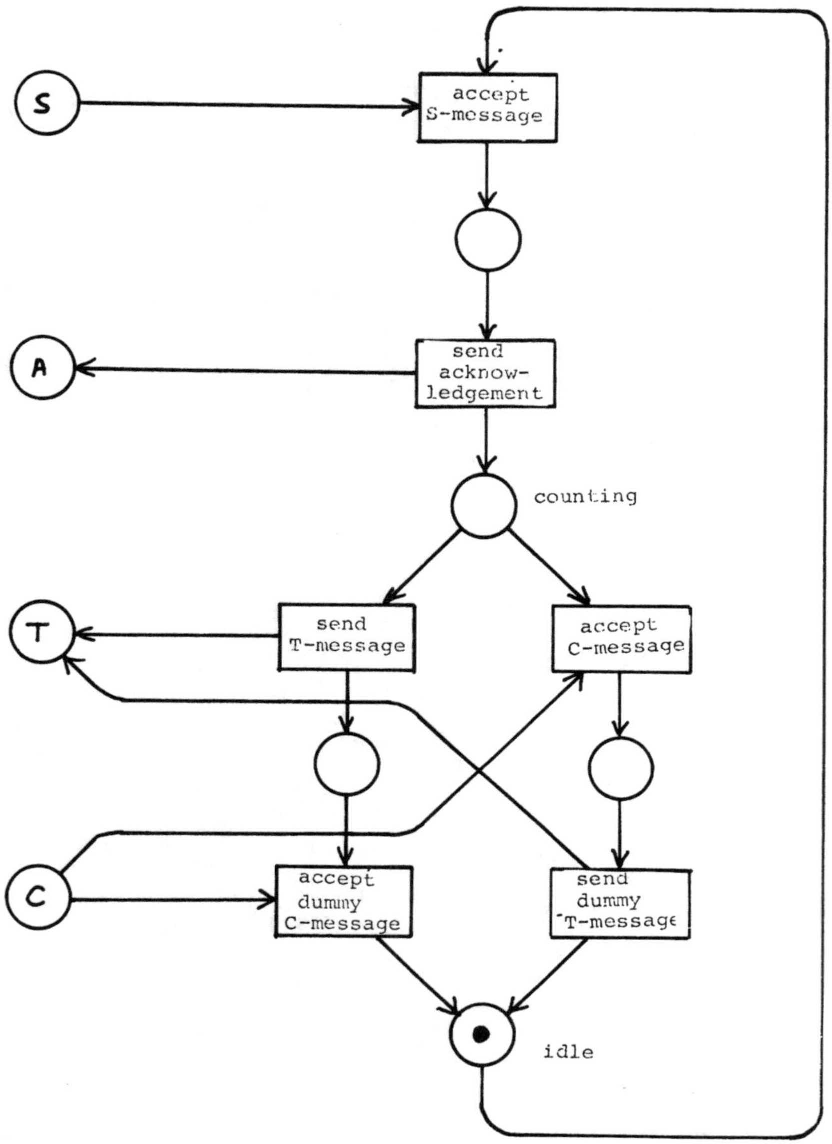

<u>Fig. 7.</u> Net representation of CK.

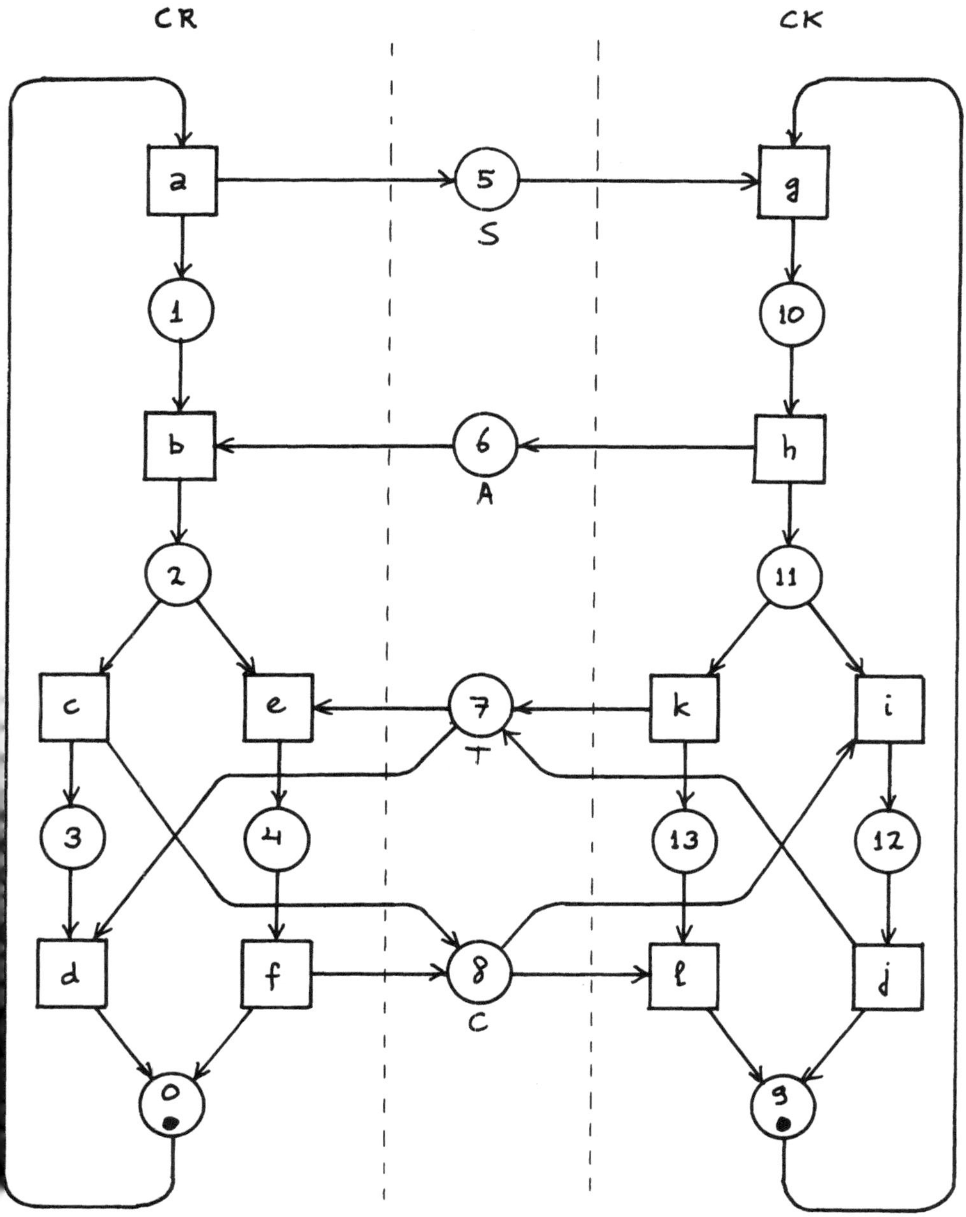

Fig. 8. Net representation of the CR-CK system.

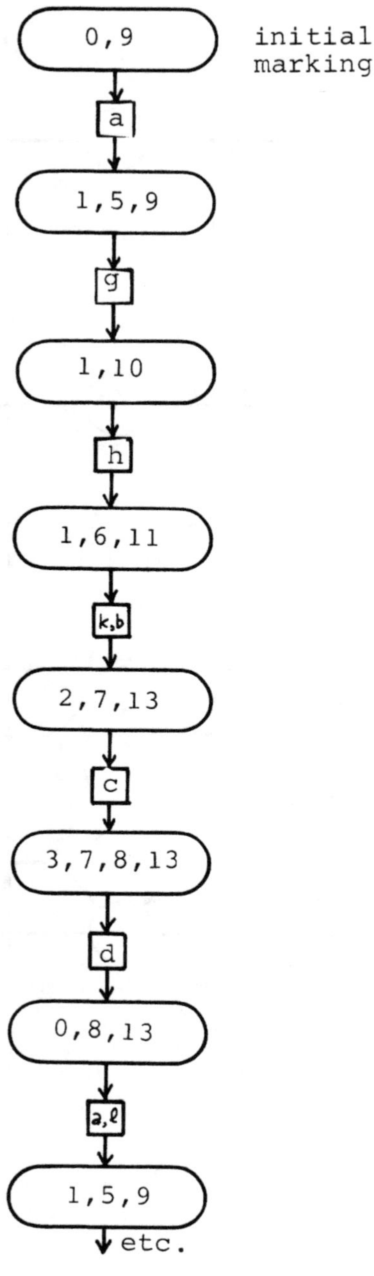

Fig. 9. Simulation of the CR-CK system.

lyses will be based on algebraic techniques. In general a lot of system properties can be proved by means of the marking graph. However, in many cases generating the marking graph is a cumbersome activity, but then the use of machines may be of great help.

E5. The net of E3 is represented in matrix notation as follows:

C =

	a	b	c	d	e	f	g	h	i	j	k	l	
0	-1	0	0	+1	0	+1	0	0	0	0	0	0	CR
1	+1	-1	0	0	0	0	0	0	0	0	0	0	CR
2	0	+1	-1	0	-1	0	0	0	0	0	0	0	CR
3	0	0	+1	-1	0	0	0	0	0	0	0	0	CR
4	0	0	0	0	+1	-1	0	0	0	0	0	0	CR
5	+1	0	0	0	0	0	-1	0	0	0	0	0	comm. path
6	0	-1	0	0	0	0	0	+1	0	0	0	0	comm. path
7	0	0	0	-1	-1	0	0	0	0	+1	+1	0	comm. path
8	0	0	+1	0	0	+1	0	0	-1	0	0	-1	comm. path
9	0	0	0	0	0	0	-1	0	0	+1	0	+1	CK
10	0	0	0	0	0	0	+1	-1	0	0	0	0	CK
11	0	0	0	0	0	0	0	+1	-1	0	-1	0	CK
12	0	0	0	0	0	0	0	0	+1	-1	0	0	CK
13	0	0	0	0	0	0	0	0	0	0	+1	-1	CK

(columns a–f: CR; columns g–l: CK)

The following S-invariants hold:

* m(0)+m(1)+m(2)+m(3)+m(4)=c0
* m(9)+m(10)+m(11)+m(12)+m(13)=c1
* m(1)-m(5)-m(6)-m(10)=c2
* m(3)-m(4)-m(7)-m(8)-m(12)+m(13)=c3
* m(2)+m(3)+m(6)-m(7)-m(11)-m(12)=c4

} for all markings m

Observing the initial marking ($m_0(0)=m_0(9)=1$ and $m_0(p)=0$ for all other places p) one may conclude that c0=c1=1 and c2=c3=c4=0. Hence the invariants become:

(S0) m(0)+m(1)+m(2)+m(3)+m(4)=1
(S1) m(9)+m(10)+m(11)+m(12)+m(13)=1
(S2) m(5)+m(6)+m(10)=m(1)
(S3) m(4)+m(7)+m(8)+m(12)=m(3)+m(13)
(S4) m(2)+m(3)+m(6)=m(7)+m(11)+m(12)

} for all markings m

Capacity analysis:

Since $m(p) \geq 0$ for each place p, we may conclude:

S0 $\Rightarrow$ $m(1) \leq 1$

S2 $\Rightarrow$ $m(5)+m(6)=m(1)-m(10) \leq m(1)$ $\therefore$ $m(5) \leq 1$ and $m(6) \leq 1$

$S0 \Rightarrow m(1)+m(2)+m(3) \leq 1$
$S2 \Rightarrow m(6)=m(1)-m(5)-m(10)$
$S4 \Rightarrow m(7)=m(2)+m(3)+m(6)-m(11)-m(12)$
$\therefore m(7)=m(1)+m(2)+m(3)-m(5)-m(10)-m(11)-m(12)$
$\leq m(1)+m(2)+m(3)$

$\therefore \underline{m(7) \leq 1}$

$S1 \Rightarrow m(11)+m(13) \leq 1$
$S4 \Rightarrow m(7)=m(2)+m(3)+m(6)-m(11)-m(12)$
$S3 \Rightarrow m(8)=m(3)-m(4)-m(7)-m(12)+m(13)$
$=-m(2)-m(4)-m(6)+m(11)+m(13)$
$\leq m(11)+m(13)$

$\underline{m(8) \leq 1}$

Conclusion: the message-holding places will never receive more than one token, hence single flip-flops may be used for communication.

E6. Consider the net of E3 and the S-invariants S0 through S4 of E5. May a total deadlock occur? Let D denote the set of deadlock markings, i.e. $D \overset{def}{=} \{ m | N(m) \}$ where N(m) stands for "no transition can fire at m", or equivalently:

$N(m) \Longleftrightarrow$	$m(0)=0$	(a cannot fire)
	$\wedge\ (m(1)=0 \vee m(6)=0)$	(b cannot fire)
	$\wedge\ m(2)=0$	(c cannot fire)
	$\wedge\ (m(3)=0 \vee m(7)=0)$	(d cannot fire)
	$\wedge\ (m(2)=0 \vee m(7)=0)$	(e cannot fire)
	$\wedge\ m(4)=0$	(f cannot fire)
	$\wedge\ (m(5)=0 \vee m(9)=0)$	(g cannot fire)
	$\wedge\ m(10)=0$	(h cannot fire)
	$\wedge\ (m(8)=0 \vee m(11)=0)$	(i cannot fire)
	$\wedge\ m(12)=0$	(j cannot fire)
	$\wedge\ m(11)=0$	(k cannot fire)
	$\wedge\ (m(8)=0 \vee m(13)=0)$	(l cannot fire)

For the deadlock markings m:

$N(m) \wedge S0 \Rightarrow m(1)+m(3)=1$	(1)
$N(m) \wedge S1 \Rightarrow m(9)+m(13)=1$	(2)
$N(m) \wedge S2 \Rightarrow m(5)+m(6)=m(1)$	(3)
$N(m) \wedge S3 \Rightarrow m(7)+m(8)=m(3)+m(13)$	(4)
$N(m) \wedge S4 \Rightarrow m(3)+m(6)=m(7)$	(5)
For all places p: $0 \leq m(p) \leq 1$ (see also E5).	
Suppose $m(1)=1$	(6)
$N(m) \wedge (6) \Rightarrow m(6)=0$	(7)
$(1) \wedge (6) \Rightarrow m(3)=0$	(8)
$(3) \wedge (6) \wedge (7) \Rightarrow m(5)=1$	(9)
$N(m) \wedge (9) \Rightarrow m(9)=0$	(10)
$(10) \wedge (2) \Rightarrow m(13)=1$	(11)
$N(m) \wedge (11) \Rightarrow m(8)=0$	(12)
$(4) \wedge (8) \wedge (11) \wedge (12) \Rightarrow m(7)=1$	(13)
$(5) \wedge (7) \wedge (8) \Rightarrow m(7)=0$	(14)

(13) is in contradiction with (14), hence the supposition that m(1)=1 is false.

Ergo $m(1)=0$ (6')
$(1)\wedge(6')\Rightarrow m(3)=1$ (7')
$N(m)\wedge(7')\Rightarrow m(7)=0$ (8')
$(5)\wedge(8')\Rightarrow m(3)=0$ (9')
(7') is in contradiction with (9').
Conclusion: N(m) is false for all m, hence D is empty. In other words, there are no deadlock states: the system is free of total deadlock.
N.B. D being not empty does not imply that total deadlock can occur. This is only the case if such states are reachable from the initial marking!

3.3. System synthesis.

Whereas system analysis techniques are to give an answer to the question whether or not the system functions properly, i.e. in accordance with the specified requirements, system synthesis techniques aim at the implementation of the specifications according to design rules which guarantee correctness, i.e. afterwards analysis is not needed. There are much more analysis than synthesis techniques. This is easily understood if one realizes that synthesis techniques do generally stem from analytical observations. In the same way net design rules may evolve from known net properties. Examples:

E7. Use basic communication structures which are known to be correct. Besides they often allow for simpler nets.

E8. Construct a net such that it is covered by "cycles" with constant token counts (S-invariants). Along with a finite initial marking this guarantees that the system is bounded.

4. CONCLUSIONS

Along with the increase of integration scales there is a growing need for tools and techniques which assist in mastering the complexity of systems to be built on a chip. A mathematical tool which may be of help is Net Theory. In particular the Petri Net formalism may be useful for describing and specifying systems, for analysing and simulating their behaviour and in the near future possibly for providing system synthesis techniques.

[1] Agerwala, T.:
"Some applications of Petri Nets".
Proc. of the National Electronics Conference, Vol. 32;
Chicago, 16-18 Oct. 1978; pp. 149-154.

[2] Michel, C.; Kwan, C.L.:
"Hardwired logic and microprocessors". I and II.
I: Euromicro Journal 3, No. 1; 1977; pp. 73-80.
II: Euromicro Journal 4, No. 3; 1978; pp. 167-180.

[3] Misunas, D.:
"Petri Nets and speed independent design".
Communications ACM 16, No. 8; 1973; pp. 474-481.

[4] Moalla, M.; Pulou, J.; Sifakis, J.:
"Synchronized Petri Nets: a model for the description of non-autonomous systems".
Lecture Notes in Computer Science, Vol. 64;
Springer-Verlag; Berlin, 1978; pp. 374-384.

[5] Peterson, J.L.:
"Petri Net Theory and the modeling of systems".
Prentice-Hall; Englewood Cliffs, N.J.; 1979.

[6] Petri, C.A.:
"Kommunikation mit Automaten".
Schriften IIM, Heft 2; Universität Bonn; 1962.
English translation:
"Communication with automata".
Griffiss Air Force Base, New York;
Technical Report RADC-TR-65-377, Vol. 1, suppl. 1; 1966.

ARCHITECTURE AND DESIGN OF VON NEUMANN MICROPROCESSORS

F. ANCEAU

Computer Architecture Group, IMAG Laboratory
BP 53X, 38041 GRENOBLE cédex, France

1. INTRODUCTION

Microprocessors are computers built on LSI technology. The constraints of this technology lead their designer to use different internal organisations as those used for computer built on SSI and MSI technology.

2. ALGORITHMIC SPECIFICATIONS

The behaviour of a computer may be described with an algorithm, like all sequential machines.
The features of the algorithmic language used for writing this specification may, or not, allow its easy translation to hardware. The easiness of this translation depends upon the adaptability, or not, of the set of primitives (available in the language) to the behavioural description of the hardware devices.

3. LEVELS OF INTERPRETATION

The translation of an algorithmic specification to a physical machine may be done through one or several steps.
Each of these steps may be considered as a level of interpretation where the execution of each instruction of the preceding level into several steps corresponding to the execution of instructions of this level.
e.g.: The execution of an I8080 instruction is decomposed into several "machine cycles". The execution of each machine cycle is decomposed into several "timing states", and a timing state is,

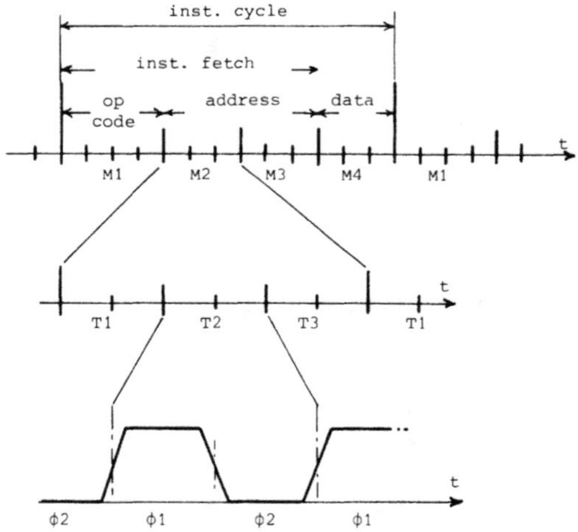

Fig.1 - Timing levels in the microprocessor I 8085

itself, decomposed into two "clock phases" (ϕ1 and ϕ2).
From this point of view the behavioural description of a processor may be viewed as a stack of interpretation layers.

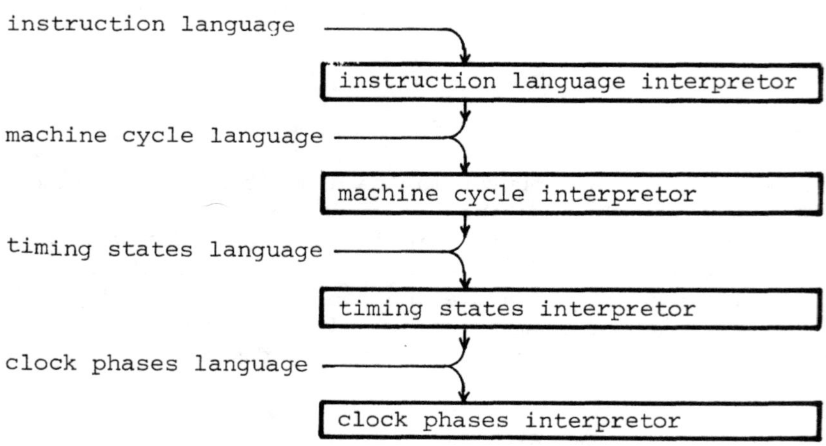

4. NOTION OF OPERATIVE AND CONTROL PART

A sequential machine may be decomposed into two sub parts :

- an operative part which performs the data manipulations and operations.
- a control part which animates the operative part in order to perform the given sequences of operations.

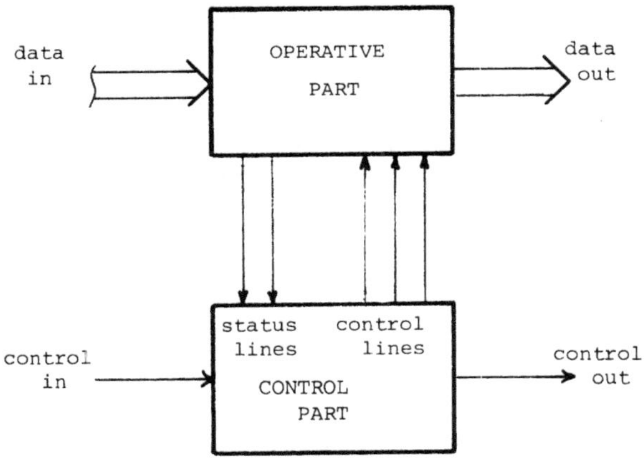

Fig.2 - Operative and control part

The control part drives the operative part through a set of control lines.
The operative part sends its status to the control part through a set of status lines.
Such a decomposition may be done from an algorithmic specification which allows an easy translation to hardware.

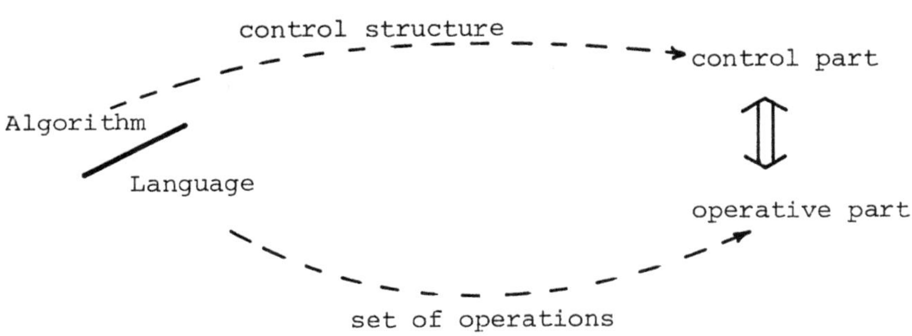

Operative and control parts may be defined at each level of interpretation, so giving :

- the <u>virtual machine</u> from the interpretor of the instruction set,
- the content of the <u>recorded sub-parts</u> of the control part of the physical machine (i.e. microprograms, PLA ...) from the intermediate levels,
- the <u>hardware</u> (operative and control parts) from the last level of interpretation.

5. TECHNOLOGICAL CONSTRAINTS

The LSI technology gives some constraints to the microprocessor designer. The main constraints are :

- data flow limitation between the chip and its environment. This constraint comes from both the limited amount of I/O pins available, and the speed of the I/O drivers on the chip. This constraint has a great influence on the external architecture (at the system level):
  . the decomposition of a system into a set of chips,
  . the multiplexing of bus,
  . the coding of the control lines.
- overall complexity of the chip. This constraint is now decreasing with the technological improvements.
  This constraint has, however, indirect effects:
  . the power dissipation of a chip increases with its size which increases slowly with its complexity. The technical limitation of a 1 to 2 w power dissipation leads the designer to choose a low power dissipation circuitry and technology (dynamic logic and CMOS).
  . the design cost,which is now close to one transistor per man per hour,increases with the complexity. The economical constraints lead the designer to use more efficient CAD tools and design methods.
  . the testability which becomes more and more difficult to achieve when the complexity increases.
- planearity of the chip which introduces geographical constraints into its global organisation. Such a constraint is due to the cost of the interconnection lines inside the chip in term of area of silicon and loose in speed. As an example, the control part organisation must provide control signals at the right place where they are needed. Such a constraint may suggest decentralized control part organisation.
- speed of the gates. The improvements of the technology have diminished this constraint. The to-day gates inside a LSI chip are so fast as discrete TTL gates.

6. OPERATIVE PART ORGANISATION

The operative part of a microprocessor is a set of memory elements (registers and local memories) and arithmetic/logical units (ALU) connected by a bus called : data path.

6.1. Bussing system

The number of buses used to interconnect the memory elements and the ALU depends upon :

- two buses allow to bring simultaneously two operands to the ALU. A simple bus uses less area of silicon to be implemented but requires more steps to move the same quantity of data.

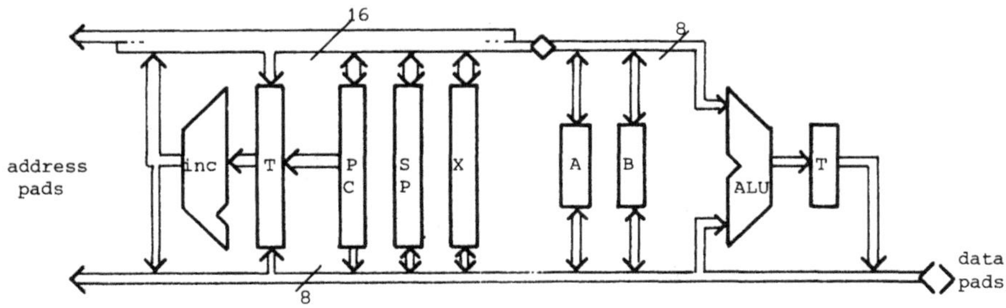

Fig. 3 - MC 6800 operative part

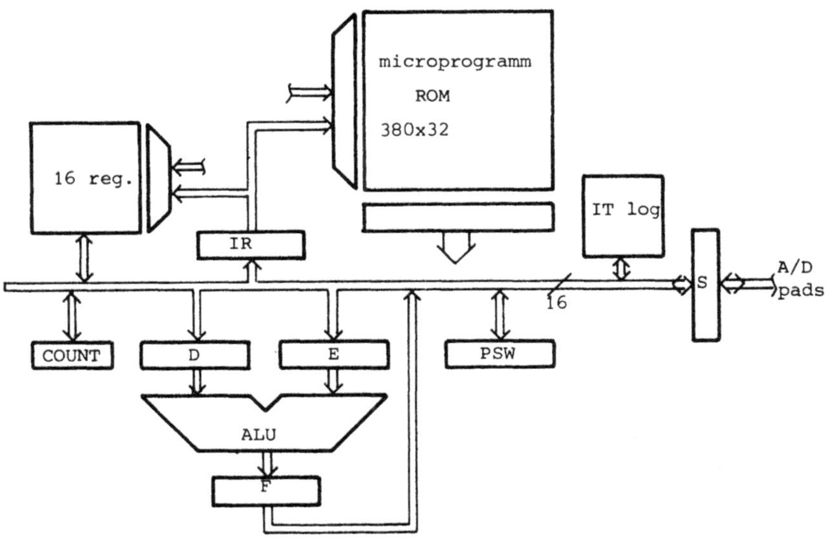

Fig. 4 - S 2752 FAST

6.1.2. The distribution of memory elements

The use of registers directly attached to the ALU allows to keep operands ready to be used and may decrease the data flow from the register bank.

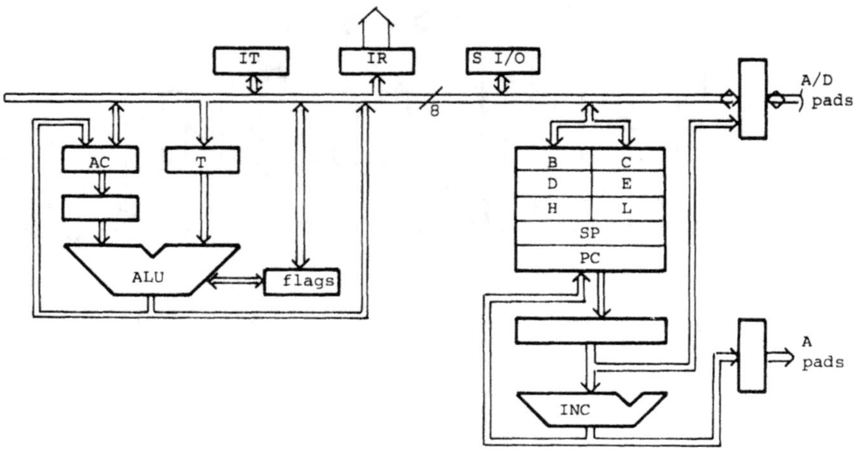

Fig. 5 - I 8085 operative part

6.1.3. The pre-load of frequently used data

The load in advance of some registers by data which are frequently used allows to perform the movement of those data when the bus is idle and reduces the data flow, e.g. pre-load of an hardware accumulator register in the Z 80.

6.1.4. Electrical realisation of the bus

A bus may be
- tri-state driven by tri-state super buffers (sometimes using boot straps). Due to electrical limitations only short buses can be used.
- precharged bus driven by open-drain drivers. The bus can be longer but is slower due to the two cycle timing(precharge, decharge).

6.2. Memory elements

The data manipulated by the instruction interpretor may be stored into a register bank or a local memory :
- a local memory needs less area than a register bank if more than 10 words are stored ;
- a local memory (single or twin ported) gives less freedom for bus

connections than a register bank ;
- a local memory has a more complex timing which requires two cycles, reading (stabilisation, read) and writing (precharge, write).

6.3. Serialisation

The serialisation of data manipulation and operation is a speed versus area of silicon design compromise. The use of a data path narrower than the logical data size reduces the area of silicon used to implement bus, temporary registers and ALU,but requires more steps to execute instructions, e.g. Z80, GICP1600, F100.

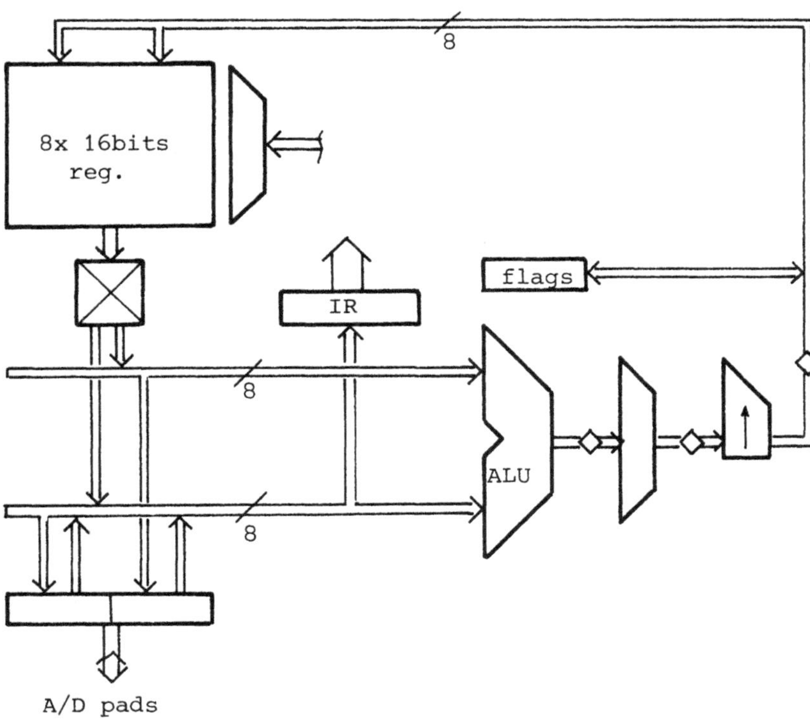

Fig. 6 - CP 1600 operative part (16 bits microprocessor)

6.4. Concurrent operative parts

The use of several sub-operative parts for parallel address and data computation permits to speed up the microprocessor but increases its area of silicon . The address sub-operative part is often limited to the address incrementation . More complex

address computation (e.g. add displacement to index), requires the address sub-operative part to be connected to the data sub-operative part by transmission switches .
These transmission switches also allows each sub-operative part to use the registers of the other sub-operative part as operand.

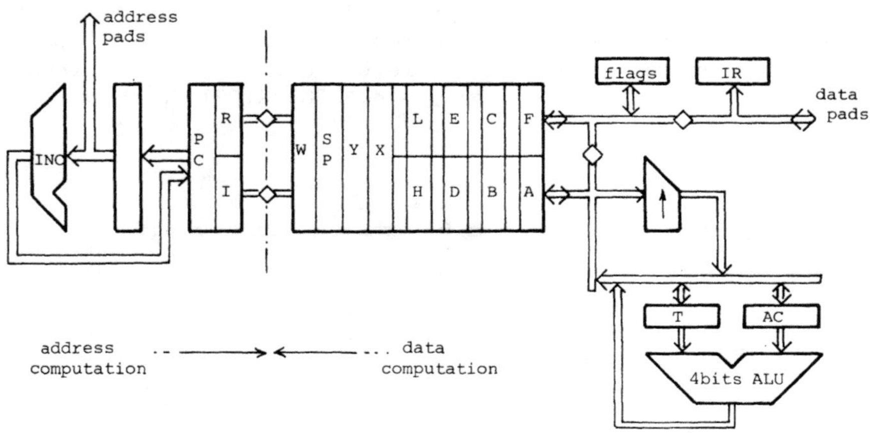

Fig. 7 - Z 80 operative part.

6.5. Operative part layout

The layout of an operative part may be organised
- as separated blocks (RAM, ALU, buses),
- as bit slice.

Bit slice organisation is based on a library of bricks (RAM elemen register element, one bit ALU ...) with a segment of bus. The bricks are designed to fit together from the topological and electrical point of view (height of the brick, fan out). The bricks are assembled to form a one bit operative part which is duplicated Some corrections must be done on this layout for non repetitive features (i.e. half carry output, carry look ahead ...).

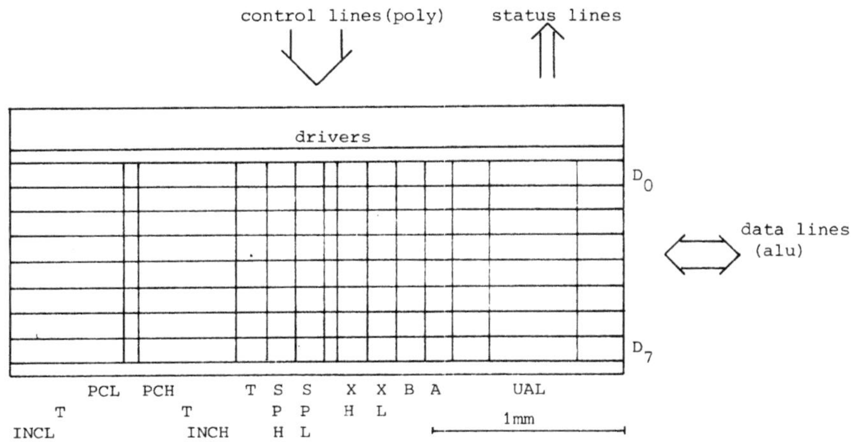

Fig.8 - MC 6800 operative part layout

The data buses are made with the aluminum layer allowing long and fast communications. The commands and states lines are made of polysilicon because they are shorter than data lines and less loaded (only one driver per command line).
Automatic tools (P1) may be writen to build such operative part from high level specifications.

7. CONTROL PART ORGANISATION

The control part of a microprocessor is designed after the operative part is implemented, in order to know the meaning and topology of the control and state lines.

7.1. PLA optimisation

Several control part organisations use PLAs. A PLA may be seen as a read only associative memory or as an "and" matrix gates followed by an "or" matrix gates.

When such a PLA is used for decoding (i.e. generating uncoded commands) the or matrix is very empty. An optimisation (P2) which consists of cutting and reorganizing the output lines of this matrix reduces its area by a factor or 2 to 3 (e.g. Z 80, Z 8000).

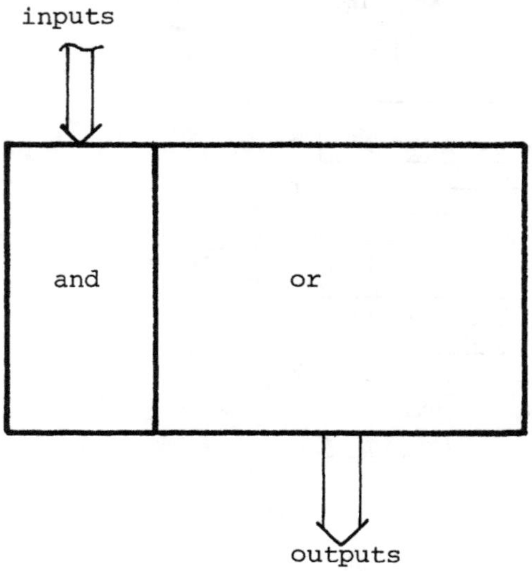

Fig. 9 - Unoptimised PLA

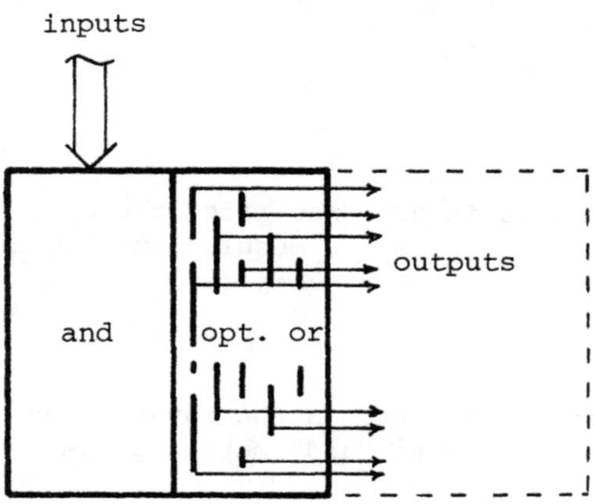

Fig. 10 - Optimised PLA

7.2. Single level of interpretation control parts

7.2.1. Wired control parts (e.g. MC 6800)

In order to design these control parts, the flow chart of the interpretation algorithm of the instruction set is directly translated into a set of gates and dynamic shift register (P4).

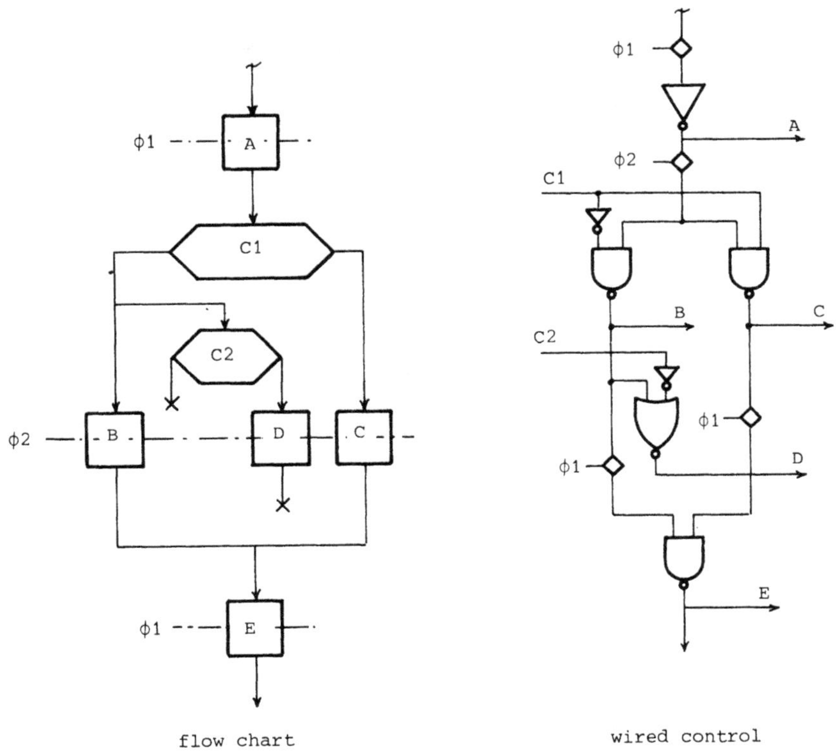

Fig. 11 - Wired control part

As a reference, the topological organisation of the MC 6800 is given in order to compare the efficiency of the different "styles" of control part organisation (P3).

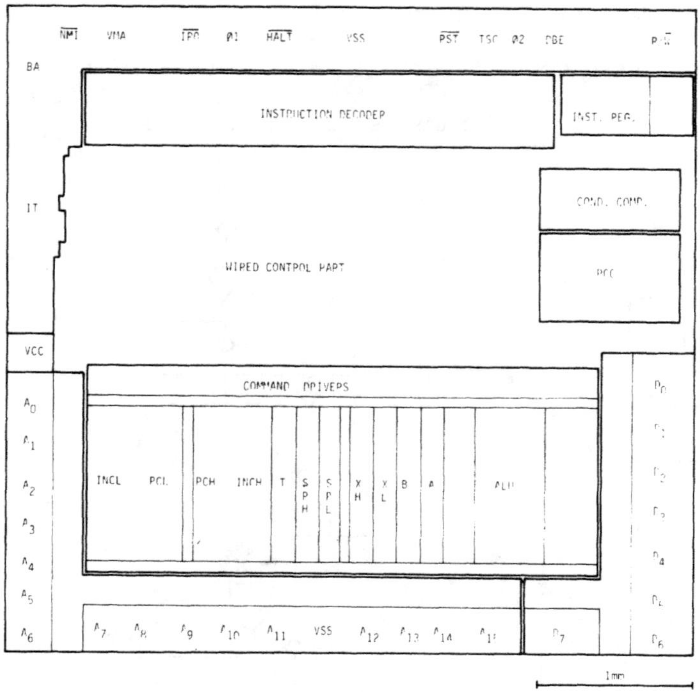

Fig. 12 - MC 6800 topology

This design technique gives a high density layout. Such a technique ,which is difficult to design an debug, may only be used for small and medium complexity instruction sets. It may be used also to implement small automata (e.g. timing generator.see7.2.5)

7.2.2. **Single PLA control part** (e.g. NS SC/MP)

Such a control part is considered as a single automaton implemented with a single PLA. The "and" matrix decodes both the executed instruction, the status of the operative part and the step register. The "or" matrix generates the next content of the step register and the commands for the operative part. The "or" matrix may be optimised.

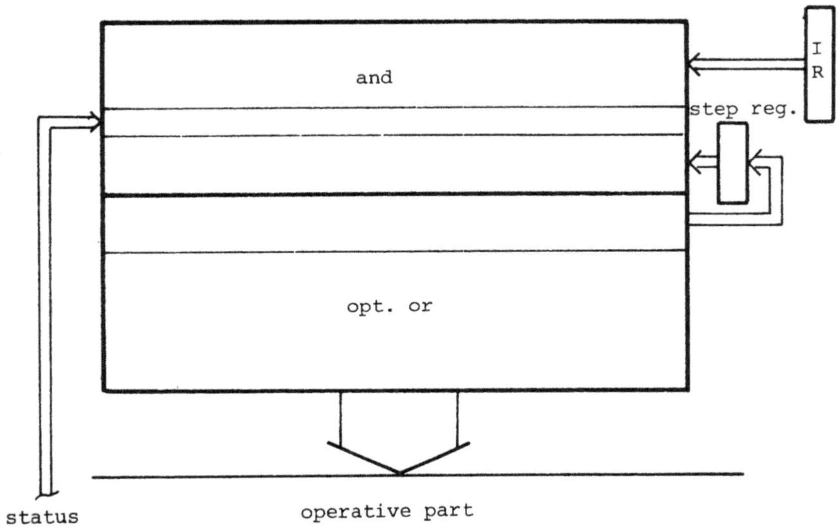

Fig. 13 - Single PLA control part

As an example, an hypothetical redesign of the MC 6800 using this technique is presented. The total area of the microprocessor is increased by about 30% (the design using the same technological rules).

This technique gives a control part easy to design and to implement. The area of the optimised PLA becomes too large when the complexity of the algorithm of interpretation increases. Such a technique may only be used for small complexity instruction sets.

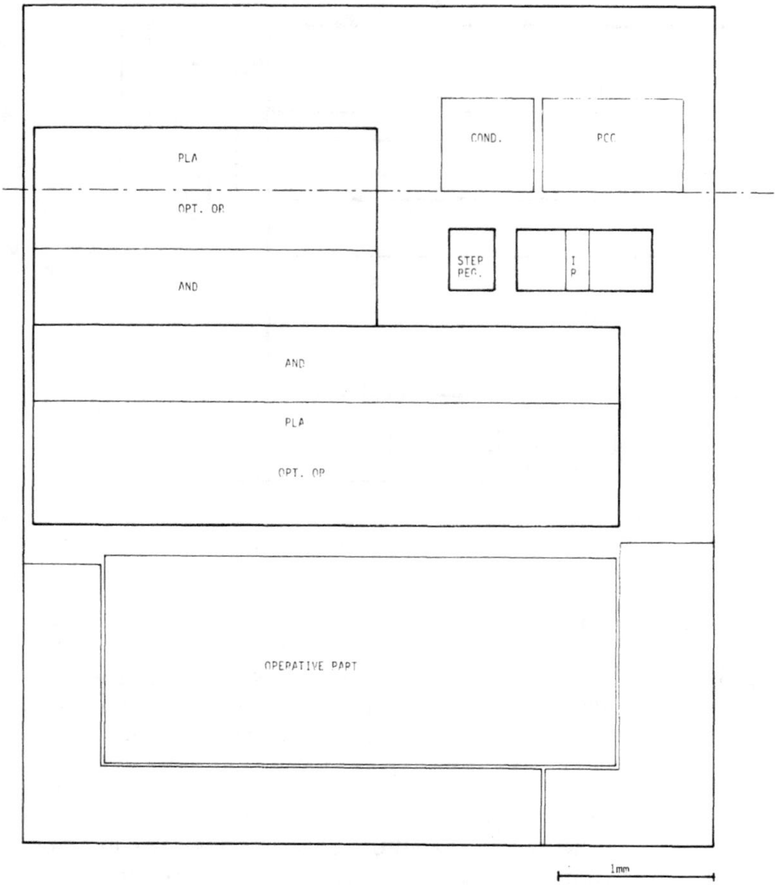

Fig. 14 - Redesign of MC 6800 using a single PLA control part

7.2.3. Multiple PLAs control parts (e.g. HPMC2, GICP 1600)

With this technique, each function of a control part is realised with a separated PLA :

- sequencing,
- command generation (or selection),
- parametrisation which is a direct generation of static commands. They can be pre-defined (e.g. addition for address computation) or directly computed from the current instruction. Selection lines coming from the preceding PLA may switch static commands and commands computed from current instructions or command lines.

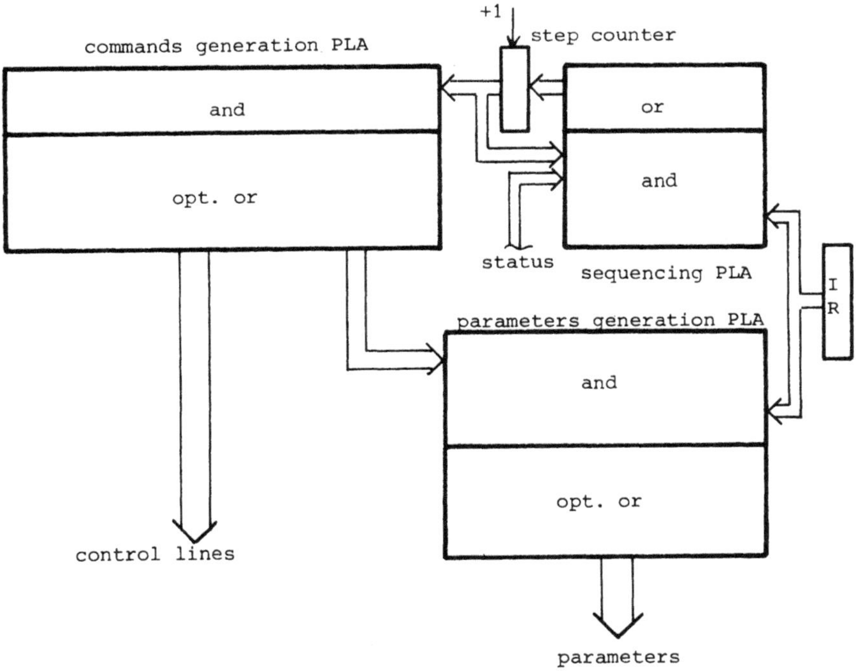

Fig. 15 - Multiple PLAs control part

As an example, an hypothetical re-design of the MC 6800 using this technique is presented. The total area of the microprocessor is not increased.

This technique allows the building of well optimised control parts easy to design and to implement. The command generation PLAs and the parametrisation PLAs may be optimised. Such a technique allows the building of control part for medium complexity instruction sets.

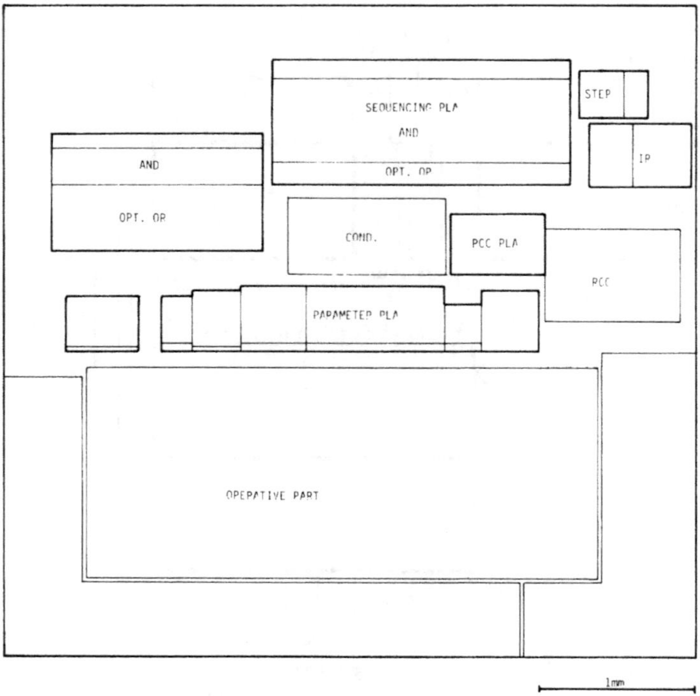

Fig. 16 - Redesign of MC 6800 using multiple PLAs control part

7.2.4. **Microprogrammed control part** (e.g. S2752 FAST, MC 68000)

This well known technique consists of recording in a ROM a microprogram which describes the sequence of control actions to be performed for interpreting the instruction set. The next address is recorded in each microinstruction. The complex decoding and branching operations are performed by an extra PLA connected to the word lines of the ROM. The microinstruction may be decoded by a PLA to generate the commands for the operative part. An uncoded microinstruction format allows the optimisation of the ROM by considering it as an "or" matrix of a PLA.

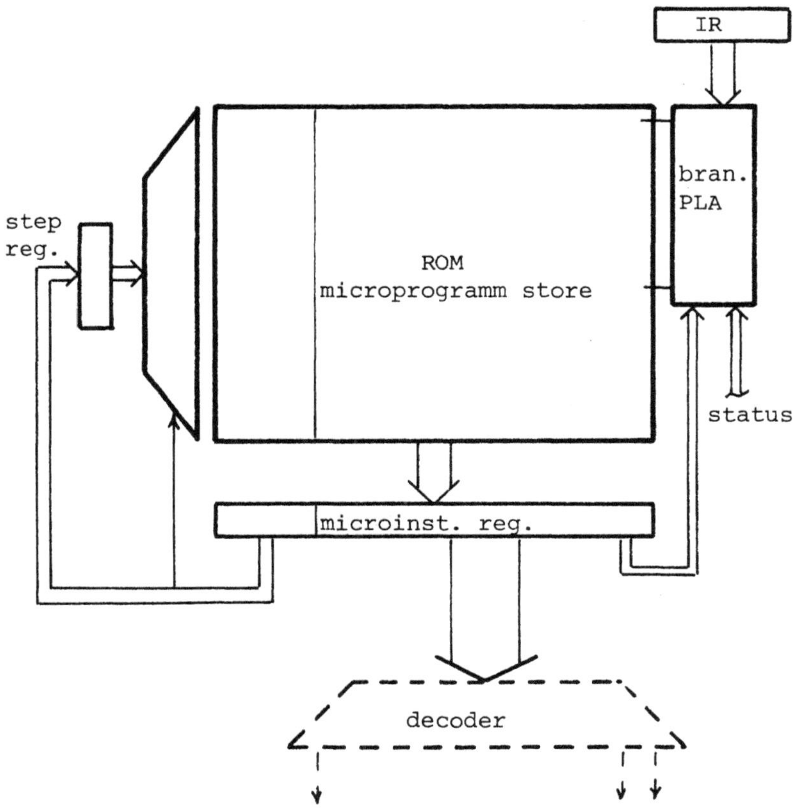

Fig. 17 - Microprogrammed control part

As an example, an hypothetical redesign of the MC 6800 using this technique is presented. The total area of the microprocessor is only increased by 20% roughly.

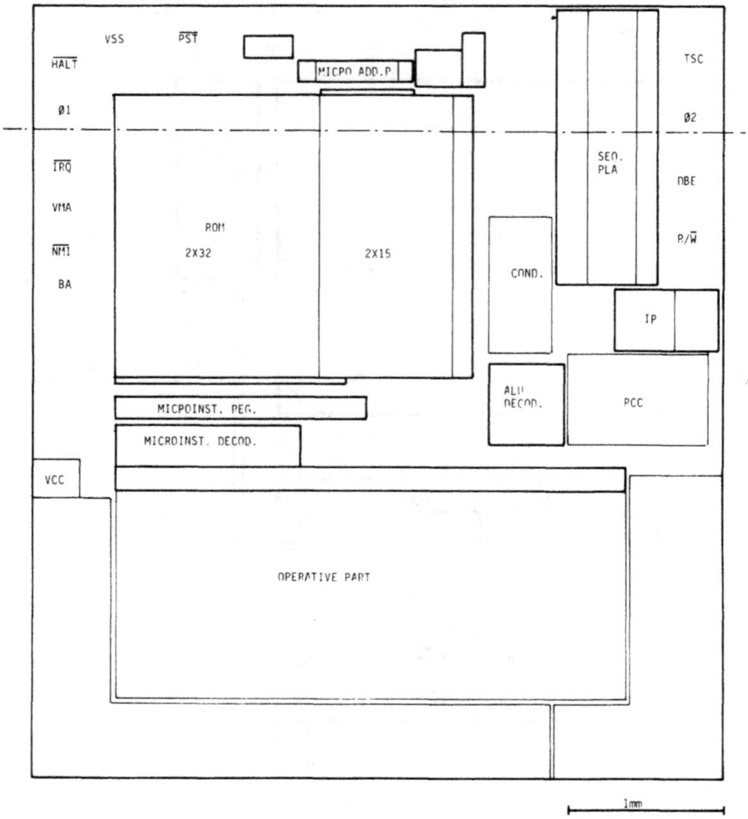

Fig. 18 - Redesign of MC 6800 using microprogrammed control part

This technique allows the building of optimised control parts, easy to design and to implement for complex instruction set.

7.2.5. Control part using timing generator (e.g. I 8080, I 8085, Z 80, Z 8000)

With this technique, command lines are generated through the validation of lines (coming from the decoding of the instruction register), by timing lines (coming from a timing generator). The timing generator is a small automaton which represents the "squeleton" of the instruction execution ; it is controled by the outputs of the instruction decoder. The validation of the commands by timing lines may be done by :

- and or gates (Z 80, Z 8000)
- a PLA (I 8085).

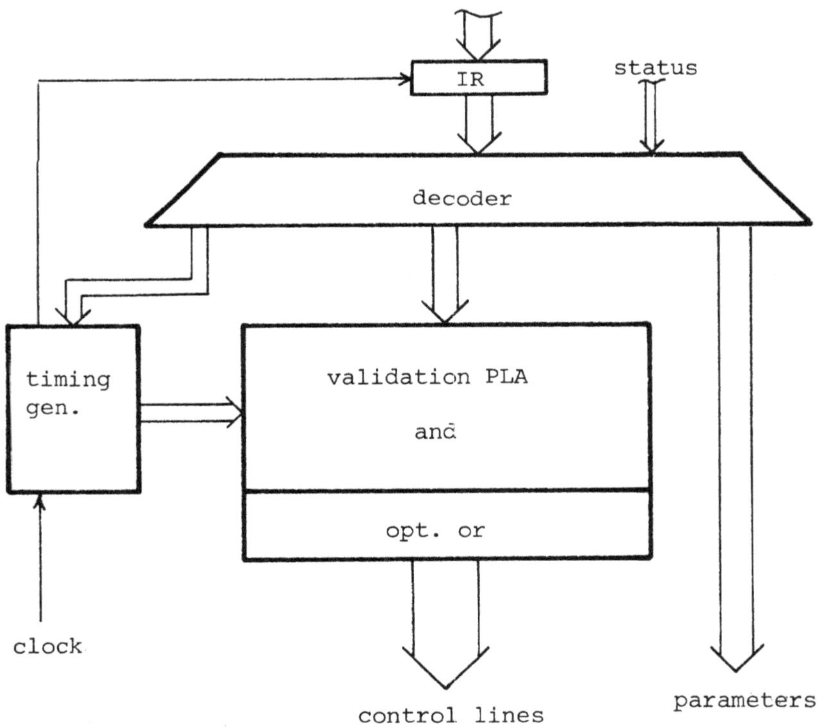

Fig. 19 - Control part using timing generator

As an example, an hypothetical redesign of the MC 6800 using this technique (PLA for validation) is presented. The total area of the microprocessor is not increased.

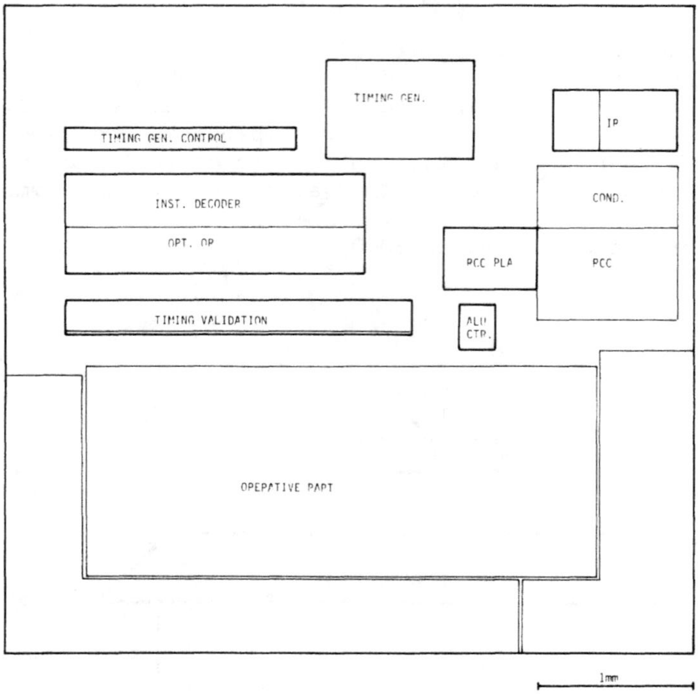

Fig. 20 - Redesign of MC 6800 using timing generator

This technique allows the building of well optimised control parts. The use of a PLA for validation needs more area of silicon but it is easy to implement. Such a technique allows the building of control parts for medium complexity instruction set.

7.3. Control part using several levels of interpretation

This technique consists of stacking several single level control parts. The commands lines coming from the upper one are considered as instruction inputs for the just lower one.

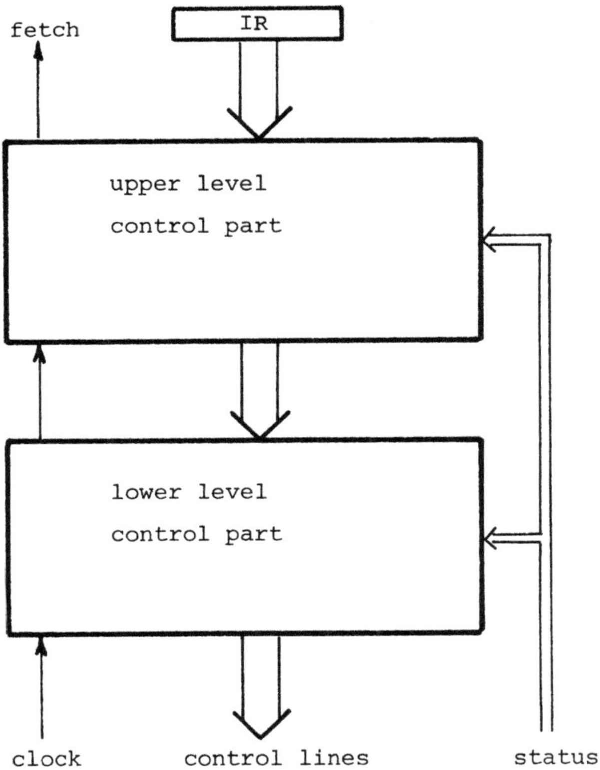

Fig. 21 - Two levels control part

The lower level of interpretation must provide a signal to the upper one to indicate that its interpretation cycle is finished. The corresponding signal coming from the upper level of interpretation indicates that the interpretation of the current instruction is finished (fetch signal).

The notion of parameter may be generalised to the commands coming directly from the upper level of interpretation.

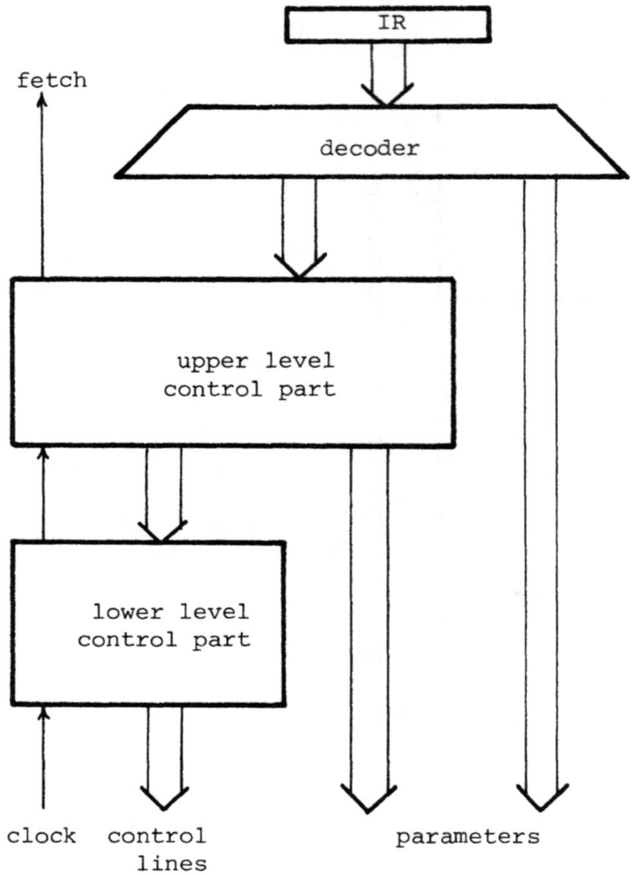

Fig. 22 - Generalisation of the notion of parameters

7.3.1. Realisation

Such a technique may be used to stack any kind of single level control part. In the microprocessor world we find only the stacking of control parts using timing generators (I 8080, I 8085, Z 80, Z 8000). The reason why is :

- these kind of control part use a validation network whose inputs (coming from an instruction decoder) are of the same kind as its outputs.

- this technique reduces the total number of timing lines by the factorisation of the timing instants (e.g. machine cycle x timing states).

In these microprocessors, the validation of the two levels are realised by a single set of and-or gates (I 8080, Z 80, Z 8000) or by a single PLA (I 8085).

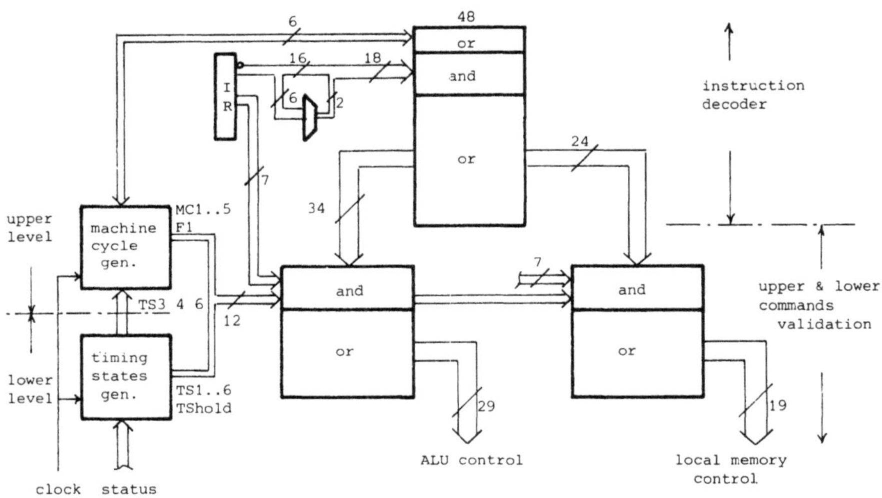

Fig. 23 - I 8085 control part

8 - OVERLAPPING IN THE INSTRUCTION EXECUTION

In order to speed up microprocessor execution, a parallelism between instruction executions may be realised by well known techniques in computer architecture :

- The last part of the execution of an instruction may be overlapped with the fetch of the next one (if the overlapped part of the first instruction does not contain any reference to memory) (e.g. MC 6800).

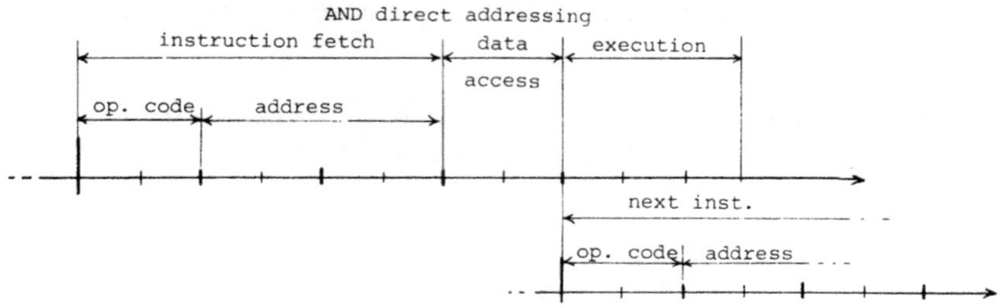

Fig. 24 - MC 6800 instruction execution overlapping

- An input queue of instructions may be managed in the microprocessor (I 8086). It permits to avoid almost all memory delays in instruction fetching. The content of the queue must be removed and reloaded when a branch instruction occurs.

- A two stage pipeline organisation may be used. The internal structure of the microprocessor is decomposed into two blocks connected by a queue :

  . the first block performs
    - the addressing of instructions and data,
    - the fetch of instructions and data,
    - the decoding of instruction.

  . the second blocks performs
    - the instruction execution.

None of the existing microprocessors has this internal architecture

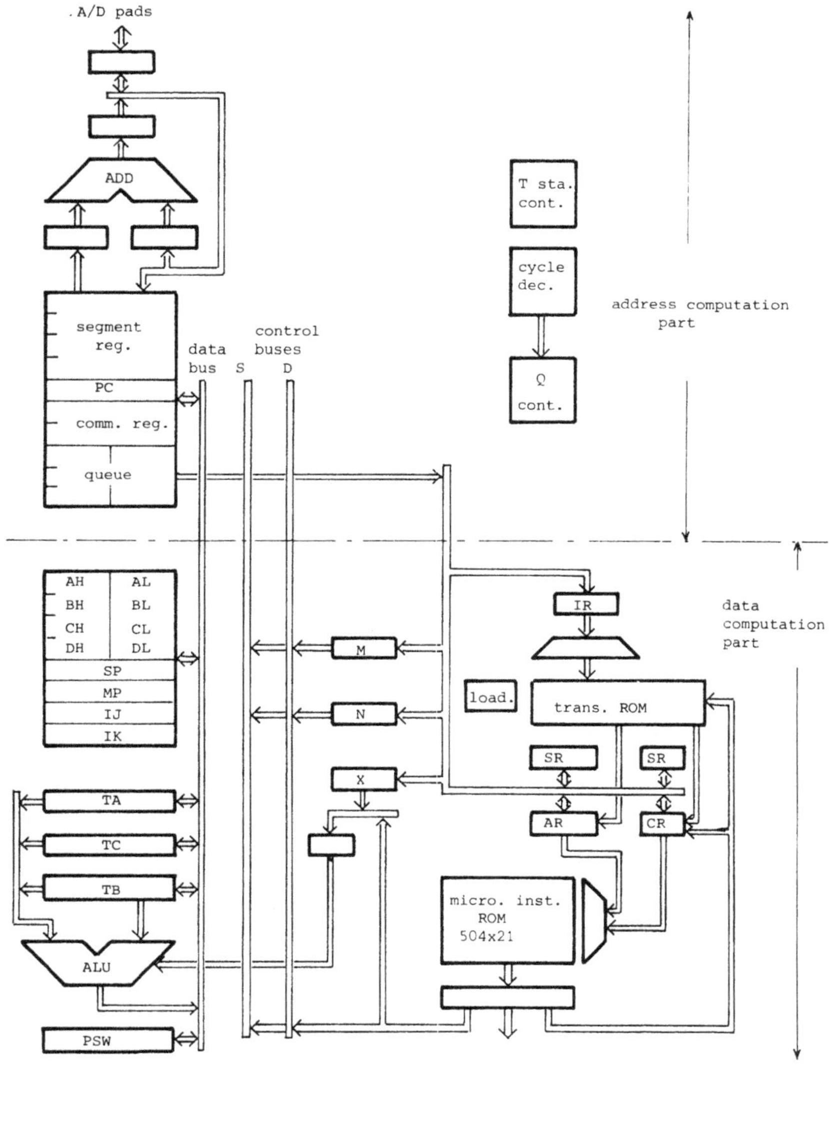

Fig. 25 - I 8086 internal architecture

The result of an economical choice in the design of the I 8086 is presented [7].

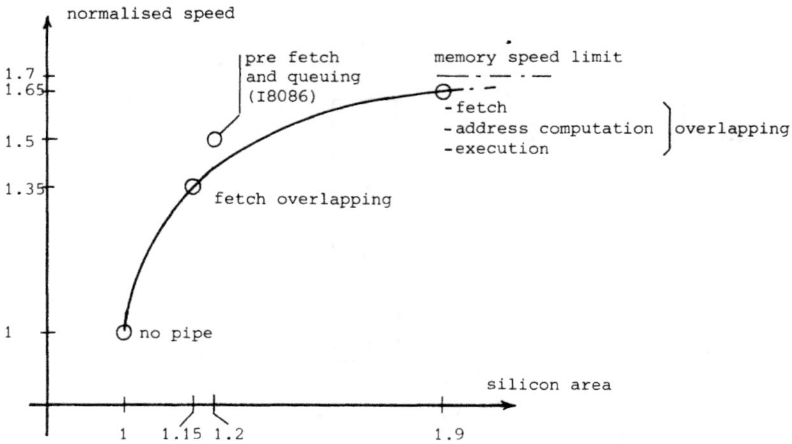

Fig. 26 - Speed/complexity trade off for I 8086

9 - CONCLUSION

This paper presents the main choices for the internal architecture of Von NEUMANN microprocessors. We have not yet investigated every possible organisation but the main rules to optimise the design of a small or complex (second generation) microprocessor seems to us, very clear. The technological evolution will decrease the optimisation requirements on small products in the favour of CAD methods decreasing the design cost and increasing the testability.

10 - ACKNOWLEDGEMENTS

This paper includes some early results of the CAPRI research project of the computer architecture group of the IMAG Laboratory, including the following sub-projects :

P1 - Building blocks system for operative part design
A. SUZIM

P2 - Optimisation of PLA by rearranging their OR matrix
T. PEREZ SEGOVIA

P3 - Comparison of the efficiency of control part organisation
M. OBREBSKA

P4 - A formalism to describe control part behaviour
M. NEMOUR , R. ETIENNE

P5 - Top-down design methodology of processor design, J.P.SHOELLKO

11 - BIBLIOGRAPHY

1. Lam Giang Dang, P.B.Ashkin, Randall Yee, M.O'Brien, A CMOS/SOS 16 bits parallel micro CPU, ISSCC Conf., 1977.
2. C.Erickson, H.K.Hingarh, R.Moeckel, D.Wilnai, A 16 bits monolithic I^3L processor, ISSCC Conf. 1977.
3. D.Staurn, D.Budde, B.Morgan, A single chip, highly integrated, user programmable microcomputer, ISSCC Conf., 1977.
4. G.Lowie, J.Wipfli, A.Ebright, A dual processor serial data controler chip, ISSCC Conf., 1977.
5. J.Beekmans, P.Danielse, C.Fernandes, L.Queré, F.Schiereck, A.Vermeulen, A complete 16 bits microprocessor on a chip, EUROMICRO Conf. 1978.
6. Masatoshi Shima, Demystifying microprocessor design, IEEE spectrum, july 1979.
7. J.McKevitt, J.Bayliss, New options from big chips, IEEE spectrum, march 1979.
8. F.Anceau, M.Marinescu, M.Henry, J.P.Potet, MSQ: a microsequencer as a monolithic component, EUROMICRO Conf. 1975.
9. L.Queré, F.C.Schiereck, An experimental custom LSI-mini computer in LOC MOS, EUROMICRO Conf. 1976.
10. A.Guyot, M.Henry, M.E.Vergniault, The MOM 400 single chip microcomputer, ESS CIRC 1979.
11. R.Hoshikawa, H.Kikuchi, S.Baba, S.Sato, K.Kawato, N.Inui, O.Wada, A 10000 gates CMOS LSI processor,ISSCC Conf.1980.
12. J.P.Schoellkopf, A top-down approach of design and fonctional description of hardware, Workshop on CHDL Dormstadt, august 1974.
13. J.P.Schoellkopf, Microprogramming: a step of a top-down design methodology, MICRO 7, Palo-Alto, october 1974.
14. A.Feller, S.E.Oza, Th.J.Lombardi, CAD VLSI design techniques and microprocessor applications, ISSCC Conf. 1978.
15. D.Johannsen, Bristle blocks: a silicon compiler, 16th Digital automation Conf., 1979.
16. A.D.Lopez, H.F.Law, A dense gate matrix layout style for MOS LSI, ISSCC Conf. 1980.
17. F.Anceau, Microprocessor world: programmed logic or microcomputer science?, EUROMICRO Conf. 1976.
18. D.Queyssac, Projecting VLSI's impact on microprocessors, IEEE spectrum, mai 1979.
19. F.Faggin, Trends in microcomputers,EUROMICRO Journal,11/1979.
20. B.Lattin, VLSI design methodology: the problem of the 80's for microprocessor design,16th Digital Automation Conf., 1979.
21. J.G.Posa, Microprocessors and microcomputers, ELECTRONICS, october 1979
22. F.Anceau, Evolution des microprocesseurs et son influence sur l'architecture des systèmes, Real Time Data Conf., october 1979.

A VLSI CHIP ASSEMBLER

J. Craig Mudge, Carol Peters, Gary M. Tarolli

Digital Equipment Corporation,
Massachusetts, U.S.A.

I. INTRODUCTION

A chip assembler is a data collection, manipulation, and storage program that aids designers of very complex (100,000 transistor) custom VLSI chips. The designers specify a chip as a hierarchical assembly of elements. These elements are called function blocks and have geometrical, electrical, and functional characteristics.

Such hierarchical assembly of data is the key component of the design style, or design methodology, which we have developed to deal with the exponentially increasing design time discussed in the first lecture (Figure 1). A chip assembler is simply a tool to support this particular design style.

The first section of this lecture revisits the methodology, describes its roots and singles out those aspects of design, principally intra-chip interfaces, for which the tool provides bookkeeping help.

Succeeding sections discuss interface management, provide examples of chip assembly, and describe progress in evaluating the chip assembler concept.

Structured chip design has three basic principles:

a. It is hierarchical

b. It emphasises the physical, not logical, domain

c. Designers, not CAD programs, place chip elements and their wiring

The physical hierarchy of a chip represents the geometric block nesting as shown in Figures 2-5. The functional hierarchy contains such items as ALUs, Datapaths, Programmable Logic Arrays (PLAs), and Register Files (Figure 6). A key component of our methodology is that physical and functional

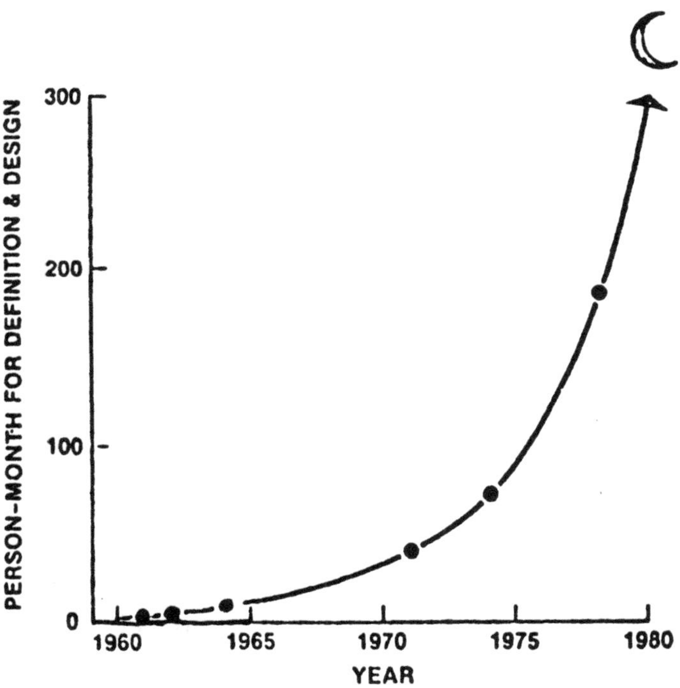

Figure 1.

(from Gordon Moore, CALTECH CONFERENCE ON VLSI, January 1979)

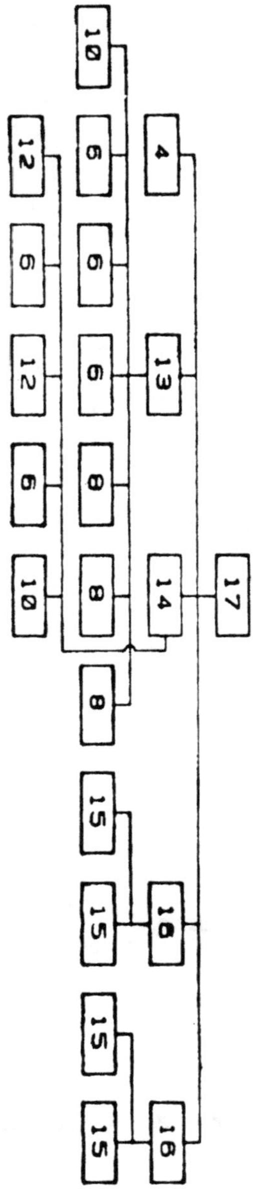

figure 2
Tree representation of physical hierarchy for chip
in Figures 3-5

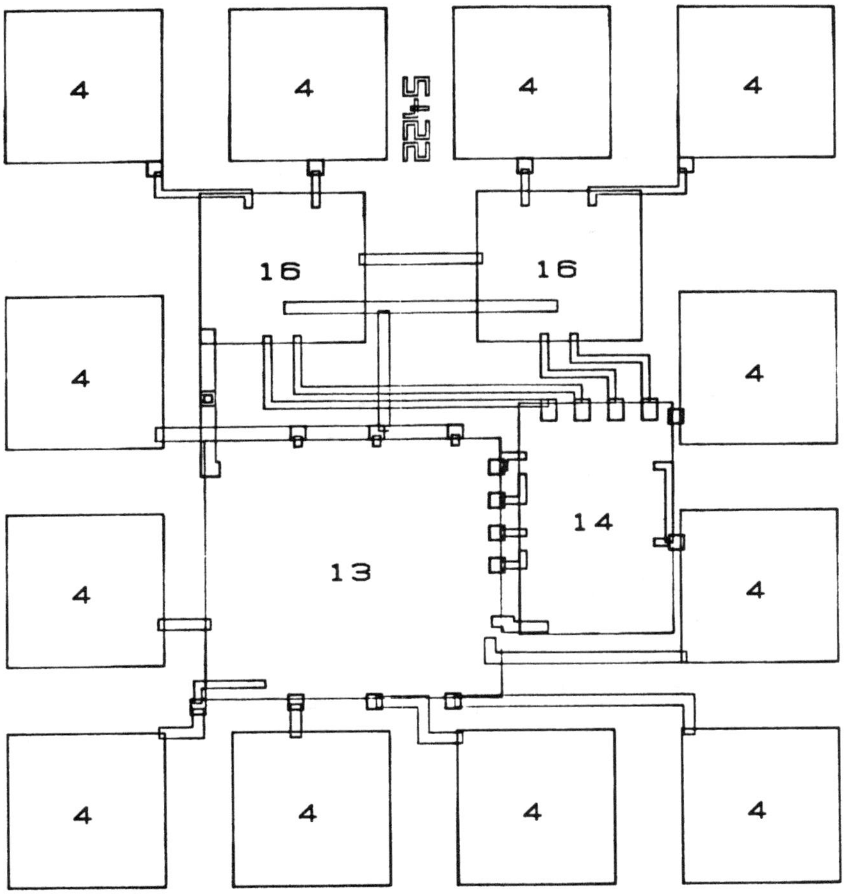

Chip plotted showing one level
of symbol nesting

Figure 3.

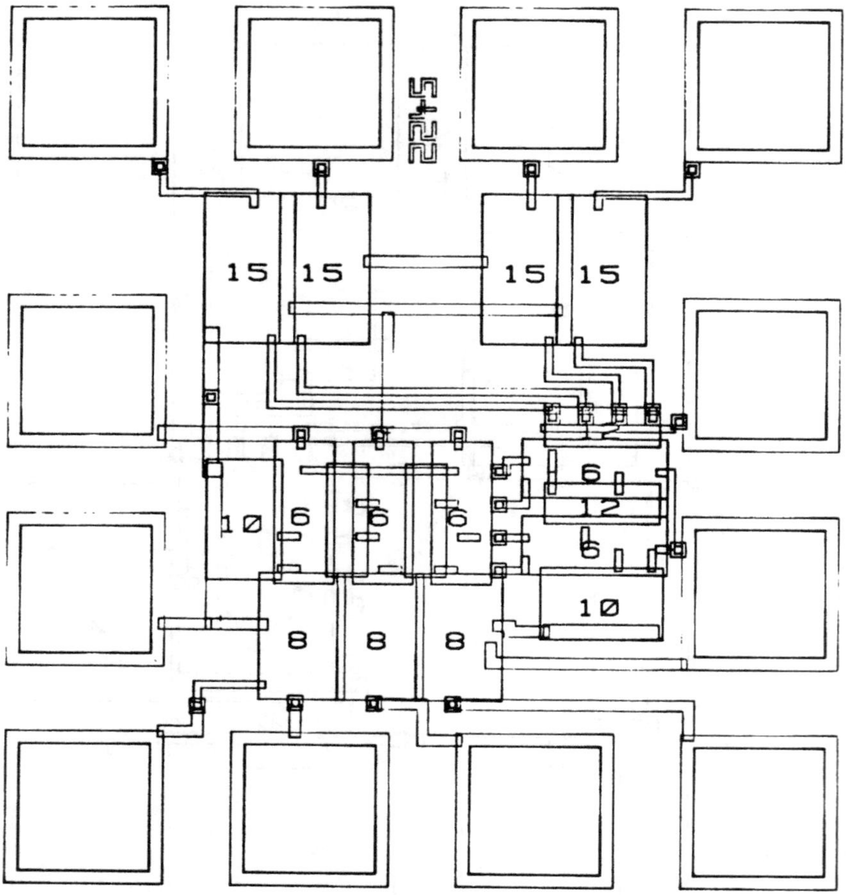

Chip plotted showing two levels of symbol nesting

Figure 4.

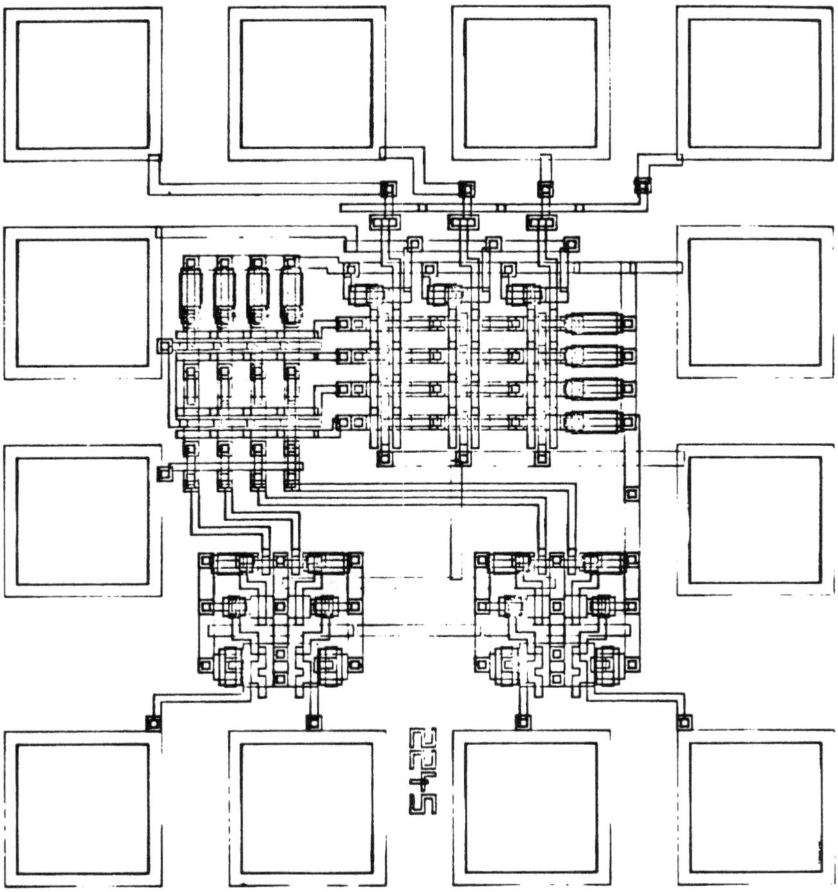

Fully instantiated plot of chip

Figure 5.

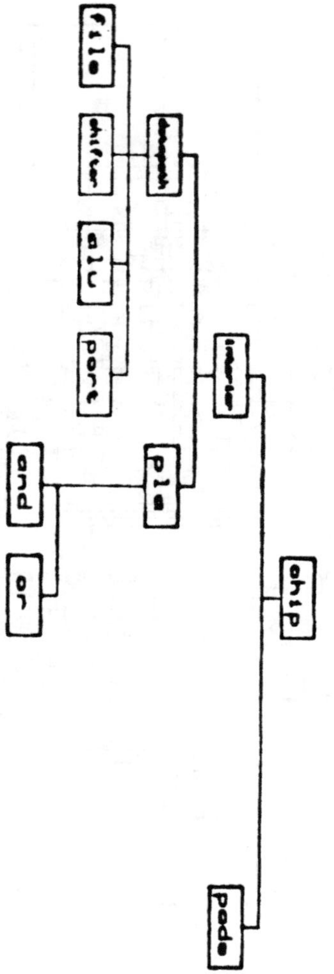

Tree representation showing functional Hierarchy of a chip

FIGURE 6.

hierarchies are identical.

Since the hierarchies are identical, well defined interfaces can be imposed between interacting blocks. This simplifies the design and conveniently partitions it so that different engineering teams can work independently on different sections of the design. The chip assembler supports structured VLSI design by taking advantage of the fact that the physical and functional hierarchies are identical. It integrates the functional and physical data for a function block into one database. This avoids problems that arise when two databases describe the same chip. One such problem is verifying physical and functional agreement. Another problem with two discreet databases is that create, update, delete operations are more complex.

We also strive for a high degree of regularity. In such design, the ratio of placed transistors to drawn transistors is very high. Thus a regular chip of 100,000 transistors may only have 5,000 drawn transistors: a ratio of 20:1. We are currently trying to achieve ratios between 50:1 and 75:1. Several techniques are used. We plan the wiring stategy first, then design the logic beneath the wires. We reserve low resist mask layers like metal for data, control, and power buses; and run these orthogonal to each other. We use structures such as PLAs and finite state machines, which are highly parameterized and can be computer generated. Computer programs translate logic equations into PLA layout geometries. Repeated geometries such as Register Files and wide datapaths also increase this ratio. Finally, we use microcode to supply control functions wherever possible.

Most of these principles have their roots in the work of Carver Mead at Caltech. The best example of the Caltech design ideas is OM2, a complete 16-bit microprocessor (externally microprogrammed) used as a non-trivial example of regular chip design in [Mead and Conway, 1979].

We are proceeding from the Caltech starting point, but see the OM2 as too limited a problem to demonstrate feasibility in the industrial environment:

a. OM2 is too small an example (9ØØØ instead of 1ØØ,ØØØ transistors).

b. We implement functions that are much less regular than the OM2 instruction set.

c. The OM2 architecture does not meet our performance goals. We design with more complicated circuit structures -- separate precharging of AND and OR planes of a PLA, for example.

d. Our product design environment permits many local optimizations and, hence, as the design is refined, many iterations of a chip's layout occur. The chip assembler addresses the iteration problem directly.

e. Total chip aspects are not addressed.

Our need to cope with functional irregularity and with a large variety of chip sections (datapath, PLA, microsequencer, RAM, clock generation, testability snakes) demands a flexible design support tool.

The universality of a programming language gave us the flexibility we required. However, because many of our hardware designers were not comfortable with programming, a simple command language has been implemented. For the (small number of) difficult parts of the layout where the command language fails, a programmer is available to the hardware designer to encode layout procedurally.

Moreover, because everything is represented as drawings and textual descriptions in a database, we have a common representation of the geometrical, electrical, and functional chip data no matter how a chip section has been designed and digitized. The designer requests various representations of this data in order to examine layout, analyze electrical characteristics of circuits, reconsider functional partitioning, verify the design, etc. The chip assembler produces the data in formats appropriate to its use: plotting, simulating, etc. Adding interfaces to new CAD tools is a straightforward programming task of reformatting output data or interpreting a new form of input data.

The chip assembler as a bookkeeper allows chip designers to iterate an evolving design many times. A chip always has layout data in the chip-assembler database - from the first plot (the floor plan showing only bounding boxes and interface bristles) through to the last check plot (the fully verified chip just prior to PG tape generation). Chip designers iterate in the following ways:

a. inserting detail in the bounding boxes as design continues: the design technique of step-wise refinement

b. intentional deferred sizing, e.g., width of power and ground buses

c. design refinement, e.g., modification of buffer sizes as parasitic capacitances are better understood

d. correction of mistakes (e.g., logic errors, circuit errors, layout errors, design rule violations, or chip overflow)

We concentrate our design-aid effort on interfaces because:

a. interfaces are the essence of physical hierarchy

b. interfaces are crucial to multi-person designs (and to complex single person designs)

c. interfaces are wires first and function second

The term "chip assembler" expresses two independent meanings -- one procedural and the other logical. The first refers to the procedure of assembling the parts of a chip. The second distinguishes it from "compiler"; it is a much lower-level aid (the designer makes more decisions) than a silicon compiler, such as Bristle Blocks and gains, thereby, tremendous flexibility.

II. INTERFACE MANAGEMENT

The chip assembler manages interfaces between rectangular function blocks. For example, we manage the interfaces between, and within, function blocks of the type shown in Figure 7, where the blocks are a RAM, a sense amplifier, a shift array, an adder, and two PLAs. There are about 2Ø to 3Ø such function blocks on a chip. Blocks nest, of course: each function block consists of many smaller blocks, the highest level block being the whole chip.

Each function block has four interfaces, one on each side (north, east, south, and west). Each interface consists of a number of bristles. The following data fully describes each bristle:

1) name

2) type and direction

3) color (mask layer)

4) width

5) capacitance

6) delays (estimated, minimum/maximum accepted, computed ...)

When assembling blocks with the chip assembler program, the designer responds to prompts that request the direction, alignment, and ordering of adjacent blocks. The chip assembler places the blocks and matches and connects interfaces. Figures 7 and 11 are examples of such an assembly.

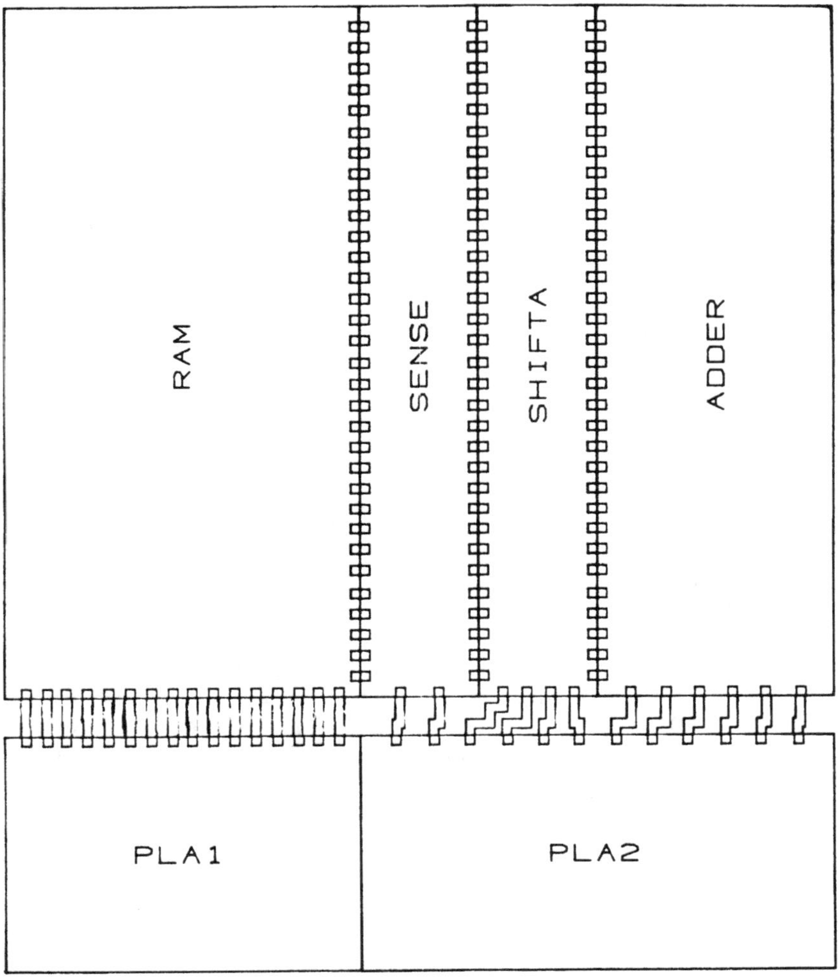

Function blocks assembled using chip assembler

Figure 7.

Interfaces are matched, i.e., connected, by one of four methods.

1. Manual pitch matching of bristles

Bristles with fixed coordinates, e.g., those between a ROM output and a microword latch, are matched by hand. In this case, the chip assembler only verifies that connecting bristles are of compatible material and function.

2. Automated stretching to match bristle pitches

In a datapath section each cell is stretched by a stretch coefficient determined by its neighbors. In the case of Y stretching (stretching along the Y dimension) in the RAM or shifter, the stretch coefficient is set equal to the cell having the greatest Y dimension so that all horizontal signal wires are straight lines. Figures 8 and 9 show a section of a datapath that has been stretched.

The control lines for the adder and the shifter originate in a PLA and run vertically. The PLA outputs (and associated buffers) are spaced horizontally by X stretching so that their pitch matches the pitch of the control inputs. Figure 10 shows an AND plane of a PLA that has been X stretched.

To maintain circuit integrity, only wires are stretched, i.e., lengthened; gate regions remain constant.

3. A routing subroutine

Whereas we use stretching to match blocks that are similar in size, we match large blocks that are disparate in size, or whose inter-bristle spacing is disparate, by a simple routing subroutine. To date the router is no more than a simple connector with no provision for cross-overs.

4. Routing by hand

For certain connections, the routing problem is too hard for the simple router to handle. Instead

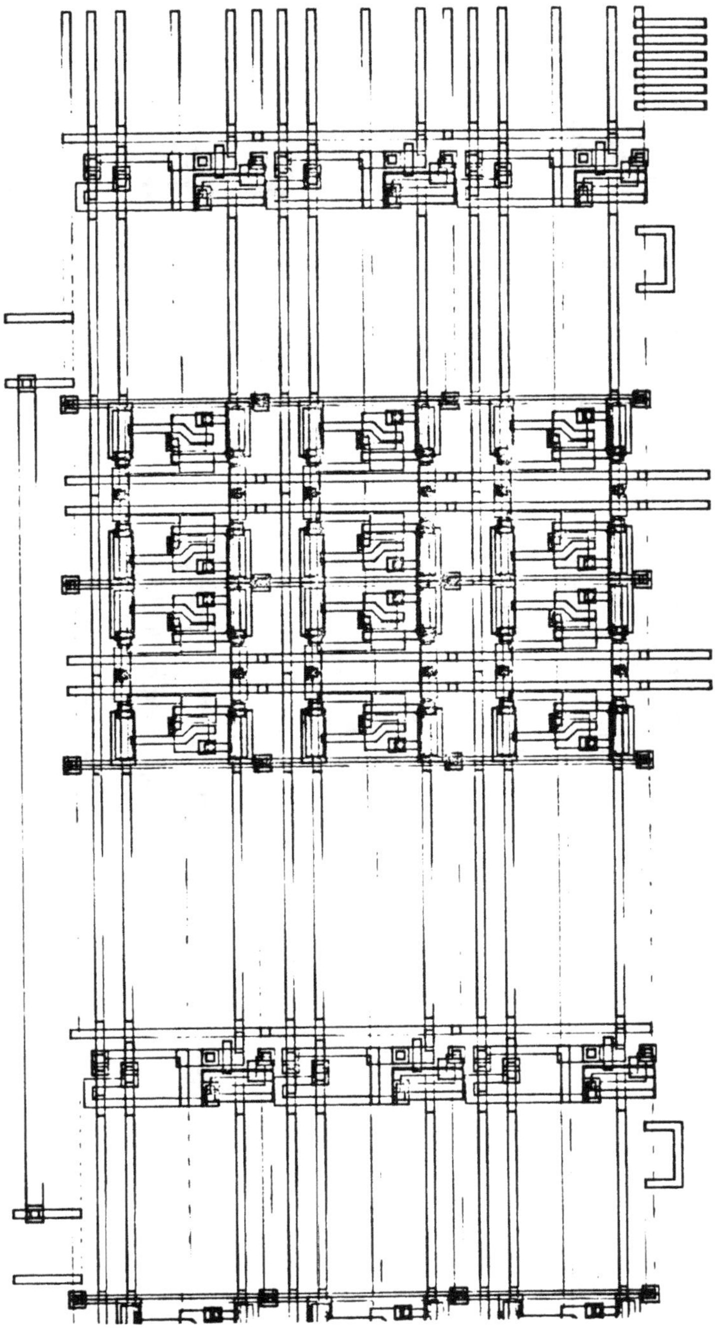

Unstretched datapath

Figure 8.

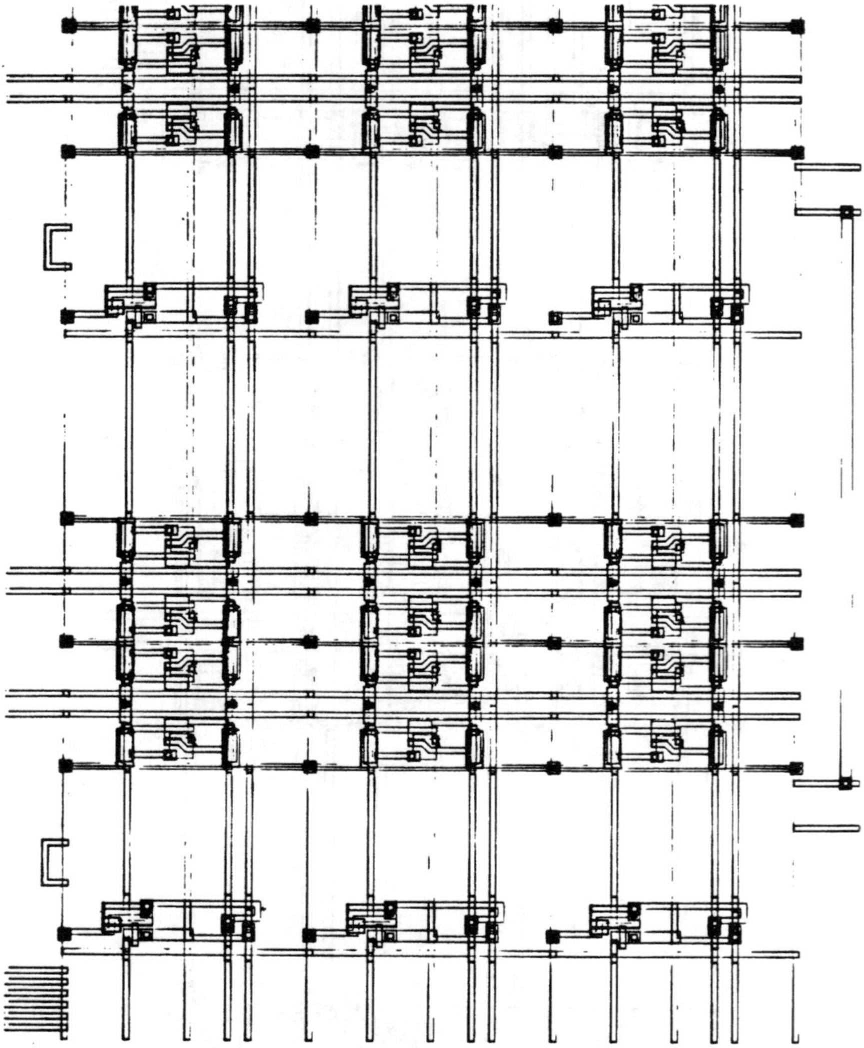

Stretched datapath

Figure 9.

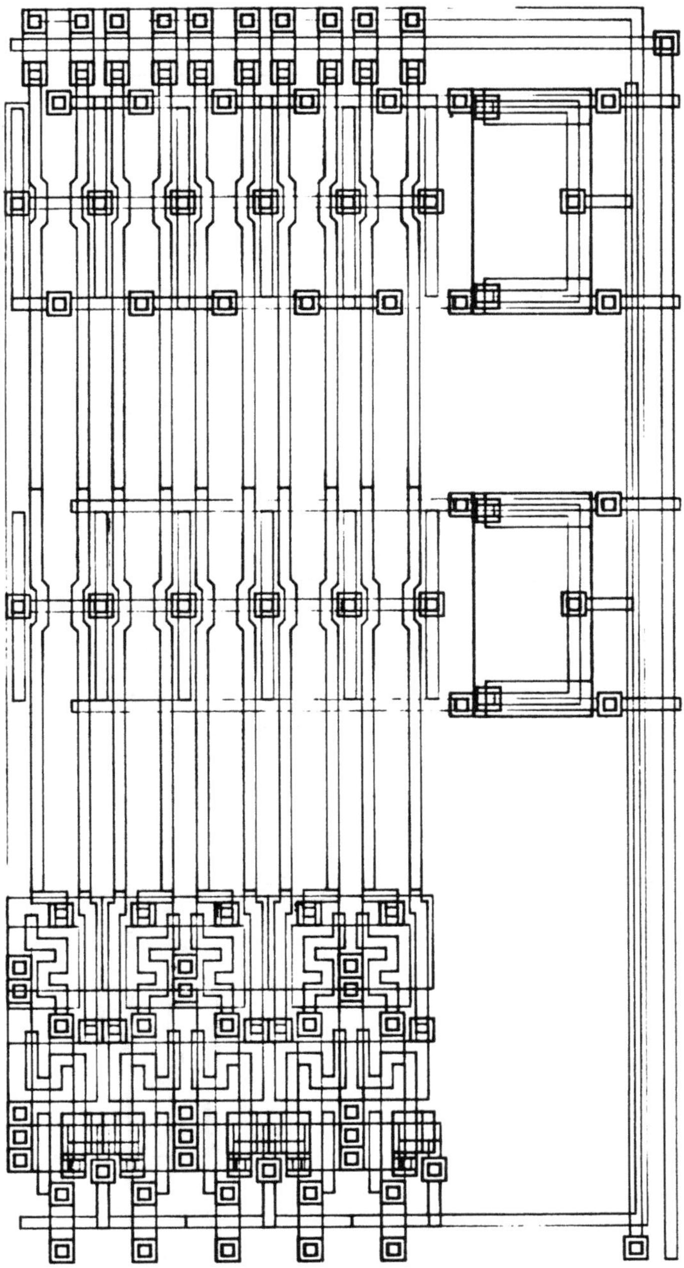

Stretched AND plane output

Figure 10.

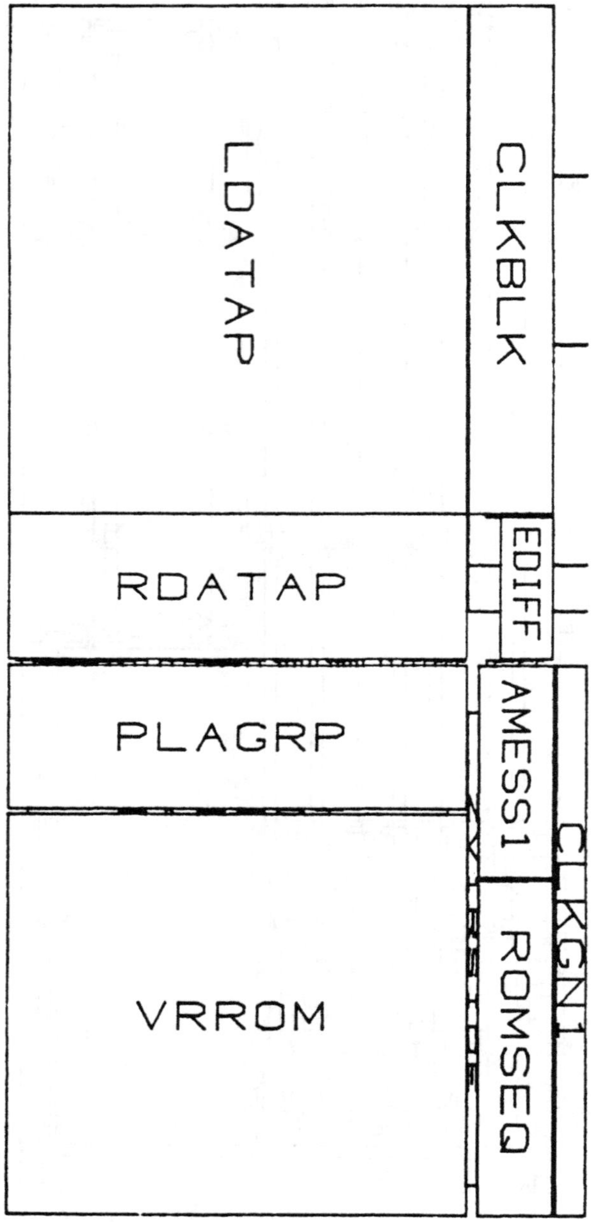
LDATAP
CLKBLK
RDATAP
EDIFF
PLAGRP
AMESS1
CLKGN1
VRROM
ROMSEQ

Figure 11.

of writing a general purpose router, we resort to manual routing, and represent the result as a block. This block, just like any other block, has a name, a block type (routing), and layout, electrical, and functional data. In the future we anticipate integrating a multi-layer routing program into CHAS to automate complex routing.

The concept of an interface manager is process independent. The number of layers of interconnect, the feature size, and the speed/power characteristics are vital to a particular chip design, but irrelevant to the concept of interface matching.

III. THE DESIGNER INTERFACE

Ideally, to preserve conceptual integrity, a chip would be designed by one person. This person would be able to architect floor plans, design circuits and their associated layout, express a design algorithmically, write well-engineered software, and document the result.

Since people who span such diverse disciplines are rare, we envisage a team that has four skills represented: circuit design, layout design, chip assembler programming, and chip architecture.

We suggest that each chip team would have one chip architect, one chip assembler programmer, and that we partition the circuit design and layout design so that several designers can work in parallel.

To encourage the chip design team to use the chip assembler, we have worked hard on creating a very approachable designer interface. The current chip assembler is a program named CHAS. The program is an interactive, menu-driven data collection program that assists the designer in specifying the layout and electrical properties of a chip. CHAS allows the chip design team to describe a chip floor plan of any shape and complexity. The designer specifies all details of the chip in a simple syntax composed of common English words. CHAS collects and stores this data for the designer in its database.

The central concept of CHAS is that the designer can describe a custom VLSI chip as nested rectangular blocks. Each block supplies one chip function. For example, the FPP block supplies the floating point processor function; the datapath block (a subblock of FPP) supplies the major operational function within the chip; the control block (a subblock of FPP) supplies the datapath control function.

To design a chip, the designer specifies rectangular blocks and then assembles these together to form larger blocks. The designer specifies the order of assembly and the orientation of placement. Blocks are placed in one of the following directions: North, East, South, or West. In any of the directions, there are two edges on which to line up the blocks. For North and South assemblies, you can line up the East or West edges of the blocks; for East and West assemblies, you can line up the North or South edges of the blocks.

After the designer has instructed the Chip Assembler to assemble the blocks, the Chip Assembler connects each block-pair together, creating the new layout and electrical data for the assembly. The Chip Assembler checks the bristles on each pair of edges that are to be connected. If the bristles match pitch, then the Chip Assembler physically abuts the blocks. Otherwise, the Chip Assembler routes between the two interfaces.

Following the discipline of regular-structure, the custom VLSI chip designer confines each functional block to a single physical segment of the chip. The implementation of logically-isolated functions does not spread throughout the entire physical space of the chip.

IV. AN EXAMPLE

The designer can generate each function block specification in a number of ways. For example, it could be created:

* in an interactive dialogue via chip assembler
* by extraction from a library
* by drawing the cell on an Applicon or Calma
* with a Sticks editor and compaction program
* by supplying parameters to a parameterized function block generator, e.g., PLA generator

The following chip assembler dialogue was used to create Figure 11:

```
@CHAS ! run the chip assembler program, first CHAS
      ! clears the screen, then CHAS displays  the
      ! following command menu



CHAS commands are:

    append    - to append data to a block
    assemble  - to assemble blocks
    create    - to create a new block
    exit      - to exit from CHAS
    help      - to display CHAS menu
    list      - to list a block's data
    make      - to make CIF or SLIC files
    plot      - to plot a tree of blocks
    read      - to read a block from a file
    save      - to save a block to a file
    setup     - to setup the I/O switches
    show      - to show the blocks defined so far
    status    - to display the I/O settings
    view      - to view blocks on a color terminal
                or plotter

CHAS:  create       ! begin left datapath creation
block name: ldatap ! CHAS displays the CREATE menu

CREATE commands are:

    exit        - to return to CHAS
    help        - to display CREATE menu
    layout      - to specify layout data
    simulation  - to specify simulation data
    sketch      - to sketch the rough outlines
    text        - to write a textual description

CREATE:  sketch ! we only want to sketch it,
                ! CHAS displays the menu
```

```
SKETCH commands are:

    electrical   - to sketch the electrical outline
    exit         - to return to CHAS
    help         - to display SKETCH menu
    layout       - to sketch the geometric outline
    text         - to write a textual description


SKETCH:  layout ! we want to sketch the layout
please enter a description of ldatap
width:  3128
height:  2352
north bristle count:  2
east bristle count:  20
south bristle count:  2
west bristle count:  20

SKETCH:  exit    ! all done, now lets assemble the
                 ! chip

CHAS:  assemble
name of combined block:  datap  ! the datapath
which dir. do you want to build (N/E/S/W):  e
align at north or south boundaries (N/S):  s
enter function block names, or type EXIT.

next block name:  ldatap        ! just created
next block name:  rdatap        ! stored on disk
reading block from file rdatap.fb ... done.
next block name:  exit
placing blocks horizontally ... done.

CHAS:  assemble
name of combined block:  ltop   ! the top left
which dir. do you want to build (N/E/S/W):  e
align at north or south boundaries (N/S):  n
enter function block names, or type EXIT.

next block name:  clkblk
reading block from file clkblk.fb ... done.
next block name:  ediff
reading block from file ediff.fb ... done.
next block name:  exit
placing blocks horizontally ... done.

CHAS:  assemble
name of combined block:  lside  ! the left side
```

```
which dir. do you want to build (N/E/S/W):  n
align at east or west boundaries (E/W):  e
enter function block names, or type EXIT.

next block name:  ldatap
next block name:  ltop
next block name:  exit
placing blocks vertically ... done.

CHAS:  assemble
name of combined block:  chip   ! the whole chip
which dir. do you want to build (N/E/S/W):  e
align at north or south boundaries (N/S):  s
enter function block names, or type EXIT.

next block name:  lside
next block name:  rside     ! assume we aready did
                            ! this and saved it
reading block from file rside.fb ... done.
next block name:  exit
placing blocks vertically ... done.



CHAS:  show

ldatap           rdatap        datap        clkblk
ediff            ltop          lside        rside
rside            chip

CHAS:  save
block name:  chip        ! now save it
saving block chip ... done.
saving block lside ... done.
saving block datap ... done.
saving block ltop ... done.

CHAS:  EXIT     ! all done
```

V. THE NEXT STAGE OF IMPLEMENTATION

We intend to concentrate on the following topics during the next 3 months of the FPP experiment (the experiment is detailed in the following lecture):

* expand the chip assembler's command capabilities:

  add expression evaluation and relative size/location specification to the chip assembler's interactive command language

  simplify and augment designer access to the existing chip database

  develop alternative graphical views of chip layout, such as circuit schematics of cells, interconnect-only plots, selective masking/display of block internals

  collect, store, and compute power consumption and timing statistics

  record inter-block layout dependencies in the database so that the designer can change one block's layout, and ask the chip assembler to recalculate the layout of all geometrically dependent blocks

* verify that layout obeys process design rules, meets timing requirements, and fulfills interconnect intentions:

  verify that routed or abutted bristles share the same physical material and bristle type

  compare early timing estimates with simulation results and cross-chip delay calculations to assess whether the chip will work

  produce network lists of bristle connections between blocks, and use these lists to verify that the chip's physical implementation matches its logical specification

* integrate existing CAD tools into the chip assembler:

  accept cell layout from Applicon or Calma systems, and send appropriately translated block layouts to such systems for graphical editing and resubmission to the chip assembler

produce circuit simulator input files, for individual blocks and for assembled blocks, with designer control over the electrical detail supplied for each block

integrate designer-synthesized circuit simulation output into the chip assembler database

run simulations, design rule checks, critical path analysis, etc, under direct control of the chip assembler

VI. EVALUATION

It will be several months before we complete the FPP experiment. Therefore, data on design time, silicon area, chip performance, power, and CHAS performance is not yet available. However, we can report on some user experience to date.

So far in the chip experiment, the designers have used CHAS in the following manner:

* Specified outer blocks such as clocks, datapath, PLA, ROM, microcode sequencer; for each such block, the designers estimated size, shape, and a number of connections (called bristles) on each block.

* Assembled these outer blocks, in a designer-specified order, into an outermost block named FPP; CHAS abutted or routed between blocks depending on whether adjoining blocks had equal or unequal bristle pitches.

* Replaced large blocks such as the datapath or PLA with assemblies of inner blocks, such as, in the case of the datapath, shifters, sense amps, RAM, etc.

* Recombined blocks in various patterns, sometimes adding or subtracting blocks, to improve space utilization and to minimize wiring.

* Defined lowest-level blocks to represent individual circuits (called cells): for each lowest-level block the designers specified as much geometric and electrical detail as was available at the time.

* Assembled lowest-level blocks (representing cells) into blocks consisting of rows of cells, and reassembled outer blocks by replacing roughly-defined blocks with the new detailed cell-row blocks.

CHAS checks pitch matching between connecting bristles during each block assembly, and chooses block abutment or inter-block wire routing as appropriate. All routing is single-layer river routing, and the routing wires are placed according to 2-micron process design rules.

As the design progresses, the designers ask CHAS for video or hard copy plots of all or part of the chip. The plots include optional block labels to aid in block identification. The designers use plots both as visual records of current layout, and as assists in evaluating chip layout, wiring, functional breakdown, etc.

The designers occasionally ask CHAS for hard copy plots of a tree-shaped representation of the block nesting on the chip. These tree plots give the designers a simple view of the chip's functional partitioning.

The FPP layout designers are unenthusiastic about using cell stretching. They are extremely successful in achieving maximally compacted cell layouts, and believe that when a cell layout does not match the space requirements, that they should relayout the cell to meet those requirements. This approach sounds too time-consuming, but we have a good group disagreement on the issue, which should provoke testing of both viewpoints.

ACKNOWLEDGEMENTS

The use of Y stretching for interface matching in a datapath chip section comes from Dave Johannsen's Bristle Blocks system. Discussions with Ivan Sutherland helped us see that the unifying theme of the chip assembler is interface management. As chip assembler users, Bill Herrick and Marge Tait have made invaluable suggestions.

REFERENCES

Mead, Carver and Conway, Lynn, "Introduction to VLSI Systems", Addison-Wesley, August, 1979

A SINGLE-CHIP FLOATING-POINT PROCESSOR
- A Case Study in Structured Design

J. Craig Mudge, William V. Herrick, Hank Walker

VLSI Advanced Development
Digital Equipment Corporation
Hudson, Massachusetts USA

A SINGLE-CHIP FLOATING-POINT PROCESSOR
- A Case Study in Structured Design

J. Craig Mudge, William V. Herrick, Hank Walker

VLSI Advanced Development
Digital Equipment Corporation
Hudson, Massachusetts USA

I. GOALS OF THE EXPERIMENT

We have undertaken the design of a 100,000 transistor floating-point processor chip as an experiment to test our emerging structured chip design methodology, the chip assembler which supports it, and our process technology.

We chose floating point for three main reasons:

1. An FPP chip can be self contained.

   In a multi-chip-set implementation of a conventional computer, a reasonable chip-set partitioning places floating-point arithmetic in a self-contained chip. We assumed such a partitioning.

   Design decisions for the experimental chip could then be made independent of global chip partitioning issues such as instruction fetch strategy and chip-set control strategy.

2. The task is a real, not a toy, problem.

   Floating-point instruction sets deal with a complicated data type and usually have 30 to 40 instructions. Typical MSI implementations have used four or five hundred chips.

3. Floating-point arithmetic is inherently irregular

   Because of its inherent irregularity, we felt that floating-point arithmetic was the hardest test of our structured design methodology.

Three examples suffice to show irregularity: (1) The datatype has four parts: exponent, fraction, sign of the exponent, and sign of the fraction. (2) Floating-point addition is a complex operation: it requires an exponent subtraction to determine the amount of alignment preshift and after fraction addition, normalization postshift generally occurs. (3) To add further complexity, we chose to exploit several algorithm optimizations possible in floating point addition: frequency distribution of floating point operands show that both pre-shift alignment and post-shift normalization are highly skewed towards low numbers of shifts, and preshift behavior can predict post-shift behavior.

The chip is being designed to use a 2 micron n-channel Si Gate MOS process. The goals of the experiment are to measure improvements in design time and penalties, if any, in chip area, power, and performance from the use of our structured design methodology.

Our experiment is aimed at understanding VLSI complexity. The increase in complexity in going from TTL board-level to chip-level systems is at the circuit and layout-level, and not the logic level. Consequently our CAD tools work very well on logic correctness in chip designs, but not so well on circuit and layout correctness. For this reason the chip assembler concentrates on these two areas while integrating existing logic simulators and checkers. Since one of the goals of our experiment is to test our tools, we emphasize layout and circuit correctness over logic correctness. In fact, no logic simulation will be done, nor will most of the microcode be written.

II. CHIP ARCHITECTURE

The chip implements all instructions of the VAX-11 floating- point instruction set except for MOV, MNEG, CLR, and parts of POLY and ACB. All datatypes, including single, double, extended double, and quadruple precision, are implemented.

The chip acts as a self-contained execution unit, with its own on-chip microcode and all other logic necessary to execute each instruction once given the operands. A master chip in a hypothetical chip set loads operands into the temporary registers, and an instruction into the FPP instruction register (FIR) and then signals the FPP to start execution. When the instruction has completed, the FPP signals that it is done, and the master chip unloads the results.

One design alternative considered was to keep a copy of the general-purpose registers on the FPP. This avoids the delay of loading and unloading operands for register to register operations, which compose half of all floating-point operations. This delay represents as much as 40% of the time for addition and subtraction. A register file copy requires either a separate bus to keep the copy updated whenever another chip modifies its registers, (which is difficult because of pinout limitations) or a smaller bus to mark registers invalid. This second alternative is feasible, but only results in a performance increase if the same registers are used for sequences of floating-point instructions that immediately follow one another. Intervening integer instructions would invalidate the FPP registers, requiring an off-chip fetch for the next floating-point operation. Unfortunately simulations show that this is often the case. A problem with maintaining the VAX registers on-chip is that they have a different format than the FPP internal floating-point format. This requires a conversion to take place whenever they are used. A comparison between the conversion time, off-chip operand fetching, and instruction frequencies and lengths leads to the conclusion that the potential performance advantage is not worth the added complexity. Moreover, pipelining techniques can hide most of the costs associated with moving operands between chips.

The FPP function blocks can be seen in Figure 2. The control structure consists of a ROM and microsequencer which control PLAs which generate datapath control signals. The PLAs handle the data-dependent operations, and can be thought of as subroutines called by the microcode. Off-chip microcode can be inserted at the ROM output if the 32K bits allocated is insufficient. The datapaths contain multiplexing logic for format conversion between on-chip and off-chip formats; a register file for temporary and constant storage; two right/left shift registers for accumulators, multiplication and division shifting, and normalization; a 4-bit shifter for alignment during addition and subtraction, and multiplication; an ALU for addition and subtraction for all algorithms. The exponent path uses its shift registers only as accumulators. The latch replaces the shifter. It is used during quadruple precision operations.

The chip is designed to be easily testable. All internal states are made easily visible by requiring all flip-flops to be parallel load shift registers [Eichelberger and Williams, 1977]. These shift registers allow the ROM, microsequencer state, control registers, temporary register files, shift register accumulators, and PLA contents to be dumped off-chip. These pieces of the chip can be tested independently, greatly reducing the combinatoric explosion of test vectors during low-speed functional test. The connections of the test shift register are detailed in the section on chip layout.

III. ALGORITHMS

The algorithms for addition and subtraction are quite similar. The two operands are loaded into the shift registers from the file. The exponents are subtracted. The EDIFF PLA looks at the result. The exponent difference could be so large that the preshift would shift away all significant bits. The larger operand is the result stored. The difference could be within the range of significance, but require shifting more than 4 bits. This is done by multiple cycles through the shifter. The last possibility is that the difference is less than 5, so the preshift alignment can be done in one step. The smaller operand is shifted, and then added or subtracted to the other operand. This is the case two-thirds of the time. The result may have to be normalized, since all results are stored in normalized form. The exponent is adjusted accordingly. The fraction is then rounded. Renormalization may be needed. This algorithm, along with guard bits, guarantees that the result will be accurate to half a least-significant bit.

Multiplication is done using a modified Booth's algorithm with a two-bit fixed shift. This multiplies the fractions together two bits at a time. The multiplicand is loaded into a temporary register. The multiplier is loaded into a shift register. The two least-significant bits plus the previous bit of the multiplier are examined to determine whether the multiplicand should be added or subtracted to the partial product, which resided in the other shift register, and whether the multiplicand should be multiplied by two. The partial product is shifted with the shifter. The shift register shifts the multiplier in two cycles. Rounding and normalization occur at the end of the multiplication loop just as in addition and subtraction.

Division is done by nonrestoration. This retires one bit per cycle, with no losses due to wrong guesses. The divisor is loaded into a temporary register. The dividend is loaded into a shift register which will contain the partial dividend. The divisor is subtracted from the partial dividend and a one is recorded in the least-significant bit of the quotient register, which is the other shift register. The sign of the previous result determines whether an addition or subtraction will occur on the current cycle. This also determines what bit will be recorded in the quotient register. Normalization and rounding occur at the end of the division cycle.

The above algorithms are very similar for single, double, and extended double-precision datatypes. Quadruple precision operations require that the fraction datapath be double-pumped, because it is only 67 bits wide instead of the 116 bits required.

IV. PERFORMANCE

Estimates of the chip's performance place it somewhere between the FP11-C and FP11-E floating-point accelerators used on the PDP-11/70 and PDP-11/60. The datapath can perform a register to register 64-bit ALU operation in about 140 ns. PLA propagation time is about 70 ns. This results in an add time of about 1.4 us. This performance is comparable with floating-point accelerators containing 400 74S-type integrated circuits. This is substantially better than the 8087 floating-point coprocesser.

V. STRUCTURED LAYOUT DESIGN

The simplified first-pass layout plan is shown in Figure 1. One immediately notices the large chip aspect ratio of 2 to 1. The chip was more than 400 mils long. Packaging limitations precluded this. Notice also that the fraction and exponent paths reside side by side with their control signals coming from the PLAs. The signals for the fraction path must run over the exponent path. Simulations showed that poly control lines were unacceptably slow for such long paths. This led us to experiment with materials of much lower resistivity. Sample cell layouts were also done for this floorplan. These were used to generate density estimates and rules of thumb for parasitics. They also helped identify which process features should receive higher priority.

Primarily because of chip aspect ratio problems, a new floorplan was generated, and is shown in Figure 2. The following remarks apply to this floorplan.

The floorplan shows the exponent and fraction datapaths separated, with the PLAs in between. This shortens the control lines and avoids running fraction control lines over the exponent path, and places the PLAs in a more convenient position with respect to the ROM microcode output. Some of the busing area, such as the microcode bus, is placed on top of the PLAs. Combining the PLAs, rather than separating them on the exterior of the chip,

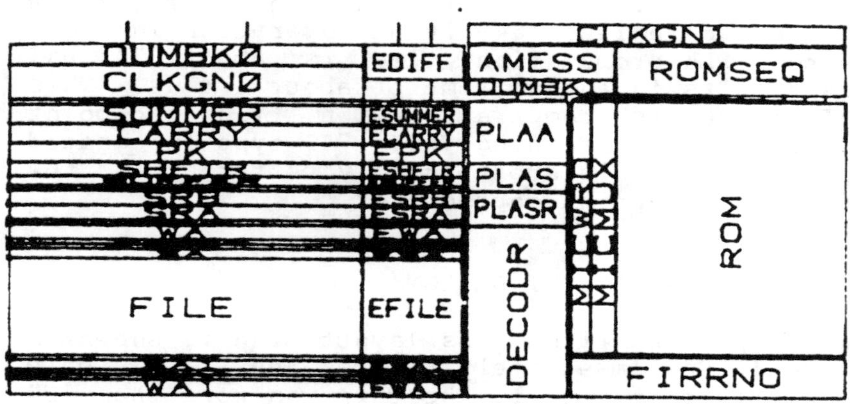

Figure 1

Initial layout plan for floating-point processor chip. 83-bit wide datapath (67 for fraction, 16 for exponent) with data bus wires running vertically occupies the left side of the chip, microcode ROM and control PLA's the right side. Sequencing logic, machine state, and clock logic are at the top of the chip, data multiplexing and formatting on the bottom. Control lines in the datapath run horizontally form the PLA's. Several factors contribute to the high degree of regularity: fixed pitch cells in the datapath, control residing in PLA's and ROM, and all PLA's instantiations of one general-purpose primitive. Random wiring is restricted to 3 interior channels and pad routing.

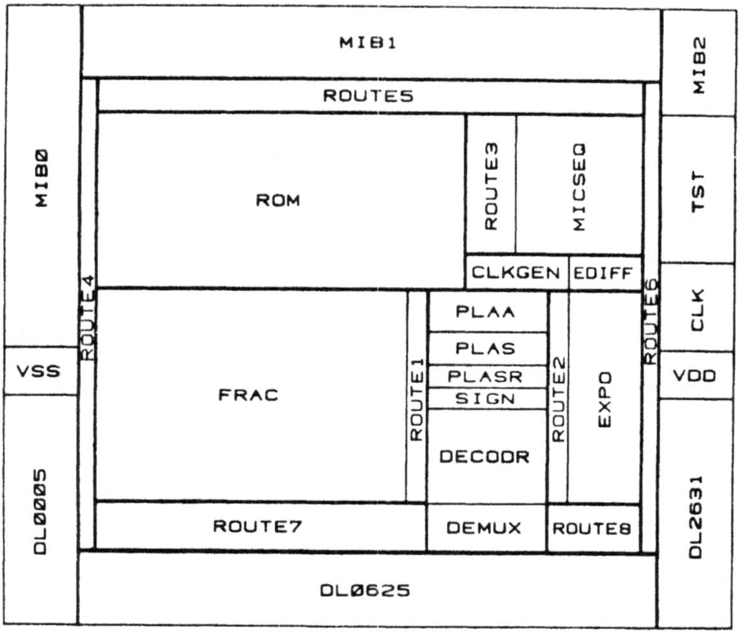

Figure 2

Final layout plan for floating-point processor chip. 83-bit wide datapath, split into a 67-bit fraction and 16-bit exponent path, with data running vertically. Datapath control PLA's reside between the two halves of the datapath along with 2 routing areas. The microcode bus runs above the PLA's. The microcode ROM, routing area, and microsequencer occupy the top of the chip. The core is surrounded by the pad ring and routing area. Separating the datapaths simplified control routing to the datapaths, from the ROM to the PLA's, from the MIB pads to the ROM, and from the datapath to the DAL pads.

allows us the option of combining some terms used by both the fraction and exponent paths.

Pad locations were defined. The MIB bus carries instructions and microcode, so it was placed at the top of the chip near the ROM and microsequencer. The DAL bus carries data to and from the datapaths, and so was placed on the bottom of the chip next to the multiplexing logic. The external clock pins were placed on the right side of the chip, not far from the clock generators. The test pins were placed next to the clock pins since their logic is associated with them.

The clock logic was moved to be in a central location, close to the PLAs and ROM, where most of the chip control signals originate. The ROM size was reduced to 32K bits to reflect a more accurate size estimate. The sign processing PLA was explicitly shown.

The radical revision of our initial floorplan provided a first test of the chip assembler. Note that most of the cells in the datapaths, ROM, PLA's, and pads remain unchanged. Because of this, little new layout was actually required. In one day, a technician took our blackboard drawings of the new floorplan and wiring strategy and had entered it into the database and generated a new plot.

Several techniques have been used to obtain a regular design. We describe the four most important below.

1. Global Wiring Strategy

   All data signals run north-south and all power and control signals run east-west in the datapaths. Microcode from the ROM is carried to the PLAs by a bus that runs above them.

   All interfaces between blocks is done by abutment (defined in the previous lecture on our Chip Assembler) except for four routing areas. These four areas can be seen in Figure 2. A section next to the ROM carries the ROM outputs to the PLAs, and wiring from the microsequencer to the OM address decoder, the clock generators, the PLAs, and the datapaths. Two sections next to the PLAs carry routing from the PLAs to the datapaths, and to the microsequencer, clocks, and inter-PLA signals. The last section surrounds the chip core and carries wiring from the pads to the datapaths, ROM, microsequencer, and clock generators.

The routing in these four channels is done automatically using programs similar to those used for PC board routing. Global wiring is often a source of error during layout. By confining this wiring to a few areas, the problem is simple enough that a program can generate the layout automatically once given the geometrical constraints and netlist.

Datapath cell interfaces are matched by abutment, maintaining a constant cell pitch throughout the datapath. Layout designers find this fixed width constraint as easy to work with as freer form styles. Some of the PLAs are designed so that they also abut, such as the EDIFF PLA.

The test shift register runs as one long shift register through the chip. It begins on the right side of the chip, goes through the microsequencer registers, the ROM output pipeline register, down the left side of the PLAs to the fraction shift register, through the shift registers, back out to the PLAs, down to the bottom of the PLAs, up the right side of the PLAs, into the exponent shift registers, back out to the PLAs, and up to the EDIFF PLA, and back out to the right side of the chip.

2. Parameterized Control Primitive

Control signals for the datapaths originate from the four PLAs between the exponent and fraction datapaths: PLAA, PLAS, PLASR, and DECODER. To increase the chip's regularity, each of these blocks is an instantiation of a parameterized, general control primitive shown in Figure 3. The PLA labelled EDIFF, at the top

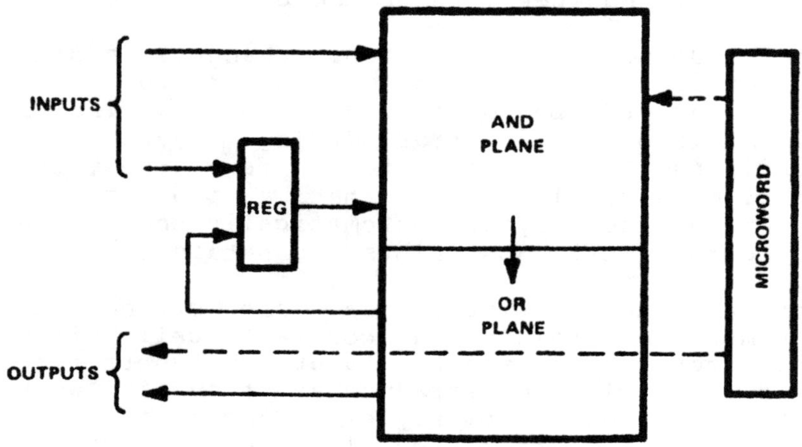

Figure 3

Parameterized PLA control primitive. Basic PLA used to generate all control signals. Parameterized to adjust number, locations, and timing of inputs, outputs, feedback terms, and microcode inputs. Presence and positions of AND and OR planes is also programmable.

of the exponent datapath, generates control state from the exponent difference of two floating point numbers; this is also an instantiation of the general control primitive.

The general primitive has two parts: a PLA with optional feedback register and microcode signals from the chip's ROM. Inputs to the AND plane come from up to four sources: datapath; microword; feedback register; or from elsewhere, e.g., from the EDIFF section. Control outputs come from the OR-plane or directly from the microword (the dotted lines in Figure 3 which pass over the PLA). Note that the PLA in this figure was originally designed for the first layout plan, so the microcode inputs are shown coming from the right. In the actual layout, the microcode bus runs above the PLAs, and connections are made directly into the interior of the PLA from the bus whenever necessary. This represents a minor change in the primitive's design.

The parameters to this structure are the number of inputs, outputs, feedback terms, microcode inputs, product terms and geometric spacing. Locations of these are also specified. Each of the five instantiations is different. EDIFF has no register, and its pitch matches that of the EALU. PLASR has a register which holds a shift count for multiplication and division. The decoder is just the AND-plane of the array. PLAA, controlling the ALU sections of the datapaths, has no feedback register.

3. Regularity by discipline

One example is forcing the exponent and fraction parts of the datapath to use identical cells. The only differences occur at the edges of the datapaths, such as carry-ins, or examining different bits of some registers. We are also using a simple clocking strategy that may cost us some performance, but which reduces the amount of clock logic required. This logic usually results in very random layouts. The control structure was designed as a two-level combination of ROM and PLA microcode. The data-dependent operations are separated out into the PLA's, which reside close to the datapaths. This code is called as subroutines by the ROM microcode. This structure retains the performance advantages of horizontal microcode and the ease of programming of vertical microcode.

4. MICROSEQUENCER, a catch-all section

The ROM microsequencer and miscellaneous control logic which could not be easily be mapped into the preceding regular structure, such as machine state, is collected in this section. Some of this logic includes condition-code settings, floating-point datatype, error flags, and multiplexing of some of the micro-branch conditions.

One measure of the regularity of a layout design is the number of actual to drawn transistors. Structures such as ROM's, RAM's and PLA's have a high degree of regularity because arrays of thousands of cells only contain one basic drawn cell. This measure tends to favor designs that are microcode intensive. For example, if the amount of microcode on our chip increased from 32K to 50K bits, then the regularity would increase according to our measure. Would this design be any "better". This is obviously not so since more area and power is required to accomplish the same function. The measure of drawn transistors also breaks down on chip wiring. The routing areas of our chip occupy a significant amount of the area, but they contain no transistors, so they do not enter into the regularity measure. Conceivably a chip could be designed with only a ROM surrounded by a huge amount of wiring area. This chip would have only a few drawn transistors, but is obviously an irregular design. The addition of a density measurement, and total chip area measurement will detect these types of irregularity. Our chip will have an actual to drawn transistor ratio of about 75. We will use about 100k transistors to accomplish the task which takes about 50k of microcode plus datapaths in TTL implementations. Our density figure will be about 1.2 transistors per square mil (about 500 square microns per transistor).

When discussing structured design, hierarchy is often mentioned as a useful tool for managing complexity. Our experience shows us that at current chip transistor counts, the number of levels in the hierarchy is quite small, as can be seen in Figure 4. The figure only shows the upper levels of the tree. For example, the box labelled "frac" represents the fraction path. The next level down in the fraction path would be the register file. The level below that would be a row in the file, and then a RAM cell, so there are three more levels which are not shown, for a total of eleven. There are several reasons for this. The first is that the bottom-level cells are not as small as they could be. The minimum-size cell is something like a RAM cell. This has to do with the number of concepts that humans can comfortably deal with at any one time. There are about 11 levels of hierarchy on the whole chip because the branching ratio of the tree is high, varying between 2 and 1000, but probably averaging about 20. The higher branching ratios occur at the bottom of the hierarchy, such as in ROM or RAM arrays, while the top levels of the chip only branch by factors of three or so. If this branching rate continues for more complex chips, then chips containing hundreds of millions of transistors will have fewer than fifteen levels of hierarchy.

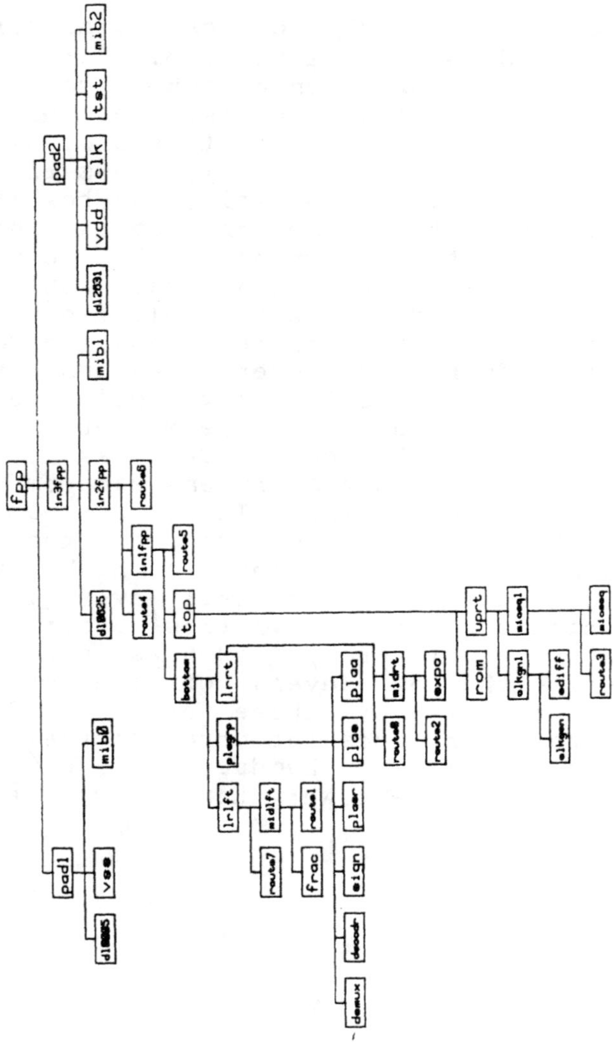

Figure 4

Hierarchy tree of floating-point processor layout. The tree starts at the top with the entire chip, and then breaks it down into pads and core, and then further into datapaths and PLA's. Not shown is the bottom of the tree such as datapath registers and RAM cells.

REFERENCES

[Eichelberger and Williams, 1977] Eichelberger and Williams: "A Logic Design Structure For LSI Testability", Design Automation Conference Proceedings, 1977.

GENERAL PURPOSE V L S I COMPONENTS FOR DISTRIBUTED COMPUTING SYSTEMS

Carlo H. Séquin

Computer Science Division
Electrical Engineering and Computer Sciences
University of California, Berkeley, CA 94720

ABSTRACT. *Current trends in the design of VLSI chips are analyzed to explore what kind of general-purpose components might be useful for the construction of very large, modular, digital computing systems in the mid 1980's. One such building block is a general-purpose communications component that permits to link together a large number of computers in closely coupled networks of arbitrary topology. Computation and communication are handled by separate processors, so that both can be performed in parallel.*

1. INTRODUCTION

Over the last few decades computing demand has always stayed ahead of available computing power, and there is no reason to assume that this trend will end in the near future. Similarly, individual integrated circuits become ever more powerful, but there will always be customers who want to have computing power substantially beyond the capabilities of a single VLSI chip.

If a system is to be extended over several integrated circuits, the question arises how it should be distributed onto the various chips. This partitioning is crucial since the connections between chips represent inherent communications bottlenecks. Firstly, the number of interconnection points on a VLSI chip is limited and increases with the years at a much slower rate than the number of gates in the circuit. Secondly, with present-day packaging technology, the bandwidth through those pins is limited. The physical paths leading from one chip to the next possess parasitic capacitances that are one or two orders of magnitude larger than that of on-chip signal paths; an impractical amount of power would be required to achieve external bit rates comparable to those that can be obtained on internal interconnections.

It is thus important to define proper functional blocks to be integrated onto individual chips. For instance, the partitioning must not cut through the high-bandwidth path between main memory and the data processing circuits. Thus memory and processor should stay tightly coupled on one and the same chip. VLSI technology in the mid 1980's, will make it possible to put a sophisticated processor and a substantial amount of "main" memory onto a single integrated circuit.

2. GENERAL PURPOSE V L S I COMPONENTS

No matter how fast, convenient, and sophisticated VLSI design tools will become, standard off-the-shelf parts will always be in strong demand for people who want to put together experimental systems or who want to bring a product to market quickly. They cannot afford to start from scratch and to determine in all details, how the optimum chip for their purpose could be constructed. For small and medium volume production, it is far more cost effective to pick a component from a limited selection of existing parts. If these parts have already been debugged and if applications manuals are available, the user need not worry about all the critical details of internal timing and interconnections. It normally pays to choose a somewhat overdesigned component, using more circuitry, power or silicon area than absolutely necessary, thus guaranteeing that the requirements of the application will be safely met.

The following questions thus arise: Is the concept of general purpose components still viable in the realm of VLSI? What are the circuits that can be placed on the shelf and will prove useful in a large number of applications? Within the framework of digital computing systems, what kind of chip will take on the former role of NAND gates, flip-flops or registers? Simple extrapolation of existing trends may not be the right answer for general purpose components with a hundredfold greater functional capability.

Memory chips with four times more storage capacity now appear on the market about every four years. Certainly, a 64k-word by 1-bit chip is a nice part to have, but will a 1M-word by 1-bit chip be equally attractive? - Or would most customers rather use a 64k-word by 16-bit part? Also, from the point of view of optimal systems partitioning, discussed in the previous section, a simple memory chip is a very poor component. It has virtually no internal interactions, and every time it is used, information has to go in and out through the package pins. These shortcomings should be remedied by adding processing power onto the memory chip itself.

Looking at the field of random logic, the current trends are more obscure. There is a certain evolution from the NAND gate to more complicated Boolean functions such as multi-bit logic functions, parity checkers, arithmetic functions, or priority encoders. However there is a limit to the size of useful logic functions that are commonly used. The trend is to add input or output registers to the logic functions to permit easy assembly of synchronous machines. Here again, we find a mixture of processing and storage elements.

The most complicated parts available today are the single-chip microcomputers with a certain amount of on-chip memory. But can these chips really be regarded as general-purpose, modular building blocks in the sense of a NAND gate? While one can prove that all computing systems could be built from nothing but NAND gates, it is rather doubtful that we can interconnect many, say, INTEL 8048 and produce a computing system of superior power ! What is missing is an answer to the question how several such components should be interconnected and how they should interact. While successful TTL parts have been designed so that many of them can be combined to form a system with more capabilities than the sum of its parts, the interconnection issue was definitely not a primary concern in the construction of the first single-chip microcomputers. In the design of modular, general-purpose VLSI components for digital computing systems, the communication issues have to be given full attention.

Standardizing the communication between computers is a considerably more complicated issue than standardizing interconnections between Boolean logic chips. In addition to voltage levels and load considerations, which apply to logic chips, interconnection of computers also involves issues such as link allocation, timing, message formats, addressing and routing, all of which must be considered in the design of truly modular, general-purpose VLSI components. Hardware for effective support of these features has to be incorporated on the chip. The case for such *computer modules* with proper consideration of the communication issues was first made by Bell *et al.* [1973].

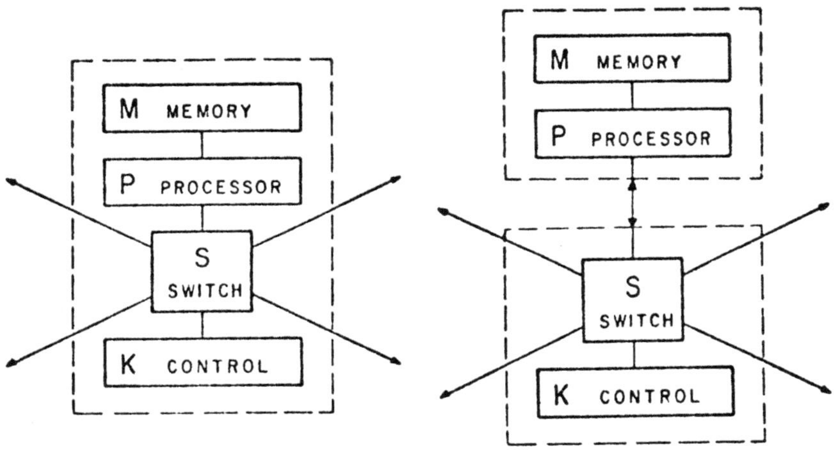

Figure 1. X-NODE, a modular VLSI building block of the mid 1980's (a) and its partitioning into a general-purpose switching component and a computer (b).

3. PROJECT X-TREE

In 1977 a research project to investigate these questions was started at the University of California at Berkeley. One specific issue was to define a modular component from which general purpose computing systems of arbitrary size could be built. Based on the considerations discussed in the previous section, the block diagram of such a potential building block has been defined (Fig.1a). It must contain at least a processor, a substantial amount of memory, a communication switch and its controller. These components can be connected in many possible ways to form large computing systems. A particularly attractive class of interconnection topologies for general purpose computing systems, as will be shown later, is based on tree-structured networks. This led to the project name X-TREE [Despain 78]; correspondingly the proposed single component, from which such computing systems are constructed, will be referred to as X-NODE.

So far research has mainly focussed on communications issues using a top-down approach [Séquin 78]. Investigations of the architecture of the main processor within X-NODE have been started [Patterson 79]. The anticipated time frame for a potential realization of a complete X-NODE on a chip is the second half of the 1980's, when VLSI chips with about a million devices will be commercially feasible. However, a recent study [Patterson 80] indicates that even a million transistors may not be sufficient to implement a good processor running a distributed operating system plus a satisfactory amount of main memory (>100K bytes).

Studies have therefore been initiated to determine how X-NODE could be partitioned into smaller chips without violating the guidelines developed in Section 2. A natural place to partition the circuit shown in Fig.1a is between the processor and the switch. The communication format at this point is still the same as that on the links between adjacent nodes; introducing a chip boundary in this branch will thus not result in a dramatic loss of communication bandwidth. The price that one has to pay is an increase in latency time since messages have to make two additional "jumps" to pass the two extra links between any two processors. As an advantage, this partitioning of X-NODE permits the definition of a more general, stand-alone switching component. The development of the communication network can thus be separated from the design of new processor architectures, permitting more specialization and optimization for particular applications. The resulting general-purpose switching parts will be discussed in more detail after a review of some of the results obtained from the studies under project X-TREE, giving the necessary understanding of the overall systems concepts of such modular multiprocessor architectures.

4. MESSAGE BASED COMMUNICATION

In an extended system with distributed control, it is more natural to communicate through messages rather than through synchronized signals. With proper autonomy and buffering in the message transport system, processors don't have to be in a state of constant alert, either actively listening for signals or ready to be interrupted at any time. Incoming messages are queued, possibly sorted by priority, and are processed at the convenience of the processor. The overhead of packaging a piece of information, that must be sent to another processor, into the proper message format is typically too expensive for small systems. In future systems with many computers, message based communication is well worth its overhead, in order to obtain reliable and processor independent communication.

A message based communication system also permits a clean separation between the transport mechanism for the messages and the interpretation of the actual message content by the receiver. This paper is mainly concerned with the first issue and will describe general building blocks for the construction of such a communication system.

4.1. Network Branching Factor

The communication system to be discussed is aimed at local installations where most computers are within the same building. One extreme is the case where a collection of .single-chip computers takes on the role of a high-throughput single-user machine, running computation-intensive jobs. A closely coupled, high-bandwidth network is thus considered.

Each node has a fixed upper limit for the total bandwidth available for communication with the outside world. This limit arises from the maximum power dissipation tolerable on a single chip and from the assumption of a fixed capacitance per lead leaving the node. Consequently, the more ports are given to a single node, the lower the bandwidth per port will be. One goal is to maximize the bandwidth per link, so that it is available when two nodes need to work cooperatively on a hard problem; thus there should be only a limited number of ports. On the other hand, redundant paths are required as a basis for a fault tolerant network. A practical compromise is a network with 4 or 5 ports per node. Three ports are a minimum number for the construction of a branching network. One or two additional ports permit the construction of effective network topologies, which have a low average number of links between any two nodes, good fault tolerance, and simple routing mechanisms. Examples are the halfring or fullring binary tree [Séquin 78] and the Hypertree [Goodman 80].

4.2. Multiplexing

The desired high degree of computational concurrency will lead to a large number of messages traveling through the network at any one time. Many of these messages may be routed over the same link. An obvious design goal is to prevent long or stalled messages from blocking a link for all other traffic. One solution is to use bidirectional links transmitting bytes alternatingly in both directions, and to multiplex messages in each direction in separate *virtual transmission channels.* Since messages move in bursts, and since switching from one message to another is associated with some overhead, a round-robin allocation scheme between all virtual channels is inappropriate. The link is assigned on a demand basis, and the message channels thus must be identified every time the assignment is changed.

Each piece of a message is preceded by a special tagged control byte NEXT#, identifying the corresponding virtual transmission channel by the number (address) of the physical buffer through which the message passes. These channel numbers are assigned at the time the virtual channel is set up and have local meaning only. In each subsequent node through which the message passes, separate channel numbers are created by the local routing controller. Once they are assigned, these channel numbers as well as the identification of the proper output port for each message are stored in small translation tables at the input ports for direct use by the switching hardware. Subsequent pieces of messages on an established virtual channel can then pass through the node without the help of the routing controller.

4.3. Routing

The main role of the communications system is then simply that of a basic routing and switching system for messages, i.e. establishing and controlling communication channels between processors. Higher level tasks, such as destination process identification, message numbering and acknowledgement, or even CRC generation and checking, are left to the users of the communication system, thus providing the most flexibility at the expense of some overhead for the processors wishing to communicate.

When alerted by a control byte NEW, the routing processor in a node will examine the beginning of the message header, to obtain the address of the destination node. Based on the relative position of the target node address with respect to the current location, the routing controller will select the proper direction, i.e. the proper port, through which to forward the header, in order to build a virtual channel towards the destination. The necessary routing algorithms to obtain an optimal path are a strong function of the chosen network topology and have been discussed in [Séquin 78] and [Goodman 80].

It is thus clear that at least the format of the header part that carries the destination address has to be defined for the switching hardware. For the X-TREE prototype it has been decided that the target node address will follow immediately after the control byte NEW and that it be represented as a byte oriented, variable length address [Séquin 78].

4.4. Message Format

The exact format of the messages to be handled in this network is beyond the scope of this paper, and will follow the usual format: a header containing destination and origin information, followed by the actual message content, and terminated with a cyclic redundancy check (CRC). However, the communication system will not be concerned with the exact format or the contents of the message itself. The message as a whole will be surrounded by two special control bytes. The control byte NEW will indicate that a new message starts and will initiate special action to set up a virtual message channel from the originating node to the destination. All bytes of the actual message remain intact during the transmission between the two communicating nodes. The control byte END will cause the removal of the established virtual channel. These tagged control bytes thus build an additional outer shell which encapsulates one or more messages and which opens and closes virtual channels as requested by the originating node (Fig. 2).

Bytes	Explanation
NEXT#	Switch to channel identified by #.
NEW	Set up a new virtual channel to
Target NA	target node address (NA), which may extent over several bytes,
NA cont.	and is in principle unlimited in length.
Data	Beginning of first submessage.
Data	
: :	Here, one or several other messages may be interspersed on the same link.
NEXT#	Switch to channel identified by #.
Data	Continuation of a previous message; data may consist of one or more submessages
Data	which may be individually checked
Data	and acknowledged to the originator.
: :	Here, one or several other messages may be interspersed on the same link.
NEXT#	Switch to channel identified by #.
Data	End of last submessage in this channel, may include crc check remainder
Data	and end of transmission character.
END	Remove virtual channel and free up corresponding slot.

Figure 2. Basic message format showing use of various control bytes.

5. COMMUNICATION PORTS AND LINKS

In order to provide a high-bandwidth path between adjacent nodes, it is important to restrict a node to only 4-5 ports. Assuming the availability of packages with 60-80 pins, one can allocate 12-20 pins per port and thus use a byte parallel communication protocol. The tradeoffs between parallel and serial data formats are not at all obvious. While byte parallel formats permit the implementation of a rugged, low-level protocol, in which each byte is individually acknowledged, serial communication protocols will typically have to deal with larger "data quantums", and thus require larger buffers, and more complicated data encoding, flow control, and acknowledgment schemes. Nevertheless, if individual processors get separated by more than a few hundred feet, serial communication would typically have to be used, because of the cost of the communication link hardware, and because of the excessive delay times involved with a handshaking protocol at the byte level.

Without trying to solve the above question one way or the other, we will discuss modular switching hardware using the example of a byte parallel data format; the possibility of an extension to a serial protocol is implied. The main purpose of the following chapters is to demonstrate the concept of modular switching components that can be put together in the way the typical logic designer assembles TTL logic parts: By obeying a few simple rules one can ignore details of voltage levels, rise times, and impedance matching, and focus one's attention at higher-level issues. Similarly, the goal here is to provide message switching parts that can "simply be plugged together" into a wide variety of network configurations without concern for the low level aspects of the communication protocols.

Since construction of arbitrarily large networks should not be excluded, different nodes must clearly be able to run on separate clocks; this requires asynchronous interfaces between nodes.

5.1. A Parallel Link

In an initial TTL prototype [Laurent 79] a byte parallel, asynchronous approach is used. Eight wires carry data, three more wires are used for parity, tag and acknowledgment, and two signaling wires handle the handshaking protocol. The tag is used to distinguish control bytes from data. Every time the link is switched to the disposition of a different virtual channel, a control byte NEXT# is sent, identifying the corresponding channel number on that link. This control byte is of significance only to the switching hardware. It tells the port at the other end of the link to what message the subsequent bytes belong. The number transmitted in this control byte NEXT# is used to select a matched pair of FIFO buffers at both ends of the link. Input and output port contain each a block of 16 such buffers which serve as queues for the arriving or departing bytes and enhance the throughput of the system by providing the necessary elasticity [Goldberg 79].

For each arriving byte parity is checked. If no violation is detected and if there is room to store the byte in the input FIFO, an acknowledge bit is sent back to the other node simultaneous with the subsequent byte transmission in the reverse direction.

Link allocation and timing is accomplished with two unidirectional lines AtoB and BtoA. With these two lines a two-cycle signaling protocol is implemented, in which a rising edge indicates that data has been placed onto the link and a falling edge that data has been received (Fig. 3). This change from the original design [Séquin 78] has been made in order to speed up the data exchange. The original 4-cycle signaling had to wait for two additional link delays.

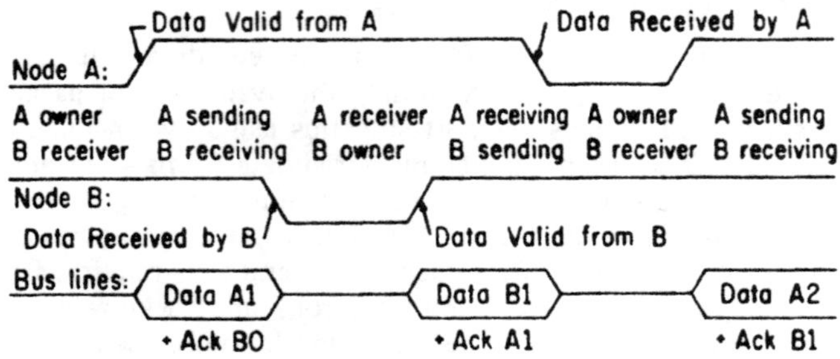

Figure 3. Two-cycle signalling handshake protocol using two unidirectional lines.

5.2. Input and Output Buffers

Before and after transmission over a link, messages are buffered in FIFO queues. These buffers are necessary to provide some elasticity for the virtual message channels. Simulation studies [Goldberg 79] have shown that there is a marked increase in average throughput if a single register is replaced with a double buffer. Further improvements in throughput are noticeable until the FIFO depth reaches about eight.

The number of parallel FIFO buffers required depends on the number of separate virtual message channels that need to be supported simultaneously across any one link. More channels are not necessarily better. With too many channels provided, it will take correspondingly longer for each channel to get serviced. With fewer channels average completion time will be reduced at the expense of increased latency. The above mentioned simulation studies seem to indicate, that a good compromise can be achieved with 16 multiplexed virtual channels per link, each with its own individual buffer at either end of the link.

A certain drawback may be that some short message, such as a confirmation to go ahead with a particular task, may have to wait until a channel is freed up. The wait time may be unduly long, compared to the length of the message. This could be remedied by reserving a few channels for short messages only.

5.3. Link Controller

Associated with each port are two finite state machines which control the interaction of the FIFO buffers and the link. An incoming control byte NEXT# will set the input multiplexer to the proper buffer, where it will remain until a control byte with a different channel number is received. Bytes in that channel are transmitted until the sending buffer is empty, the receiving buffer is full, or a parity error occurs. In the the last two cases a NACK signal is sent back, the sender switches to a different channel, and issues a control byte NEXT# identifying the corresponding channel number.

Timing on the link can be kept simple and robust. To simplify link allocation among the two attached nodes, the direction of transmission alternates after every byte. At start-up or reset time, each link has a designated direction for transmission which can be derived from the node addresses in the network. If a node has no data to transmit it simply sends a control byte DUMMY, which gets automatically acknowledged. Each link runs at its own speed given by the propagation delays across the link and through the two asynchronous finite state machines at either end. These asynchronous link circuits interface to the FIFO buffers, which run at the speed of the node clock, through the FIFO input and output registers.

6. SWITCHING COMPONENTS

So far the link hardware and the lowest level communication protocol has been defined, and thus the interface to a computer can be specified. In order to build networks of non-trivial connectivity, switching components are needed that can send messages in more than one possible direction. In the previous section we have established the desirability of nodes with four or five communication ports and we will thus discuss the construction of message switches with 5-6 link connections, according to the block diagram in Fig.1b.

6.1. Message Switch

Internal to the communication module, hardware must exist which permits it to transmit bytes between any two ports. For a switch with more than three link connections, this function can best be implemented with a time multiplexed bus, called the S-bus. This is a parallel bus which transmits simultaneously one data byte, parity, tag, and ACK bits, as well as the required addressing information, i.e. output port number and channel number [Griffin 79]. Since the time-multiplexed switching bus and all the communication ports will be contained on a single chip, the bus will not be loaded down by the external parasitic capacitances and can thus operate at chip-internal speeds, which can be more than an order of magnitude higher than external signals. This bus can then easily keep up with eight communication ports by serving them in a round-robin fashion. A more detailed diagram is shown in Fig.4.

The major cycle in which all input ports get served is subdivided into 8 minor cycles. Each port owns the S-bus for one minor cycle and may transmit one byte of data and the corresponding addressing information to any output port or to the routing controller. In addition, during all minor cycles, each port has to listen to the switching bus and pick up bytes addressed to its own port number.

A closer analysis of the kind of information which needs to be transmitted within each switching node, led to the conclusion that control information should be handled by a separate bus. The information required to update the translation and routing tables, to determine which channel numbers are free on a particular link, and the handling of various exception cases could cause significant interference with the transmission of actual data over the S-bus. It would also require a more complicated bus and additional tags to mark the different possible destination registers. It was thus decided to introduce a separate control bus for all this routing information, for the handling of exception cases, and for interaction with the operating system. While this approach may be somewhat of an overkill, it provides the required flexibility during the development stages and the necessary room for expansion of the functions of the communication node. Since the routing controller in the first TTL prototype is implemented with a Zilog Z80, the control bus has the form of a somewhat reduced Z80 system bus.

6.2. Routing Controller

The routing controller is the nerve center of the switching node. It handles all but the lowest level byte forwarding and acknowledgment tasks. Specifically it sets up new message channels, checks all channels for stalled messages, reclaims unused slots after a channel is torn down, and provides a general purpose interface to the operating

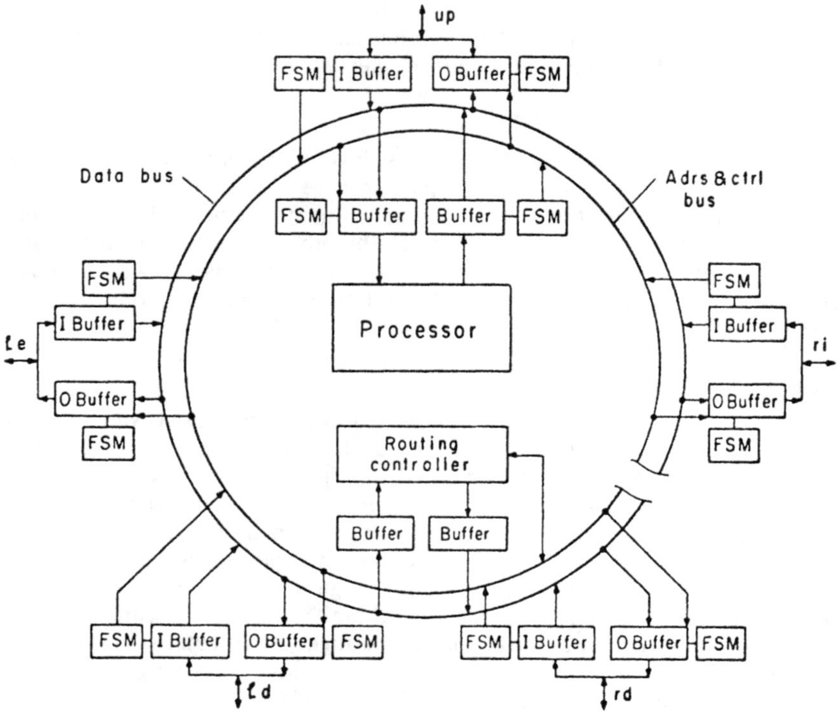

Figure 4. Expanded view of the message switching component.

system of the multi-processor system. A separate intelligence for these tasks is necessary in order to protect the actual processor at each X-NODE from the constant interruptions which would occur every time the status of a message channel through this node changes. The idea is to perform communication and computation in parallel on separate dedicated hardware.

The standard routing task involves: looking at a header, decoding the target node address, determining the proper link over which the message must be sent, finding an available channel number on that link, updating the translation table, and giving a go-ahead signal to the input port. These tasks could be done in hardware, firmware or software. The decision is based on the frequency with which the task has to be performed and on the influence that its execution speed has on the overall communication bandwidth. Correspondingly, the routing of individual bytes through each node, after the message channel has been established, should be performed entirely in hardware. On the other hand, less critical events, such as reporting the status of all links to the operating system, can definitely be handled by software.

Other decisions are less obvious. The necessary hardwired logic to calculate the relative address between target node and destination, which is required as an entry vector into the routing table for a fullring tree network described in [Séquin 78] requires about 30 MSI and SSI chips. For small networks it is more economical to use individual small look-up tables at each node, based on the absolute destination addresses. These small RAMs are loaded individually with the specifications of the proper output port for every possible destination in the network.

Furthermore, the routing controller must take care of many other tasks which are much less frequent; they can, therefore, be performed entirely in software. Such tasks are: periodic scanning of all channels for a lack of successful byte transfers since the last interrogation, tearing down of virtual channels with stalled messages, or handling of message segments with illegal channel numbers.

Overall, the functions of the routing controller are rather diverse and are best performed by a simple microprocessor. While a Z80 may be more powerful than necessary and thus is not representative of the complexity of the circuitry required on a final implementation of a VLSI version of the switching node, it is certainly worthwhile to have its power available during the development and debugging phase.

7. TREE-STRUCTURED COMMUNICATION NETWORKS

In this section we will discuss a special example of a network based on an enhanced binary tree structure. It turns out that such an interconnection scheme has many advantages for general purpose computing systems.

7.1. Advantages of Tree Structures

In an evaluation of different multiprocessor interconnection topologies [Despain 80], it turned out that most schemes have one of three disadvantages.

1) The interconnecting link constitutes a serious communication bottleneck, such as a common bus which has to be shared by all attached processors.
2) The switching hardware becomes unreasonably expensive for a large number of processors, as in fully interconnected networks or in a full crossbar arrangement.
3) The interconnection scheme is not truly modular, because the requirements for the individual components change as the system grows. This is the case in a hypercube network with the topology of a cube in n dimensions, where the number of ports has to increase with the logarithm of the number of nodes.

Networks that are truly modular and incrementally expansible comprise lattice structures in n dimensions and tree structures. Among those, trees have the further advantage that the average distance between nodes grows only logarithmically with the number of nodes.

The primary contender for the interconnection scheme in X-TREE is therefore a binary tree enhanced with additional links to form a halfring or fullring tree (Fig.5). These additional links further shorten the average path length, they distribute message traffic more evenly throughout the tree, and they provide the potential for fault tolerant communication with respect to the removal of a few nodes or links. Various placements for these extra links have been investigated, but most yield only insignificant improvement on the message traffic density and typically require significantly more complicated routing algorithms than the halfring or fullring structures.

7.2. Addressing and Routing in Fullring Tree

X-TREE is designed to be a truly modular, incrementally expansible system. To allow the system to grow to arbitrary size, neither the number of nodes nor the address space should be limited by artificial restrictions. Thus unbounded, variable-length addresses are employed throughout [Séquin 78]. All communication throughout the

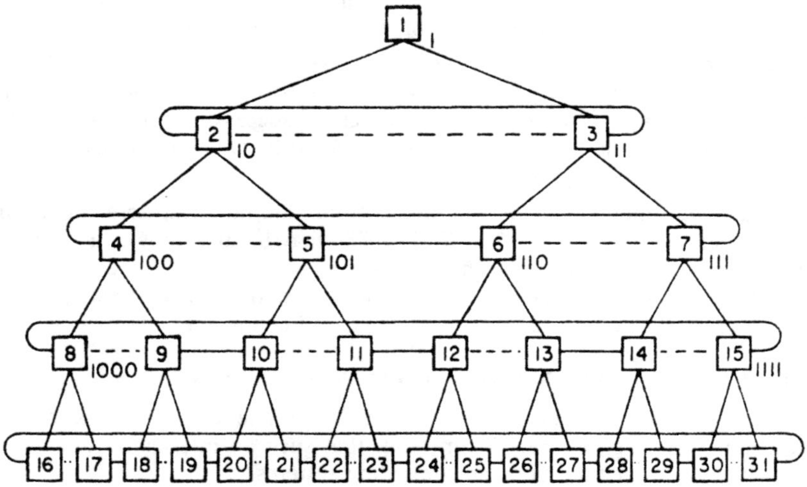

Figure 5. Binary tree with fullring connections. When the dashed branches are omitted, a halfring tree is obtained. Notice that the children of node *n* have node addresses $2n$ and $2n+1$ respectively.

tree is in the form of messages. To enable effective routing of messages, the complete address is subdivided into a node address part and a second part identifying a particular memory location if that node has any memory attached that belongs to the global address space. In X-TREE secondary memory, as well as all input/output, is restricted to the frontier (i.e. the leaves) of the tree. This is to minimize the number of ports per node, and thus to alleviate the restrictions originating from the limited number of pins. A PMS representation of an example of a small X-TREE is shown in Fig.6.

The addressing scheme is designed so that messages can be effectively routed from node to node simply based on local decisions, involving only the current node address and the destination address. In a binary tree this can be achieved in a very simple manner. The root node is assigned node address "1". The node address of a left child is formed by appending a "0", and that of a right child by appending a "1". With that scheme any node can readily be found by starting at the root and using the sequence of bits in the node address to make the routing decision at each node. To go from an arbitrary node to another node, one has to move up in the tree to the common ancestor of the two nodes, i.e. to the node where the address matches all leading bits in the target address. From there one moves down to the destination.

In halfring and fullring binary trees the routing algorithm is only marginally more complicated. The horizontal links permit a message to take a shortcut before the common ancestor has been reached. It turns out that in a fullring tree the optimal level to take the horizontal links is the one where ascending and descending path are separated by less than five link units (Fig.7). Again the routing decision can be based entirely on a comparison of the local node address and the target address.

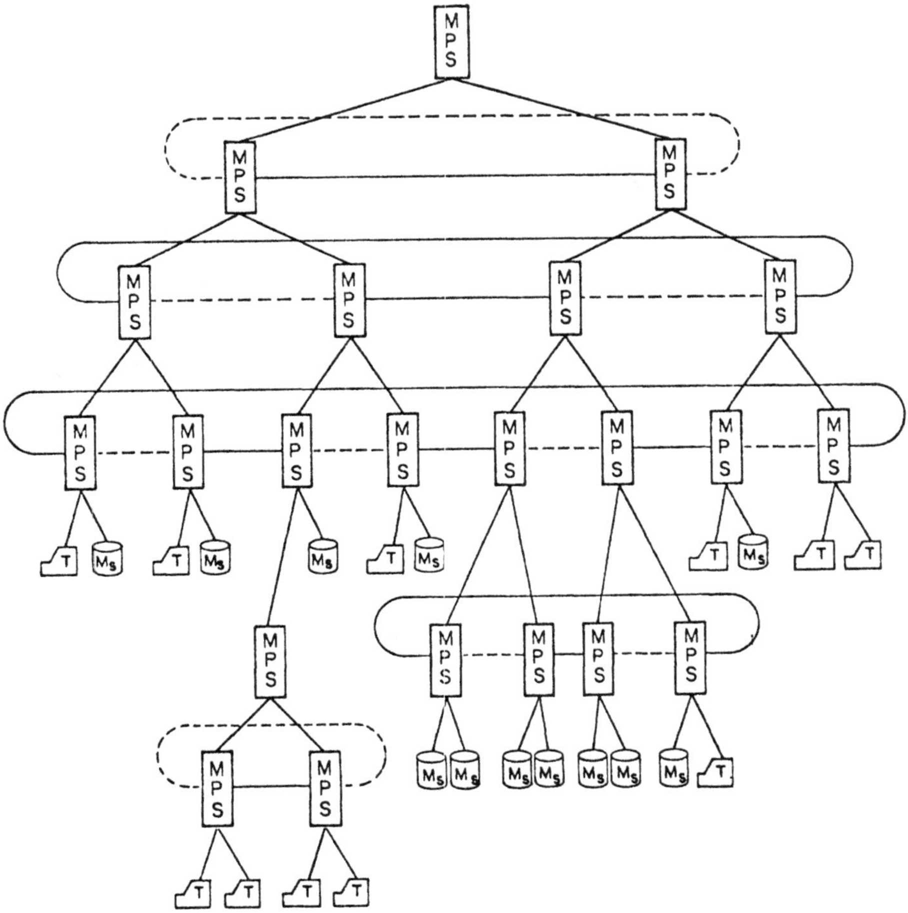

Figure 6. An X-TREE example in PMS notation.

Some fault tolerance can be achieved by listing secondary and ternary choices for the routing decision, which are taken when the primary choice cannot be followed. Such a decision table for the case of the fullring tree is shown in Table 1. The proper field is selected from a comparison of the horizontal and vertical distances between current position and target node. Within each field, first, second and third choice for the routing out of the current position are listed.

Of course there are always cases where this fault tolerant routing algorithm cannot be successful - for instance when the target node itself is missing. Special means to safeguard against cluttering the tree with worthless messages are thus required. One possible solution is to accompany each message header with a byte that counts the number of detours experienced, i.e. how often the message could not be routed in the primary direction. When this count reaches a limit, preset by the operating system, a

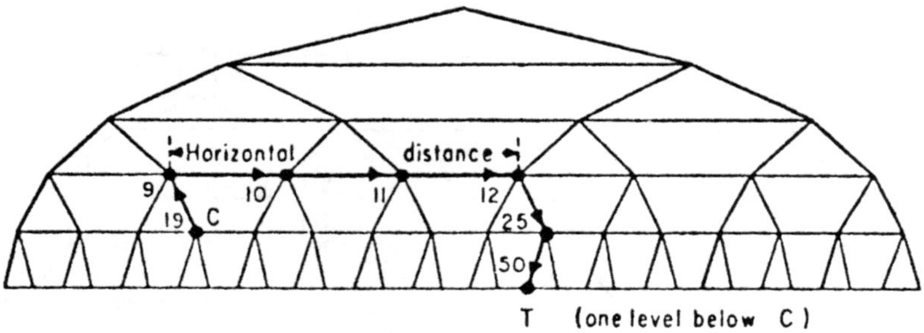

Figure 7. Shortest path in a fullring tree from current position, C, on node 19 to target, T, on node 50.

Choice of directions as a function of relative target position							
choice		far left	near left	left line	right line	near right	far right
above	1	up	up	up	up	up	up
	2	left	left	left	right	right	right
	3	right	right	right	left	left	left
level	1	up	left	proc.	proc.	right	up
	2	left	up	back	back	up	right
	3	right	ldown	back	back	rdown	left
below	1	up	left	ldown	rdown	right	up
	2	left	up	rdown	ldown	up	right
	3	ldown	ldown	left	right	rdown	rdown

Table 1. Decision table for fullring tree routing algorithm

higher-level recovery mechanism is invoked. For instance, the message could then be purged and a notice of this fact could be returned to the originator.

8. X-NODE ARCHITECTURE

In this section comments will be made about processors to be attached to X-NODE. Some general issues will be discussed that will apply to most potential processors to be used in such a network.

8.1. Link Interface

In Section 5 an example of a definition of link hardware and protocol has been given. This specifies one side of the processor/network interface. Another possible candidate for an off-the-shelf VLSI component can thus be defined: For all popular processors, interface chips which link the internal systems bus of the processor to the external, message-based communication network should be provided. While the outer part of such an interface will look very much like the port described in Sections 5.2 and 5.3, the inner part could be a DMA unit on the memory bus. Arriving messages could thus be safely stored away in a reserved area of the processor memory and only examined after they have been totally received and checked for transmission errors.

8.2. X-NODE Memory Hierarchy

Memory contained within each X-NODE is not part of the global address space. It acts only as a cache to the secondary memory contained entirely at the leaves of X-TREE. To alleviate the paging overhead, this memory should be as large as possible and thus the densest storage technique should be employed. For the mid 1980's it is anticipated that about 64k bytes could be implemented together with the X-NODE processor on the same chip. This local main memory can contain different types of information: data, descriptors, program, or microcode, each of which will be used by a different functional block. Unfortunately, typical high-density memory cannot easily be implemented as a true multiport memory, and thus it is not possible to extract four words from different locations simultaneously. Severe contentions for the memory bus can thus be expected.

An on-chip memory hierarchy can alleviate this problem. It would consist of a high-density RAM and several high-speed static caches dedicated to data, addresses, instructions and microcode, each closely tied in with the data path, an address ALU, the instruction decoder, and the microcontroller, respectively (Fig.8). Contentions for the common bus between the main memory and the four caches can be minimized by proper choice of the cache parameters. This approach combines the storage density advantage of dynamic RAM with the higher speed of static caches and yields a lot of flexibility in the allocation of local main memory space to data, programs, or microcode.

Swapping of microcode occurs through the same mechanisms as paging of programs or data from secondary memory at the frontier of X-TREE. No special mechanism has to be invoked to change the instruction set of one of the X-NODEs, and the latter can thus be tailored dynamically to best solve the current computational problem.

8.3. Expansion of Node Memory

If more memory is required than fits on the X-NODE processor chip, e.g. to run the kernel of a very sophisticated operating system, a way must be found to add more memory on separate chips without seriously reducing the critical bandwidth between the memory and the processor. One possibility is to move all main memory off the processor chip and to devote all that area to a significantly larger cache. The correspondingly larger hit ratio should then make up for the increased difference in speed due to the lower-bandwidth path through the chip peripheries.

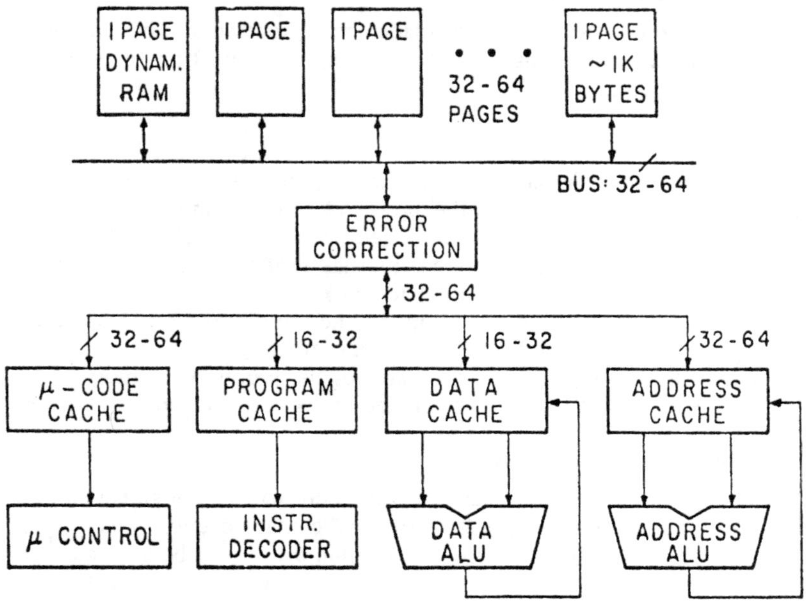

Figure 8. Memory hierarchy within X-NODE.

This approach is also compatible with the philosophy of using existing general-purpose parts as much as possible; random access memory chips should certainly not be ignored.

9. GENERALIZATIONS OF SWITCHING COMPONENTS

A minimum set of primitive switching components that permit the construction of networks with any branches at all contains: a decision element that routes incoming messages in one of two directions, and an arbiter that takes messages from two incoming branches and serializes them onto a single branch (Fig.9). These two components use unidirectional communications links.

A somewhat more complicated switching component, that works on bidirectional links, consists of a node with three branches. Incoming messages on one branch can be routed to either one of the other two links. In this case the internal switching function can be achieved with three 2-way multiplexors rather than with a time multiplexed bus.

The analogy to the family of elementary TTL gates now becomes even more pronounced: Just as there are NAND gates with a variety of numbers of inputs, there should be a group of switching components with different branching factors. Furthermore, corresponding to the different speed classes of TTL parts, a few different types of data formats and link protocols can be envisioned: High-speed byte-parallel communication links for tightly coupled networks running compute-bound jobs, and low-power serial link hardware for less critical applications, where modularity and flexibility in the network topology is the primary concern.

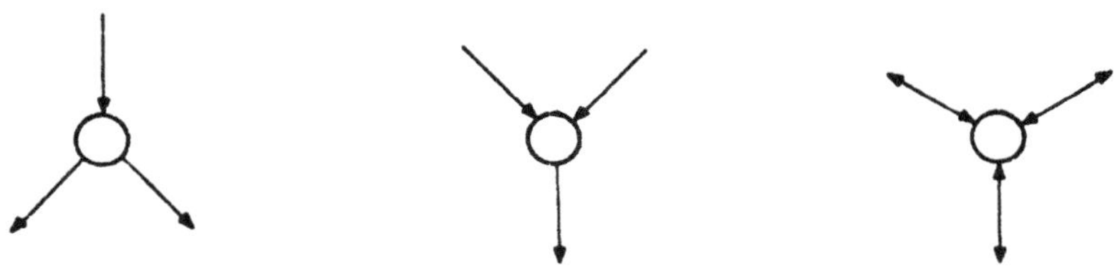

Figure 9. Simple Message Switching Components.

10. SUMMARY

In summary, there will always be a demand for off-the-shelf, general-purpose components, which can be used as truly modular building blocks for the construction of computing systems of any size. One such component has been identified and one way to organize that component into an incrementally expansible computing system has been outlined. Two crucial issues in the design of such a system are interprocessor communication and the operating system. Both issues must be addressed at an early stage of the design so that the necessary hardware support can be provided.

We felt that the issue of interprocessor communication is even more important at an early stage of a VLSI multiprocessor project. The emergence of truly modular, general-purpose VLSI components may actually depend on the definition of a standard inter-processor communication protocol and a realization of the required switching hardware on VLSI chips.

11. ACKNOWLEDGMENTS

The concepts around X-TREE have been developed in cooperation with Profs. Al Despain and Dave Patterson, and with our students in the Computer Architecture Group. The author would like to thank them all for their contributions and the many stimulating discussions.

REFERENCES

[Bell 73]
C.G. Bell, R.C. Chen, S.H. Fuller, J. Grason, S. Rege, and D.P Siewiorek: "The Architecture and Applications of Computer Modules: A Set of Components for Digital Design", *Proc. IEEE Compcon,* March 1973, pp 177-180.

[Despain 78]
A.M. Despain and D.A. Patterson: "X-TREE: A Tree Structured Multiprocessor Computer Architecture", *Proc. 5th Symp. on Comp. Arch.,* Palo Alto, CA, April 1978, pp 144-151.

[Despain 80]
A.M. Despain and D.A. Patterson: "The Computer as a Component: Powerful Computer Systems from Monolithic Microprocessors", submitted to *Comm. of ACM.*

[Goldberg 79]
W.S. Goldberg: "Communication in X-TREE: A Simulator Supported Analysis", *Masters Report,* E.R.L., U.C.Berkeley, July 1979.

[Goodman 80]
J.R. Goodman and C.H. Séquin, "Hypertree, a Multiprocessor Interconnection Topology", submitted to *IEEE Trans. Computers*

[Griffin 79]
M. Griffin, "X-SWITCH Communication Busses", *Masters Report,* E.R.L., U.C. Berkeley, Dec. 1979.

[Laurent 79]
M.J. Laurent: "Input/output ports for the X-TREE nodes", *Masters Report,* E.R.L., U.C. Berkeley, Dec. 1979.

[Patterson 79]
D.A. Patterson, E.S. Fehr, and C.H. Séquin: "Design Considerations for the VLSI Processor of X-TREE", *Proc. 6th Symp. on Comp. Arch.,* Philadelphia, April 1979, pp 90-101.

[Patterson 80]
D.A. Patterson and C.H. Séquin, "Design Considerations for Single-Chip Computers of the Future", *Joint Spec. Issue, IEEE Jour. Solid-State Circuits SC-15, No.1 and IEEE Trans. Computers C-29, No.2,* pp 108-116, Feb. 1980.

[Séquin 78]
C.H. Séquin, A.M. Despain and D.A. Patterson: "Communication in X-TREE, a Modular Multiprocessor System", *Proc. ACM 78,* Washington D.C., Dec. 1978, pp 194-203.

ARCHITECTURES FOR MULTIPROCESSOR SYSTEMS, ARRAY SYSTEM VERSUS FUNCTIONAL DECOMPOSED SYSTEMS

E. Hoerbst

SIEMENS A.G., Central Research Department, Munich

ABSTRACT. In our research laboratories, we developed two different microprocessor systems. One system is a parallel processor system, where at the moment 128 computer modules work in parallel (512 mudules are possible without any changes). With this system, the parallel processes start at the same time and stop when the last is finished. The other system is a functional decomposed system with 16 computermoduls realized (64 are possible) where the processes start and terminate independent from each other.

THE PARALLEL SYSTEM

Many problems have an inherently parallel structure (Figure 1) or can be correspondingly partitioned [1,2]. The overall problem is compised of numerous elementary tasks (ET_i) which interact. A circuit simulation will be taken as an example. Figure 1 resembles a circuit

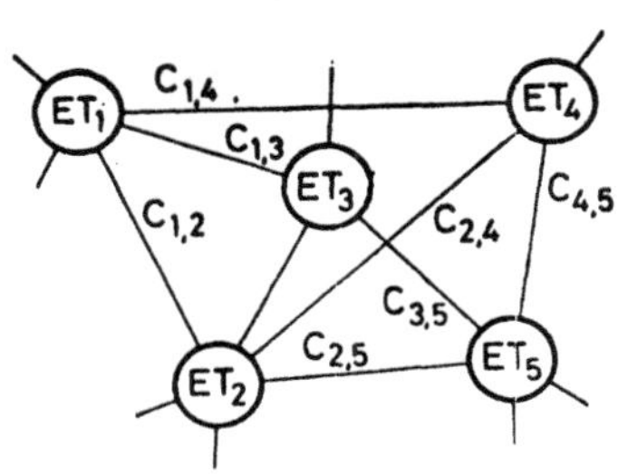

Figure 1. Structure of parallel systems

configuration in which the nodes are interconnected by electronic devices. The elementary tasks consist in the computation of the various node voltages. Further examples of this class of problem are field computations in which each elementary task represents part of the overall problem, and image processing in which the image can be subdivided into segments.

Such problems can be delegated to a parallel computer-structure as in real life. The elementary tasks can be solved by concurrently operating computer modules. Interaction between elementary problems is simulated by the exchange of information between modules. This partitioning of the overall problem simplifies programming.

1. SMS PARALLEL COMPUTER CONCEPT

A concept for a parallel-operating multi-micro-computer system has been developed with the objective of realizing high computing power for application problems of the above class. The structured multi-microprocessor system (SMS) has an easily understandable architectural and organizational scheme which avoids the principal difficulties commonly encountered with parallel computers, viz. access conflicts, the synchronization of processors and deadlocks [3,5]. The structure of the SMS parallel computer is shown in Figure 2. A number of identical modules, each consisting of a module computer with program and data memories, an input-output interface and a communication memory are connected by way of an interconnection network with a main processor.

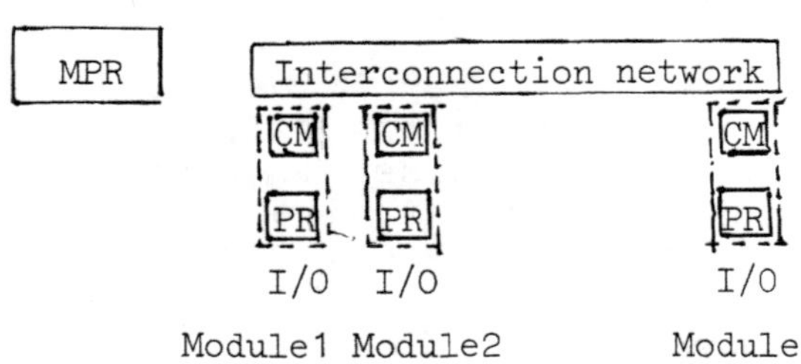

Figure 2. Structure of SMS parallel computer. MPR Main processor. CM Communication memory. PR Processor

All system components are coordinated by the main processor and operate in self-synchronizing phase cycles. In numerical calculations a phase cycle corresponds, for example, to an iterative relaxation step or time step. Since the SMS concept retains the central coordination of array computers such as ILLIAC IV [6] it therefore needs only simple organizational software. It does not however retain the fixed master-slave relationship at the instruction level. The modules can run entire programs as in the "distributed" multiprocessor systems C.mmp and Cm* [7,8]. Since the elementary tasks can be directly translated into programs for the modules, the user software can be kept simple.

2. IMPLEMENTATION OF PARALLEL COMPUTER WITH 128 COUPLED MICROCOMPUTERS

For studying the hardware and software aspects of large parallel computer systems, the SMS 201 was implemented at our laboratories with 128 coupled SAB 8080 microcomputer modules [9]. The system has been in use since the end of 1977. Its computing power is comparable with that of current main frames. The relatively slow floating-point operations of the microprocessors can be increased about 20 times by adding one specialized arithmetic processing unit (APU) per module. The principal characteristics of the system are summarized in Table 1.

Table 1. Main characteristics of the SMS 201

	SMS 201	SMS 201 with APU
Instruction stream[1]	32 $MOPS_8$	
Floating-point instruction stream[2]	0.03 MFLOPS	0.55 MFLOPS
Memory capacity	2.5 MByte	
Power consumption	3 kW	

[1] $MOPS_8$ million microcomputer operations per second
[2] MFLOPS million floating-point operations per second (memory-memory, 32 bit)

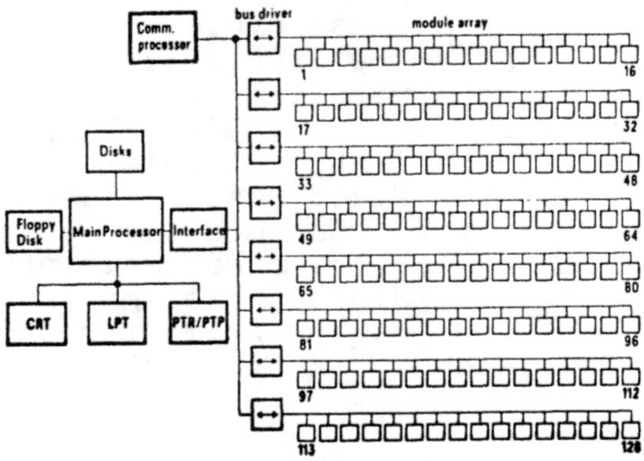

Figure 3. Structure of SMS 201 parallel computer system

Figure 3 shows the block diagram of the SMS 201. The 128 modules are arranged in 8 groups and connected to a main processor via bus drivers.

3. PROGRAMMING

The parallel programming of problems normally has to be supported by system software. Available computer languages are for this purpose supplemented by new language elements or special parallel languages may be developed. Such is not necessary with the SMS system. The overall problem is already partitioned during problem analysis and the resulting elementary tasks delegated to the modules of the parallel computer structure. Since these modules (main processor and module processors) are self-contained units, they can be programmed in the conventional manner using any available programming language such as Assembler, PL/M, FORTRAN, etc. and programming aids such as cross-software for host computers, resident software for the main processor or the microcomputer development system. The system within the the main processor is used as the operating system.

4. APPLICATIONS

The use of the SMS parallel microcomputer system in meteorology, electrical engineering and nuclear physics will be illustrated below with reference to examples.

4.1 Weather Forecasting

The first application program of the SMS 201 was a weather forecasting program for the earth's northern hemisphere which was taken over from the German Weather Service and simplified somewhat [10]. The model of the earth's atmosphere is, in mathematical terms, a finite difference approximation of partial differential equations. It divides the forecasting area into a three-level grid with 2000 grid points per level. Figure 4 shows the horizontal structure of the weather forecasting model. This is simulated with the parallel microcomputer structure as follows: each module simulates the weather in a geographical segment covering $3 \cdot 16$ grid points. Beginning with the initial values the system computes the weather development in 5-min real-time intervals. Following each interval the results are exchanged between the geographical segments in order to simulate the natural interaction of their atmospheric pressures, temperatures and other parameters.

Figure 5 shows ground-level isobars relating to a segment of the forecasting area. The initial state put into the computer is seen on the left, while the computed state after 24 h appears on the right. The 77 min computing time required for a 24 h forecast can be reduced to approx. 7.6 min by adding arithmetic processing units as described in Section 2. The remarkably short time taken for data exchange, which totals about 3,5 min, can be reduced to less than 0,5 min by intoducing a special data exchange computer.

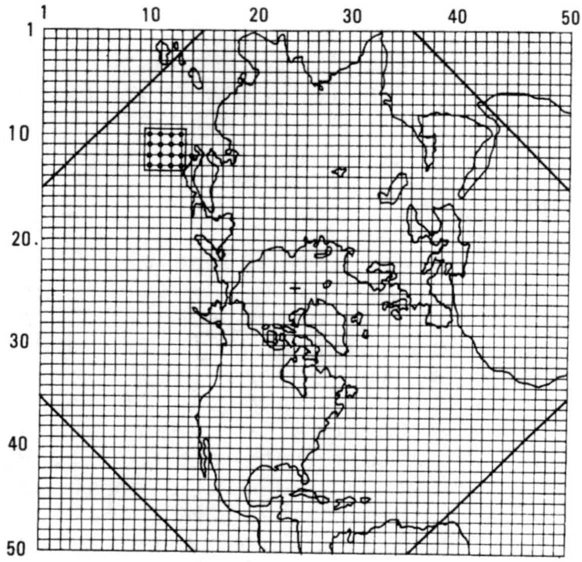

Figure 4. Weather forecasting model

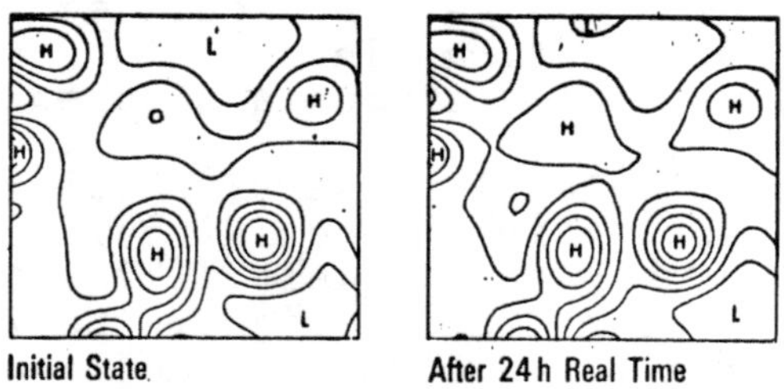

Figure 5. Ground-level isobars

4.2 DYNAMIC STABILITY COMPUTATION FOR ELECTRICITY SUPPLY NETWORKS

The dynamic response of a power distribution network is simulated with computers in order to determine the load flow and voltage variations during short-circuit and other undesired conditions. These simulations are very time-consuming, especially for large networks. Figure 6 gives a schematic representation of a power distribution network with 7 nodes to which a total of 4 generators and 4 loads are connected.

This network was simulated with the SMS 101, an 8-module predecessor of the SMS 201 5 . Each module of the para-

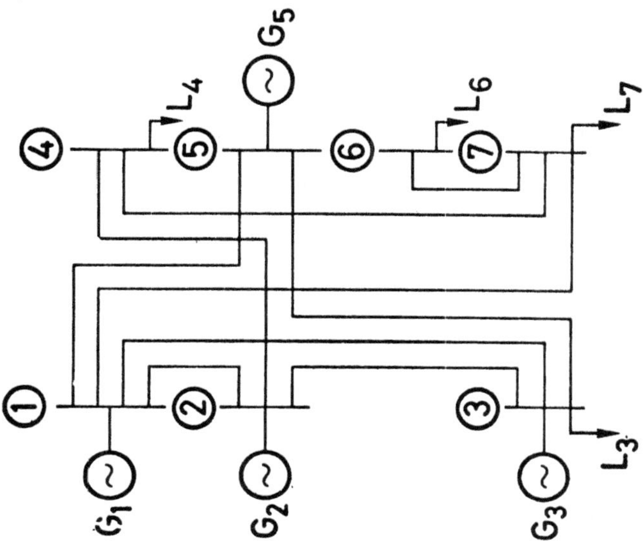

Figure 6. Power distribution network

llel system is associated with a net point. The dynamic response, e.g. the reaction to a fault, is determined in terms of time intervals. The overall microprocessor workload is about 80%. Larger network can be simulated after adding more modules. The CPU time will then increase at most only linearly with the number of nodes, and not according to square law as when using a single computer.

4.3 SIMULATION OF SPIN SYSTEMS

The macroscopic properties of meterials can be determined by simulating the behaviour of their atoms at microscopic level. The spin model shown in Figure 7 was used for investigating these magnetic properties.

The lattice sites of a crystal are occupied by atoms with a magnetic moment (spin). Within the effective field caused by the field intensity of the neighboring moments and an external field intensiy H_o the spins wobble with a precessional motion. The complicated motion of the interacting spins is computed over iterative time intervals. The result obtained is the correlation of the spins, which may be used to compute the magnetic behavior of the overall system.

A model with 128 spins was computed with the SMS 201. Each module is associated with a certain spin. The programs for the microprocessor modules and the main pro-

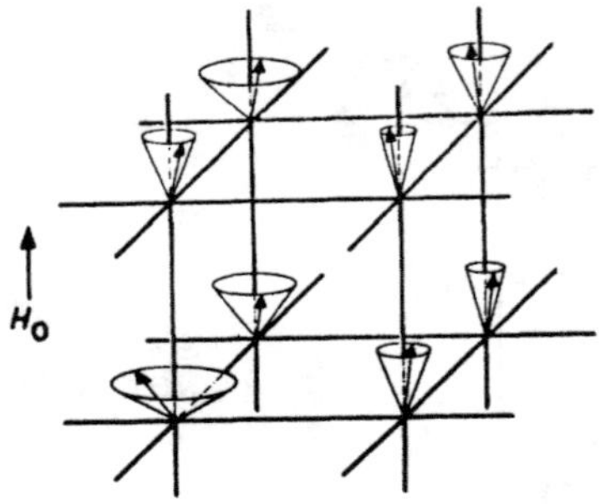

Figure 7. Spin model for investigation of magnetic properties

cessor are written in FORTRAN. The microprocessors in the modules are loaded to more than 90% of capacity.

4.4 SOLVING LINEAR EQUATIONS

Two algorithms for solving systems of linear equations are presented. They demonstrate how the parallel processor system SMS 201 is programmed and how an effective data exchange can be obtained. The efficiency of the system is discussed.

GAUSSIAN ELIMINATION

$A.X = B$,
with $A = (a_{ij})$ n.n matrix (known), $B = (b_{ij})$ n.m matrix (known), and $X = (x_{ij})$ n.m matrix (unknown), is solved by Gaussian Elimination [23].

By repeated formation of linear combinations of two rows, A is transformed into the unity matrix E:

$$A \xrightarrow{T} E,$$

i.e., the transformation T is a multiplication from the left by A^{-1}.
The same combinations are formed with the rows of B, thus transforming B into the solution X:

$$B \xrightarrow{T} A^{-1}B = X.$$

The transformation T is carried out in n steps. In the k`th step all rows (except row k) are combined with the k`th row in such a way that the elements a_{ik} become zero:

$$a_{ij} := a_{ij} - \frac{a_{ik}}{a_{kk}} a_{kj}, \; i \neq k, \; j = k, \ldots, n;$$

(1)

$$b_{ij} := b_{ij} - \frac{a_{ik}}{a_{kk}} b_{kj}, \; i \neq k, \; j = 1, \ldots, m.$$

These floating point operations consume most of the time and should be evaluated simultaneously. The following program flow indicates how this was achieved in the implemented SMS 201-program.

Initialization. The matrices A and B are loaded from a disk into the main processor. Row i of A and B is loaded into module i.

Elimination loop (n times repeated). In the k-th repetition of the loop, module k broadcasts the actual values of its row to all the other modules. This exchange phase is followed by an autonomous phase, in which all but the k`th module evaluate simultaneously the linear combination (1) of their own actual row and row k.

Normalization. After the n-th completion of this cycle, A is transformed into a diagonal matrix. In a final autonomous phase all modules divide their row by their diagonal element. Every module now contains one row of the solution X.

JACOBI`S ITERATIVE METHOD:

The system of linear equations:

$A.x = b$,

with $A = (a_{ij})$ n.n matrix (known), $b = (b_i)$ vector (known), and $x = (x_i)$ vector (unknown) is resolved with respect to the diagonal element of A [24].

$$x_i = \frac{1}{a_{ii}} \left[b_i - \sum_{j \neq i} a_{ij} x_j \right], \; i = 1, \ldots n$$

This formula is applied to improve approximations of x in the iterative steps. It uses on the right hand side the values of x from the previous step. The equation to be solved in each step for each x_i (with the super-script k denoting the iteration step) is

$$x_i^{(k)} = \frac{1}{a_{ii}} \left[b_i - \sum_{j \neq i} a_{ij} x_j^{(k-1)} \right] \qquad (2)$$

The iteration is started with $x_i=0$, $i=1,2, \ldots, n$, and repeated until the absolute value of the difference for two consecutive approximations stays below a predetermined bound for all components of x.

In the SMS 201 program for the Jacobi algorithm again every module handles one row of the system. The iterations are performed in a cycle of an autonomous phase in which every module calculates a new approximation of one component of x according to (2) and an exchange phase in which the new approximation is communicated between all modules. The cycle is repeated until all modules have set a flag indicating that the improvement in their component did not exceed the bound .

COMPUTATION TIMES

Figure 8 shows the total time for Gaussian eliminations as function of the matrix dimension. The time for one iteration step in the Jacobi program is given in Figure 9. In this case the total time would be meaningful, because the speed of convergence depends widely on the matrix A. For these investigations systems of linear equations were chosen, that could be solved on both programs: A is a matrix with only non-zero elements and a predominiant diagonal element, while B consists of a single column.

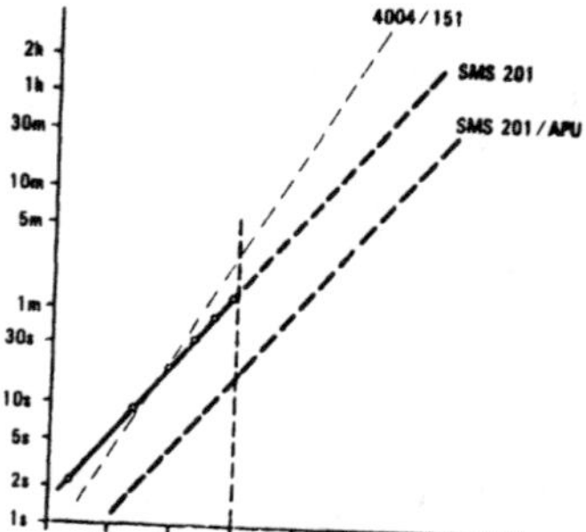

Fig. 8. Computation time for Gaussian elimination

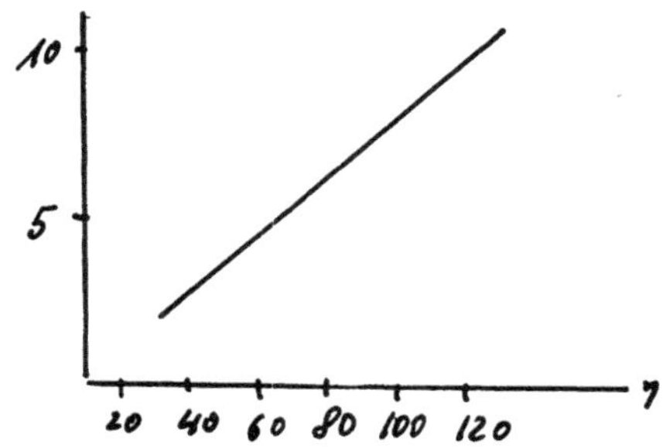

Fig. 9. Computation fime for one Jacobi step.

In order to find out how effective the program implementations are, measurements were carried out to determine the time relation between autonomous phases (t_{AUT}) and exchange phases (t_{EXCH}).

The ratio

$$n = \frac{t_{AUT}}{t_{AUT} + t_{EXCH}}$$

indicates how much parallelism was really achieved, respectively how much overhead was required for intermodule data exchange, which would not be needed in a monoprocessor.

Since in both programs the number of calculations in the autonomous phase as well as the number of data to be exchanged is proprotional to the dimension of the matrix A, it should be expected that the ratio does not depend on the dimension. This was confirmed by measurements. $\eta = 0.85$ for Gaussian elimination and $\eta = 0.90$ for Jacobi`s iterative method prove that in more than 85% of the time the modules are working parallel. That ratio will not change when the number of modules is increased above the 128 used in SMS 201.

5. OUTLOOK

The efficient processing of the described problems calls for an enormous computing power. Parallel computer structures such as the SMS system are capable of providing the capacity required. Estimates indicate that a system can already be implemented at the laboratory with present microprocessor modules (bit slices) which distinctly surpass the computer speed of current main frames for adapted problems. Further advances towards the realization of a new class of high-power computers can already be expected in the near future due to the advent of very-large-scale integrations (VLSI) [11].

THE SEQUENTIAL SYSTEM

The term "sequential" refers to the type of information processing performed in a multicomputer system ("structure"), which in the present case is composed of microcomputers. A process runs on various computers such that whereas the processing is initialized sequentially, its handling by the various computers can be concurrent. Initialization is started by the initiating computer, while coordination results from the process itself. This interrelationship will be illustrated by the following example. Process 1 is initialized in a computer in response to the input of user information at a terminal. The purpose of the process is to coordinate and interpret various userinputs from the terminal. The result is that the further processes 2, 3 and 4 are now initiated at computers 2, 3 and 4. The results flow back into process 1, where a decision is made with regard to the further running of the process.

Possible versions of such structures and specific proposals for solutions and system configurations are given in [12-14], where however predictions with reference to the usefulness of the structure are confined to selected examples.

Sequential system structures are of interest wherever the decomposition of the software for allocation to independent hardware modules is permissible. The process can be controlled either centrally or from a system module.

Typical applications are: process control (rolling mills power stations), information systems (management, manual or automatic control centers), all types of office data processing (office computer) and future high-power computer systems [15].

This type of information processing is already familiar from monoprocessor systems, where the disadvantages are due to complicated program development, servicing and maintenance and the requirements for the operating syst-

The principal advantages of multicomputer systems are:
the functional decomposition of hardware and software,
the modularity of the system and the resulting possibility of expansion,
the broad applicational spectrum.

Thus the system developed for multicomputer systems necessarily has to be structured, while the various modules can be tested largely independent of the system. The functional decomposition and modularity of the system will be illustrated below with reference to the example of a dedicated or office-oriented computer system.

6. FUNCTIONAL DECOMPOSITION OF MODULES

Office computers are interaction-oriented systems. The problem of the decomposition of functions for a multicomputer structure is assigned to a monocomputer system. Table 2 shows selected functions and their tasks.

This functional decomposition of the various hardware modules result in the separation of system functions from user functions. Thus tasks such as translation, linking and editing are user functions which are separated from the remainder of the system in the "processing" function.

These functions are now decomposed by the various ways and means and allocated to a multicomputer structure [16]. Decomposition among various levels as described in [17, Figure 6] was chosen. In the simplest case each function is introduced into a separate computer so that, if redundancies are included, a structure as shown in Figure 10 results.

At the appropriate moment a specific function at the function level is selected via user and interaction levels and accessed via the communication level, where the performance of that function is initialized. During the performance of the function the communication, interaction and user levels can be reused for other processes.

The functions need not be distributed to individual computers according to a fixed scheme but can be delegated to external memories and, if necessary, loaded into a free processor [17]. This means that a further level - a memory level - has to be introduced into the structure as shown in Figure 10, and that identical computers which are function-free when idle will be present in the function level.

Table 2. Selection of functions and their tasks with reference to the example of an office computer

Interaction functions	Mask preparation User control Input test
Processing function	Translating, linking ... Math. functions Text editing
Input/output functions for peripherals	Channel routines Text preparation
Data file function	Management Search Sorting
Diagnostic and test functions	System testing Detection of HW and SW faul Remedying of faults
Internal communication function	Internal data exchange Prevention of system failur
External communication	Interfacing of other computers Procedures Text preparation

7. MODULAR SYSTEM DESIGN

7.1 SYSTEM HARDWARE

The nucleus of the system is the communication level. The communication mechanisms have been investigated for multiprocessor systems in theory and for some practical applications [18-20].

In sequentially operating structures there is no need for strictly hierarchical systems as still used in pro-

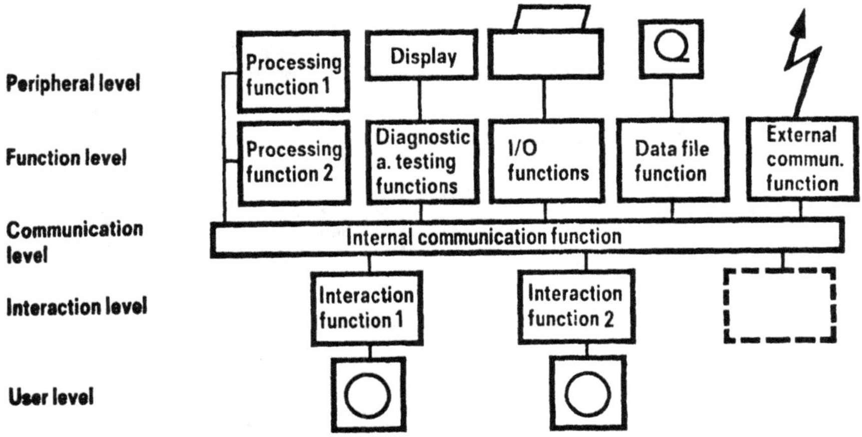

Figure 10. Decomposition of functions for allocation to hardware modules

cess control. The hardware modules of the interaction and function levels are distributed with equal status around the communication level. Possible architectures are in this case the ring structure, the star structure and the meshed structure.

Investigations of a meshed structure and a star structure in a distributed multicomputer system have shown the star structure to be far superior to the meshed structure in both expandablility and flexibility [21]. With regard to reliability the star structure should be superior to the ring structure if provided in duplicate.

If the modules of the system are operated in functional independence, the degree of modularity will be increased through the introduction of several internal communication levels. The I/O functions and the external communication function shown in Figure 10 are however omitted. Figure 11 shows a second communication level for the processing function interposed between the function level and the program memory level.

A structure similar to that for the processing function is conceivable for the data file function [17]: function level, communication level 3, database processor level, communication level 4 and data memory level.

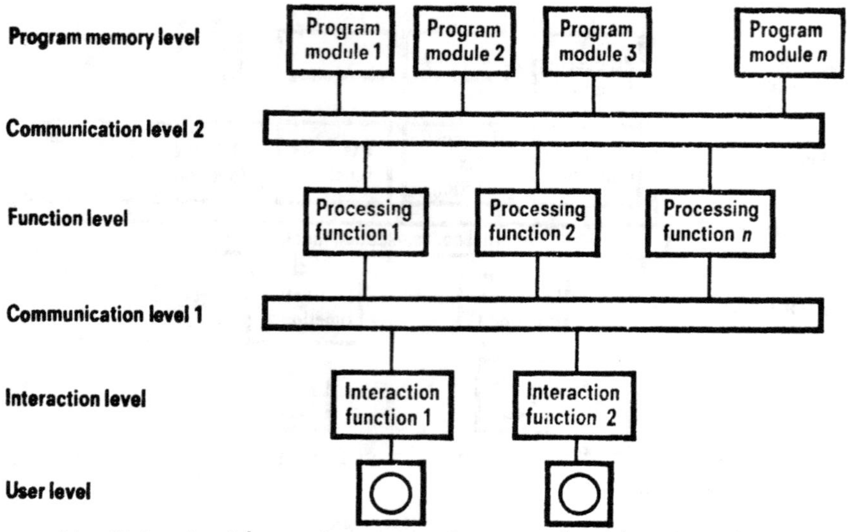

Figure 11. Introduction of a second communication level

7.2 SYSTEM SOFTWARE

The functional decomposition of the system among various hardware modules (microcomputers) means dividing the system software into functionally independent modules for incorporation in a hierarchically structured level scheme [22]. Figure 12 shows a level scheme for the office computer represented in Figure 10.

The system structure begins at the lowest level (hardware level) with the local software, e.g. device driver. Above the local software is the communication software. Since communication procedures are essentially the same for all computers, the modularity and expandability of the system are assured. The communication software takes care of internal data transfer and microcomputer synchronization. The operating system performs service routines for data transfer and program execution. The operating system level is accessed from the user level via the system interface. This hierarchical concept and the independence of the programs within any one level results in great reliability also for the system software.

Adaption to different applications and stages of expansion is implemented during system generation by means of a bootstrap ROM or by reading in the operating system from an external memory such as a diskette. The configuration can then be modified simply by changing the data medium.

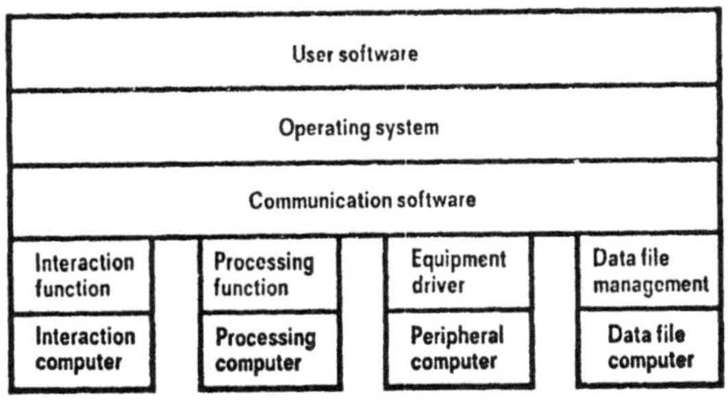

Figure 12. Structure of system software used for the office computer shown in Figure 10

8. FLEXABILITY OF SYSTEM WITH REFERENCE TO EXPANSION

Since the user accesses the system via the system interface, it is possible not only to decompose the functions for allocation to various hardware modules as shown in Figure 10, but also to adopt the reverse procedure. Thus various functions, beginning with the hardware oriented software and extending up to higher levels of the operating system (cross-reference lists, tables), can be implemented in a single computer. Figure 13 shows a configuration with an interactive computer and a date file computer in which the interactive and processing functions are implemented in the interaction computer and the data file and input/output functions are implemented in the data file computer as shown in Table 2 and

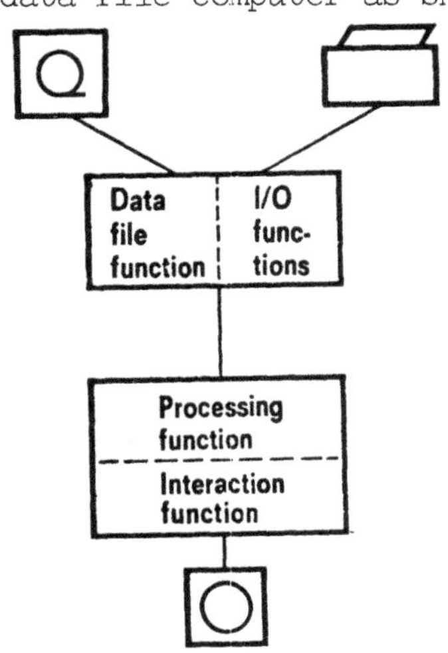

Figure 13. Modification of system configuration in Figure 10 into a minimum configuration

Figure 10.

The modular design of the hardware described in Section 9 allows for instance the connection of a printer to the data file computer

In the ideal case the operating system functions can always be automatically adapted to the modified hardware structure by, for instance, bootstrapping the system.

9. IMPLEMENTATION OF A SEQUENTIAL SYSTEM STRUCTURE

A sequential structure in the form of an office computer as shown in Figure 10 was implemented at the Siemens research laboratories.

The system hardware (microcomputers) is of modular design. Each microcomputer has an identical basic module, besides which memory modules and I/O modules for special peripherals can be plugged in according to the functions it is required to perform. The basic module is composed of the CPU group, program and data memories, parallel and serial standard interfaces and an interface for communications between the various computers.

Data are exchanged between the various microcomputers via a common bus (star node). Since software-controlled data exchange would be too slow only hardware-controlled data exchange via direct memory access (DMA) has been chosen. Bit rates up to 1 MByte/s can be realized using commercially available DMA controllers. The bus hardware consists of a central control and one interface per microcomputer.

The system and its various modules runs fully automatically. The diagnostic and testing system communicates information of the operating status of the system to the display panel. Another duty of the diagnostic and testing system is the monitoring of the various modules for failure and the communication of instances of failure to the display panel.

The functional decomposition of hardware and software for allocation to the various independent modules offers considerable advantages for the following reasons:
a structured standard system has to be developed as a matter of course,
the various modules can be tested in independence of the system,
the system can be easily expanded,
component redundancy can be provided in a simple manner,
system maintenance is easier.

The described work was done at the SIEMENS Research laboratory by Mr. Kober and Mr. Sauer who presented this work in the Siemens Research and Development Reports vol. 7, 1978, No. 6 pp. 316 - 321.

REFERENCES

1. Wirsching, J.E.: Computer of the 1980s - Is it a Network of Microcomputers? Dig. of Papers. Compcon Fall 1975, pp. 23-26
2. Kuznia, Ch.: Kopp, H.: A Model for Process Communication in Parallel Processor Systems. Proc. IMACS-GI-Sympos. on Parallel Comput. and Parallel Math. Munich, March 14, 16, 1977, pp. 339-342
3. Flynn, M.J.: Some Computer Organizations and their Effectiveness. IEEE Trans. C21 (1972), pp. 948-960
4. Weissberger, A.J.: Analysis of Multiple-Microprocessor System Architectures, Comput. Design 16 (1977), pp. 151-163
5. Kober, R.: Kopp, H.: Kuznia, Ch.: SMS 101 - A Structured Multimicroprocessor System with Deadlock Free Operation Scheme, Wuronicro Newsl. 2 (1976), pp. 56-64
6. Barnes, G.H., et al.: The ILLIAC IV Computer, IEEE Trans. C17 (1968), pp. 746-757
7. Wulf, W.A.: Bell, C.G.: C.mmp - A Multi-Mini-Processor. Fall Joint Comput. Conf. 1972, pp. 765-777
8. Swan, R.J.: Fuller, S.H.: Siewiorek, D.P.: CM* - A Modular Multi-Processor, AFIPS Conf. Proc.46 (1977), pp. 637-663
9. Kober, R.: The Multiprocessor System SMS 201 Combining 128 Microprocessors to a Powerful Computer Dig. of Papers, Compcon Fall 1977, pp. 225-229
10. Kopp, H.: Numerical Weather Forecast with the Multimicroprocessor System SMS 201, Proc. IMACS-GI-Sympos. on Parallel Comput. and Parallel Math. Munich, March 14, 16, 1977, pp. 265-268
11. Garbrecht, K.: Stein, K.-U.: Perspectives and Limitations of Large-Scale Integration, Siemens Forsch.- u. Entwickl.-Ber.5 (1976) pp. 312-318
12. Burton, T.: Multi-μP Systems Combine the Efficiency of Dedicated Microcomputers with the Throughout of Minis. Electron. Des. 25 (1977) 16, pp. 68-71
13. Reyling, G.: Performance and Control of Multiple Microprocessor Systems. Comput. Des. 13 (1974) pp. 81-86
14. Colon, F., et al.: Coupling Small Computers for Performance Enhancement. Proc. of the Nat. Comput. Conf. 1976, pp. 755-764

15. Donner, H.: Tendenzen für die Architektur von Rechensystemen. NTG-Fchber.62 (1978), pp. 21-38
16. Tibbals, H.F.: Curran, P.: Optimizing Function Distribution in a Terminal Network, Microprocessors (1977), pp. 362-368
17. Hoerbst, E.: Sauer, A.: Weiss, J.: Verkuppelt. Klassifikation der Strukturen von Mikrocomputersystemen. Micomp 2 (1977) 6, pp.40, 42, 44 and 3 (1978) 1, pp. 40, 43, 44
18. Anderson, G.A.: Jensen, E.D.: Computer Interconnection Structures: Taxonomy, Characteristics and Examples, ACM Comput. Surv. 7 (1975), pp. 197-213
19. Bauer, J-L.: Multiprocessing Systems, IEEE Trans C 25 (1976), pp. 1271-1277
20. Spetz, W.L.: Microprocessor Networks. Comput. 10 (1977) 7, pp. 64-70
21. Bellm, J.: Hoerbst, E.: Sauer, A.: Vergleich einer vermaschten Struktur mit einer Sternstruktur in ein dezentralen Mehrrechnersystem. Int. Microcomputers. Minicomputers. Microprocessors: Geneva, 1977, pp. 115-112
22. Kroneberg, A: Hierarchisch strukturierte System-Software. Siemens Forsch.- u. Entwickl.-Ber.7 (1978), pp. 349-352
23. Zurmuehl, R.: Praktische Mathematik für Ingenieure und Physiker, Springer Verlag, 1965
24. Conrad, V.: Wallach, Y.: Iterative Solution of Linear Equations on a Parallel Processor System, IEEE Transaction on Computers, Vol. C-26, No. 9, Sept. 1977

ANALOG TECHNIQUES USING VLSI TECHNOLOGY

R. W. Brodersen

Department of Electrical Engineering
and Computer Sciences
University of California, Berkeley, CA 94720

ABSTRACT The basic components needed to implement analog techniques in a MOS-LSI compatible technology will be described. Techniques will be shown which use these components to implement the interface functions of filtering, amplification digital to analog and analog to digital conversion.

1. INTRODUCTION

The primary emphasis in the development of VLSI technology has been on the optimization of the performance of digital logic circuits. This unfortunately is inconsistent with improving (or even maintaining) the performance of the analog interface circuits. It would be a considerable limitation to the usefulness of VLSI technology if techniques are not found which make it possible for the interface circuits to improve in performance along with the digital technology [1]. A review will be given of the analog techniques which have been developed which make use of MOS technology which will be followed by comments on the effect of scaling on analog circuit performance.

1.1. ANALOG COMPONENTS [2]: Elements which are of key importance in analog circuits are capacitors, resistors, switches and amplifiers. Ratios of capacitors are the precision component of A/D and D/A converters as well as determine the frequency response of filters. The near-ideal MOS switches make possible novel circuit configurations which are not possible in other technologies.

1.1.1. MOS Capacitors: The MOS transistor itself is essentially a nonlinear capacitor in which the charge induced on one plate by the applied voltage forms the conducting channel of the transistor. The dielectric of this capacitor is silicon dioxide, which is one of the most stable dielectrics known. However, the MOS transistor itself is not particularly useful as a capacitor because of its inherently nonlinear behavior. Within an advanced MOS technology, capacitors which have a sufficiently low voltage coefficient to be useful as precision passive components are generally made with two separate layers of poly-silicon which are separated by SiO_2.

These capacitance structures tend to have common characteristics. >From the standpoint of switched-capacitor filters and D/A converters the most important of the characteristics of MOS capacitors are: ratio accuracy, voltage coefficient, and parasitics.

1.1.1.1. Ratio accuracy: A key aspect of the performance of any frequency-selective filter is the accuracy and reproducibility of the frequency response and for a D/A converter the accuracy of the quantized voltage levels. MOS compatible circuit techniques have been developed which rely on a certain level of accuracy in the ratios of capacitors in order to meet these requirements.

Integrated MOS capacitors have a value which is determined by the dielectric constant, the thickness of the dielectric and the area of the capacitor. Assuming that the dielectric constant and thickness do not vary, the ratio of two capacitors made within the same integrated circuit will depend only on their area ratio. This is primarily determined by the geometrical shape of the capacitors defined by the photolithographic mask used to make the integrated circuit. Errors in the geometrically defined ratio occur because the actual definition of the capacitor is done optically, which results in errors in the capacitor edge location due to the finite wavelength of the light used and the control of the chemical or plasma etching of the metallization and in addition because the thickness of the dielectric can vary with distance across the integrated circuit. These effects can be alleviated with careful layout of the components and the errors generally get smaller as the capacitor dimensions are made larger. Generally speaking, the achievable ratio accuracies range from 1-2% for small capcitor geometries (<400 square microns) to on the order of .1% for capacitor geometries which approach the limit of economical size (40,000 square microns) [3]. This also implies that as the capacitance ratio gets larger, the achievable accuracy in a given area decreases since the smaller of the two capacitors must be decreasing.

1.1.1.2. Voltage and temperature coefficients: MOS capacitors made with heavily doped silicon plates display voltage coefficients in the range of 10 to 100 parts per million per volt. Temperature coefficients are generally in the range of from 20 to 50 parts per million per degree centigrade and are of course much lower for the value of a ratio [3]. These variations are low enough to be insignificant in almost all applications.

1.1.1.3. Parasitic capacitances: A sizeable parasitic capacitance exists from the bottom plate of the capacitor to the substrate which is the capacitance of the SiO_2 layer under the first layer of poly. Typically, this capacitance has a value of from one-fifth to one-twentieth of the MOS capacitor itself, depending on the specifics of the technology. Also, because the top plate of the MOS capacitor must be connected to other circuitry, a small capacitance will exist from the top plate to the substrate due to the interconnections. This capacitance can range from .01 to .001 of the desired MOS capacitance, depending on the capacitor size, layout and technology. These parasitic capacitances are unavoidable, and the design of filters and A/D and D/A converters must be done in such a way that they do not degrade performance [4].

1.1.2. Resistors: Resistors can be formed from the source-drain diffusions as well as by poly-silicon lines. Even though the temperature coefficient, voltage coefficient and the obtainable ratio accuracy are vastly inferior to MOS capacitors; resistors are sometimes used as precision components. The advantage of resistors are that simpler circuits are sometimes possible, since capacitors cannot supply current and thus cannot act as a voltage source. Also, capacitive dividers require some form of resetting to insure that parasitic leakage currents do not cause voltage offsets on pure capacitive nodes.

1.1.3. MOS Switches: The second principal component in analog MOS circuits is the MOS switch. This device behaves like a resistor in the on-state and in the off-state like an open circuit. The important parameters of the MOS transistor from the standpoint of an analog circuit component are the on resistance, off leakage currents, and the parasitic capacitances. Typical NMOS silicon gate technology is capable of producing switch devices with a channel length of 5 microns and assuming a length/width ratio unity, an on resistance of 5k ohms (with a gate drive voltage of five volts with respect to the source). This device would display a leakage current from the source and drain to the substrate of the order of 10^{-14} amps at 70 degrees centigrade. A value which is negligible in most applications.The parasitic capacitance from source and drain to substrate would be about .020pF (picofarads) each, and the overlap capacitance from drain to gate and source to gate would be about.005pF. The capacitance of the channel when the gate potential is five volts more positive than the source and drain is approximately .008 pF.. The values of capacitances used as passive components typically range from 1 to 20 pF. Thus the effect of the various parasitic capacitances on the behavior of the circuit must be carefully considered, as well as the flow of the charge stored in the channel of the transistor.

1.1.4. Operational Amplifiers: Until very recently, virtually all commercially manufactured operational amplifiers were fabricated utilizing either bipolar technology or a mixture of bipolar and MOS technology. Recently, the trend toward higher levels of integration on MOS-LSI chips has led to the need for all-MOS operational amplifiers to be included on such chips. A considerable amount of recent work has been carried out directed towards the realization of MOS operational amplifiers. Typically achieved levels of performance are a gain of from 60 to 80 dB, commonmode and power supply rejection rations of 60 dB, unity-gain bandwidth of 2 MHz, and power dissipation of 1-5 mW [5]. These devices typically occupy a die area on the order of 0.2 mm^2, so that the inclusion of several tens of them on a single chip is feasible.

1.2. ANALOG SAMPLED DATA FILTERS [4] The most common signal processing function is frequency selective filtering, which accounts for the variety of methods which have been developed to accomplish this task over the years. The development of low cost, high-performance monolithic operational amplifiers led to the utilization of active RC filtering techniques to replace passive RLC filters in the 1960's. These RC active filters have proven difficult to realize in completely monolithic form because stable, precise RC products cannot be achieved in standard I.C. processes.

As a result, the preferred technological approach to voice-frequency filtering until very recently has been discrete component or hybrid integrated circuit realizations of various types of active filters. Recent developments in both analog and digital filters which use MOS-LSI have made it possible to fully integrate the voice-frequency filtering function, with the usual advantages in cost, size, power and weight.

While the density advantage of MOS has led to its wide utilization for digital circuits, a second key advantage, which has only recently been recognized, has led to the increasing use of MOS technology to perform analog signal processing. In contrast to bipolar technology, MOS integrated circuits offer the ability to store charge on a node over a period of many milliseconds and to sense the value of the charge continuously and non-destructively. The former results from the high impedance of the MOS transistor in the off state, and the latter results from the essentially infinite input impedance of the MOS transistor. This inherent analog memory capability does not exist in bipolar technology.

The charge storage feature of MOS technology was first utilized in dynamic random access memory design and dynamic logic. In both instances, information storage on small nodal capacitances in the circuit allows the elimination of active or pullup transistors which are otherwise required to preserve the information content of the circuit. This results in greater circuit density. While the information stored on nodal capacitance is essentially analog in that the node voltage has a continuous range of values, in digital applications the information is usually interpreted as being only one of two logic levels; these binary levels are typically refreshed (restored) every two milliseconds.

The charge storage feature of MOS technology was first used to perform analog signal processing with the advent of the bucket brigade shift register. In this device, a time-varying analog signal is converted to a series of charge packets, proportional to the value of the input signal at the sample instants. These charge packets are passed serially through a chain of MOS capacitors, the postion of the charge being brought about by the action of appropriately clocked MOS devices. In analog filtering applications of such devices, the charge at each capacitor site is sensed non-destructively by MOS transistors. These sensed voltages can, for example, be added in a weighted summation to produce a single output, providing a realization of a sampled-data transversal filter as shown in Fig. 1 [6].

The charge-coupled device (CCD), which was developed subsequent to the bucket brigade, provides a similar function in analog signal processing applications, but with the integration of the storage capacitance with the MOS switch structure and MOS sensing device in such a way that transversal filters and other analog signal processing devices can be fabricated in a smaller silicon area and achieve higher performance. These devices provide an extremely effective technological tool for attacking filtering problems in which transversal filter structures are required, for example, as matched and adaptive filters [7].

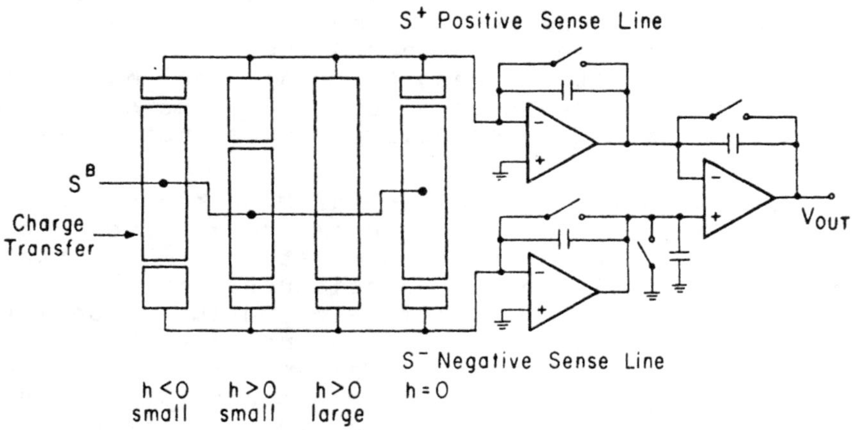

Figure 1: Block diagram of a CCD transversal filter.

However, the utilization of CCD and bucket brigade filters in large- quantity low-cost general filtering applications has been hindered by several factors. First, until very recently, the peripheral circuitry required to convert the input signal into charge packets, and to recover a low-impedance voltage signal at the output, had to be realized off- chip since this circuitry was not thought to be easily realizable on the same chip. Second, the transversal filter itself has certain limitations as a general-purpose analog filtering and signal processing technique. Among these is the fact that since the signal is time sampled at the input to the CCD, a continuous-time anti-aliasing filter is required to bandlimit the input signal to a frequency below the sampling rate. Since the sampling rate in CCD's must be relatively low (to minimize filter length and conserve silicon area) a relatively complex anti- aliasing filter is required. Also, since CCD filters are non-recursive, for transfer functions with a narrow passband which have long impulse responses, a very long CCD filter utilizing large silicon area is required.

In the last several years, the concept of analog charge storage and processing in MOS technology has been applied in new ways and has overcome some of the difficulties inherent in the transversal filter structures. First, in order to obtain highly selective filters with long impulse responses in a small silicon area, one is led to adopt discrete time, recursive filter structures. A particularly useful class of these filers have proved to be filters which are essentially discrete time simulations of active filter circuits in the limit of high sampling rates. These filters have come to be called switched-capacitor filters, and are distinguished by the fact that they utilize a capacitor and MOS switches to simulate the circuit behavior of a resistor, as shown in Fig. 2 [8].

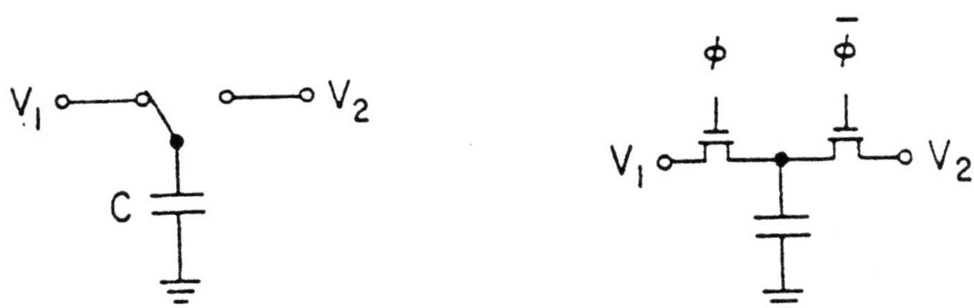

Figure 2: Switched-capacitor equivalent of a resistor.

The operation of this "resistor" is as follows: the switch is initially in the left-hand position so that the capacitor ,C , is charged to the voltage V_1. The switch is then thrown to the right and the capacitor is discharged to the voltage V_2. The amount of charge which flows into (or from) V_2 is thus $Q=C(V_2-V_1)$. If the switch is thrown back and forth at a clock rate, f_c, then the average current flow i from V_2 into V_1 will be $C(V_2-V_1)f_c$. Thus the size of an equivalent resistor which would give the same average current as this circuit is

$$R = \frac{1}{Cf_c} \qquad (6)$$

the most important advantages of the use of switched-capacitors. The 3 dB bandwidth of the conventional RC filter is

$$\omega_{3db}^{RC} = \frac{1}{R_1C_2} \tag{7}$$

The 3 dB width of the switched-capacitor filter can be approximately found by substituting the effective resistance of the switched capacitor C1 into Eq. 6 to obtain

$$\omega_{3db}^{SC} = f_c(C_1/C_2); \quad f_c >> \omega_{3db}^{SC} \tag{8}$$

(The constraint that $f_c >> \omega_{3db}^{SC}$ is imposed in order that the effect of time sampling and charge sharing can be ignored.). Since the bandwidth of the switched-capacitor filter is proportional to a ratio of capacitor values, it can be accurately defined with a high degree of stability.

>From Eq. 8 it can be seen that ω_{3db}^{SC} is proportional to the clock freuqency; thus programmability can be achieved by simply varying the clock rate (a property which is common to all discrete time filters).

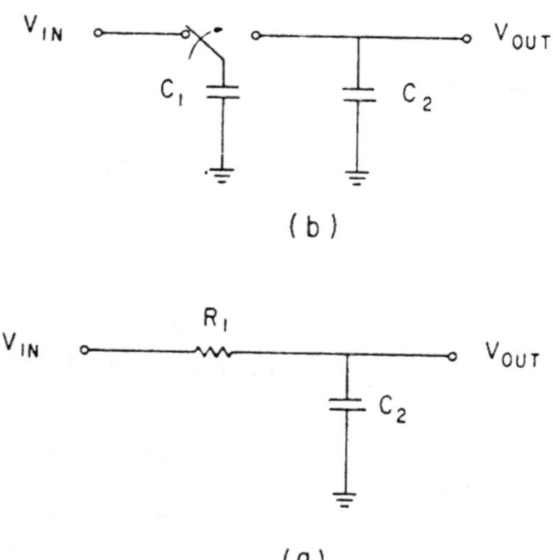

Figure 3: (a) Simple RC low-pass filter (b) Switched-capacitor equivalent.

The problem of implementing active filters in MOS technology is thus reduced to the question of what kind of active filter should be used. The filter configuration chosen is the analog computer simulation of the

If the switched rate is much larger than the signal frequencies of interest then the time sampling of the signal which occurs in this circuit can be ignored in a first-order analysis and the switched-capacitor can often be considered as a direct replacement for a conventional resistor. If, however, the switch rate and signal frequencies are of the same order then sampled data techniques are required for analysis and, as for any sampled data system, the input signal should be bandlimited as dictated by the sampling theorem.

The switched-capacitor resistors require very little silicon area to implement large resistance values. In fact the silicon area decreases as the required value of resistance increases. To implement audio frequency filters a resistance on the order of 10 megohms is needed if a monolithic capacitor of a reasonable size is to be used (~10pF). This value of resistance is easily achieved by switching a 1pF capacitor at a 100 kHz rate, requiring approximately a silicon area of .01 mm^2 (to be compared to a total chip area of 10-20 mm^2). If a 10 megohm resistor were implemented by using a polysilicon line or diffusion the area required would be at least 100 times larger.

By using this switched-capacitor resistor in conjunction with other capacitors and op amps it is possible to realize many of the circuit configurations used in conventional RC active filters. As in the case of digital filters and CCD filters, the signal in switched-capacitor filters is sampled in time so that anti-aliasing filtering is required. Fortunately, the penalty in silicon area for a high sample rate is much less in the switched-capacitor filters in comparison to the CCD transversal filters and thus the anti-aliasing filter can be quite simple.

Viewed from a conventional RC active filter standpoint, these filters have the property, which will be demonstrated below, that all the RC time constants of the filter are determined by the clock frequency which is used to drive the switches and by capacitor ratios. Thus the problem of controlling RC products to a high degree of accuracy, which is characteristic of RC active filter realizations, is reduced for these filters to the problem of maintaining accurate capacitor ratios and supplying an accurate external clock frequency. This property makes the filters much more practical for monolithic realization.

A conventional single pole RC lowpass filter is shown in Fig. 3a and a switched capacitor implementation of this filter is shown in Fig. 3b. In spite of the simplicity of this filter it demonstrates some of

equations which describe a passive doubly-terminated RLC ladder [9,10] These filters are called "leapfrog" or "active ladder" filters in the active filter circuit literature and are closely related to wave digital filters in the digital signal processing literature.

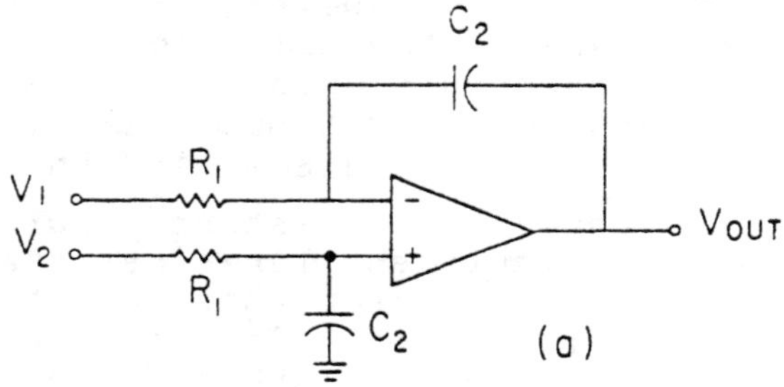

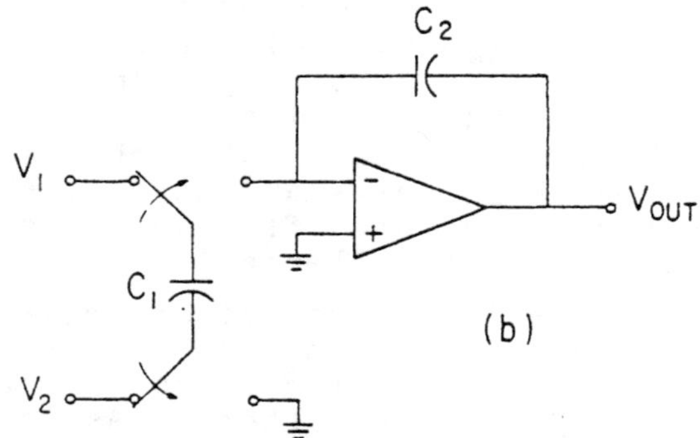

Figure 4: Switched-Capacitor integrator.

The basic building block of these filters is a differential integrator and summer which is shown in Fig. 4 (a) that is implemented in the conventional way with resistors and capacitors. The equations which describe this circuit are:

$$V_{out}(\omega) = -\frac{C_s}{C_I}V_1 + \frac{1/RC_I}{j\omega}(V_3-V_2) \tag{9}$$

Straightforward substitution of the resistors by the circuit in Fig.2 would yield a switched-capacitor circuit that is sensitive to parasitic capacitances, so an alternate method of realizing the resistors which also makes it possible to invert the sign of the signal is used and is shown in Fig. 4(b) [11]. The circuit in this figure is described by the following z-transform equation:

$$V_{out}(z) = -\frac{C_s}{C_I}V_1 + \frac{C_U/C_I}{1-z^{-1}}(z^{-1}V_3-V_2) \tag{10}$$

Appropriate interconnection of these sampled data integrators following an RC active filter prototype yields the parasitic free, fifth order switched capacitor ladder shown in Fig. 5.

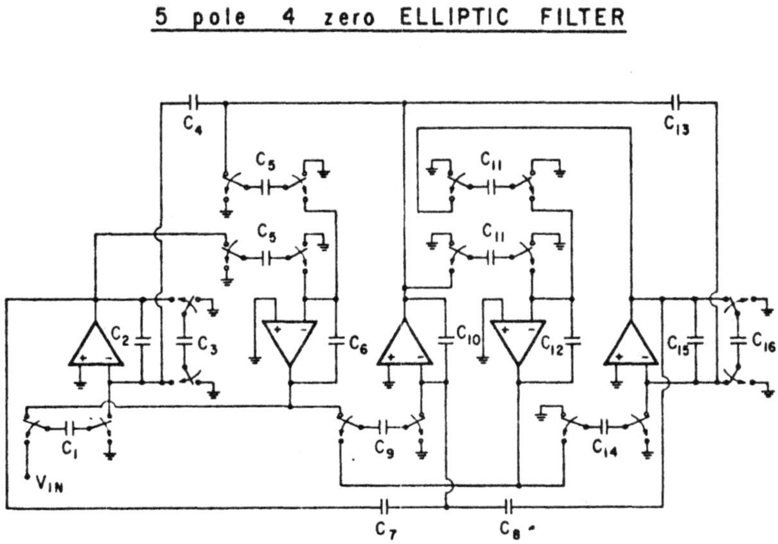

Figure 5: Ladder filter implemented using switched-capacitor integrators.

1.3. DIGITAL TO ANALOG CONVERTERS [12]

The two basic techniques which have emerged as the most suitable for precision D/A converters, which are compatible with standard MOS technology are voltage division using a resistor string and charge redistribution using MOS capacitors.

1.3.1. Resistive Voltage Dividers: The most straight forward form of D/A conversion (but not the most accurate) is to divide a reference voltage into the required number of quantized voltage levels by using a string of 2^m series resistors for an m-bit converter as shown in Fig. 6 [13]. A 1 of 2^m decoding tree which is shown can be used to tap into the desired point on the resistor string. This tree is particularly compact in MOS technology because of the availability of the MOS switch which can be merged with the resistor string, if the resistor string is fabricated using the source/drain diffusion. The area for an 8-bit converter is quite small, and 10 and 12 bit converters have been fabricated with some slight modifications of this basic technique.

1.3.2. Capacitive Charge Redistribution: Voltage division can also be obtained with the use of capacitors as the passive elements such as in Fig. 7 [14]. The operation of this converter begins by a resetting of the top plate to ground. Following this the desired combination of switches on the bottom plate are thrown to the reference voltage V_{ref}. A charge redistribution occurs which results in a voltage on the top plate of

$$V_{out} = V_{ref} \sum_i b_i 2^{-i} \tag{11}$$

Multiplying D/A converters have been fabricated using this technique. These circuits have proved to be very efficient methods for performing multiplications in hybrid analog-digital signal processing systems [15].

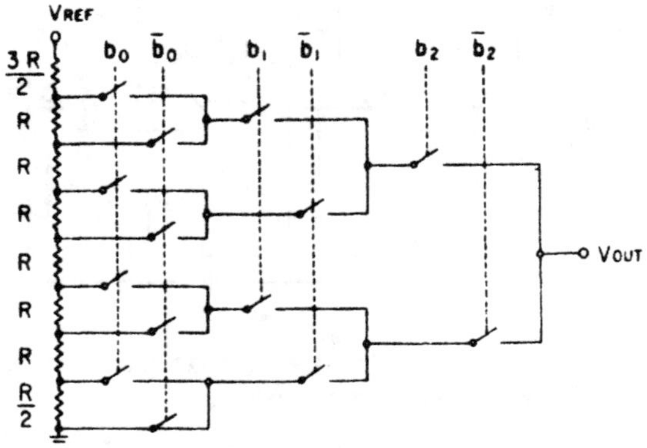

Figure 6: Resisor string digital to analog converter

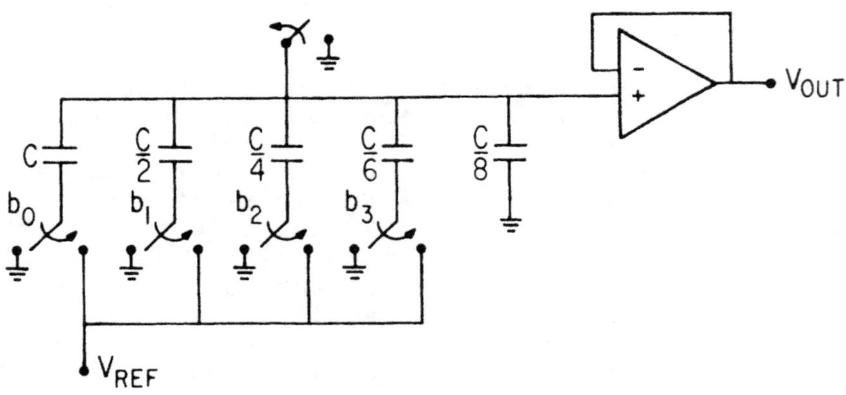

Figure 7: Circuit of a capacitive redistribution digital to analog converter.

1.4. ANALOG TO DIGITAL CONVERTERS [12]: A variety of methods have been developed to perform A/D conversion, which trade off accuracy, speed, and circuit area. The basic technique of all A/D converters is that a precise voltage is generated by a D/A converter which is then compared to samples of the incoming analog signal. Since the incoming signal must be sampled in time it has to be bandlimited before conversion. The three basic types of converters are serial, successive-approximation, and flash (or parallel).

1.4.1. Serial converters: This type of converter is for signals with low bandwidths (<100Hz) and high accuracy (> 12 bits). An example of a serial converter is shown in Fig. 8. Switch 1 connects the integrator within tha dashed lines to either the magnitude of the sampled input signal, $|V_{in}|$ or to a negative referance voltage $-V_{ref}$. The conversion begins by resetting the op amp integrator with switch S_2 and integrating until the comparator is at the threshold of switching, V_{th}. The integration is performed using the same technique as the switched-capacitor filters in which the capacitor C_0 is switched between the op amp and ground. The integrator is then switched another fixed number of clock cycles, N_{ref}. The voltage, V_{Int} at the output of the integrator is then

$$V_{Int} = \frac{C_o}{C_I} N_{ref} V_{in} + V_{th} \tag{12}$$

The next step is to integrate the negative reference voltage until the integrator output is again at the comparator threshold. The number of clock cycles (i.e.times C_o is switched), which is measured by the counter N_{out} is given by:

$$N_{out} = \frac{V_{Int} - V_{th}}{(C_o/C_I)} \tag{13}$$

Combining Eqns. 12 and 13 and assuming that the reference count $N_{ref} = 2^m$ where m is the number of bits desired, N_{out} can be written

$$N_{out} = 2^m \frac{V_{in}}{V_{ref}} \tag{14}$$

which is the desired output digital word. Since N_{out} is not a function of the comparator threshold, the absolute value of any of the capacitors or the clock rate, extremely high accuracies are potentially possible. The limitations on accuracy are charge injection by the switches and the noise level of the comparator. Since it takes 2^m clock cycles for an m bit conversion this method is inherently slow, but the small amount of analog circuitry should make serial approaches more useful as the speed of the technology is increased.

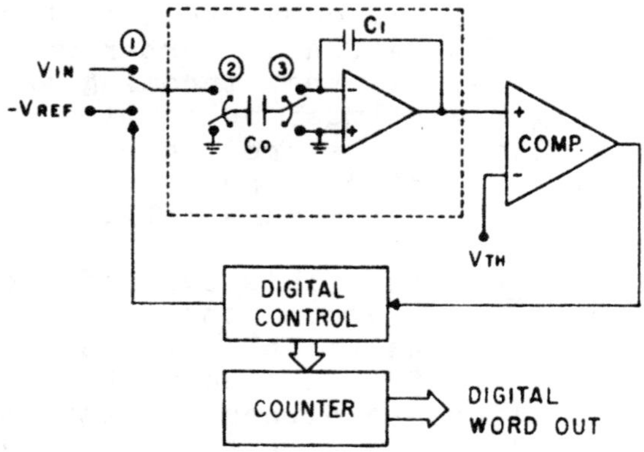

Figure 8: Circuit of a serial analog to digital converter.

<u>1.4.2. Successive-Approximation Converters:</u> The use of successive-approximation is the most important conversion technique for signal processing, since it can convert signals with high accuracy at bandwidths up to the present limits of MOS digital processors. In Fig. 9 is a block diagram of a successive-approximation A/D converter. The D/A converter in this figure can be implemented in a variety of ways (i.e. with resistors or capacitors), as long as a discrete voltage level can be generated in a few clock cycles. In the serial converter, since each voltage is generated in sequence, the tests (comparisons of the input signal with the discrete voltage levels) must also be sequentially performed. In a successive-approximation convertor a carefully selected sequence of tests are performed which allow rapid determination of the proper digital output word. The succesive-approximation sequence requires m clock cycles for an m bit conversion. The disadvantage is that the resolution of this type of convertor is determined by the absolute accuracy of the D/A converter.

The first test which is performed (after the sign of the signal is found) determines if the magnitude of the input signal lies in the upper or lower half of the allowed input range, i.e., is between 0 and $V_{ref}/2$ or between $V_{ref}/2$ and V_{ref}. This yields the most significant bit of the digital output word. The speed advantage is obtained, because if the voltage is found to be in the lower half, none of the voltages in the upper half need be tested (and vice-versa). Next, a test is performed to see if the input is in the upper or lower half of the remaning allowed range to determine the next most significant bit. Again, the result of this test allows the elimination of the testing of half of the remaining voltage levels. This sequence of tests continues until the least significant bit is found.

The bandwidth of signals which can be processed by successive-approximation converters ranges up to several MHz and accuracies as high as 12 bits have been obtained [16,17]. There are presently a number of efforts which are attempting to obtain monolithic converters in the range of 14 - 16 bit accuracies for use in consumer digital audio applications.

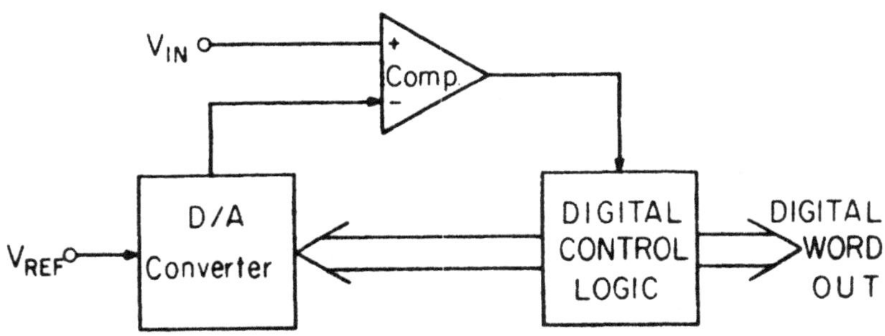

Figure 9: Block diagram of a successive approximation analog to digital converter.

1.4.3. Flash Converters: Through the use of parallelism a significant increase in bandwidth can be obtained at the expense of increased silicon area. The flash (or parallel) conversion approach is shown in Fig. 10. In this technique there are m comparators for an m-bit conversion. The resistor string can be formed by tapping a poly-silicon line or a diffusion in the silicon. The conversion time for this approach is just the time it takes for a single converter to make a comparison and thus the data rate can be extremely fast. However, since the amount of circuitry is increasing exponentially with the number of bits, the maximum accuracy at present device densities has been limited to about 6 bits (64 comparators) at 20 MHz [18]. This type of converter should be able to easily exploit the increased capabilities of VLSI technology and thus will probably be of increased importance as the technology is scaled.

1.5. ANALOG CIRCUIT SCALING: The analog circuits described in the last few sections have proved to be very useful techniques in todays technology. However, in order to assess their future importance, the effect of scaling on their performance must be investigated. The patterning process used for the passive elements is the same process which defines the features of the digital logic gates. As the device dimensions are scaled this patterning process must be improved. It will, therefore, be possible to obtain improved ratio accuracies of passive components (MOS capacitors and resistors) if their size is not scaled down or the same accuracy will be achievable in less area. Analog amplifiers will also be able to take advantage of the speed improvement which was obtained by the decreased capacitances of a scaled technology.

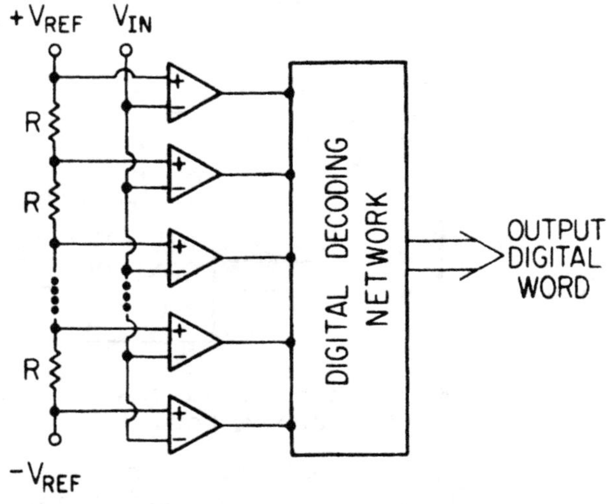

Figure 10: Schematic of a flash (parallel) analog to digital converter.

The major problem with scaling the analog circuits is the effect on the dynamic range, which is the ratio of the maximum signal which can be processed to the noise level. Since the maximum signal level decreases linearly with the supply voltage, if equivalent reductions are not obtained in the noise level when the supply is decreased then the dynamic range will be reduced. Furtunately, at this time there is sufficient margin that the supply can be substantially decreased over present day voltages (10 volts) and still retain adequate dynamic range for many applications.

To demonstrate this point it is first necessary to enumerate the fundamental noise sources and determine how they scale. There are two sources of the random noise in an MOS circuit: 1/f or surface state noise;

and thermal noise. The surface state noise has a power spectral density which increases as the frequency is reduced and is thus most important at low frequencies (<10 kHz). However, with some increase in circuitry it is possible to essentially eliminate this noise source by signal-chopping techniques such as chopper-stabilization and correlated- double sampling which significantly reduce the noise at frequencies below the chopping rate [19,20]. This is fortunate since the 1/f noise component increases as the gate capacitance is decreased.

The contribution of the thermal noise to the total circuit noise of some circuits (eg. the switched-capacitor filters) increases as the technology is scaled, since the inherent band-limiting by the circuit capacitances is reduced, if the standard scaling rules optimized for digital circuits are slavishly followed. This occurs in spite of the fact that the thermal noise itself (which is proportional to the square root of the channel resistance) is actually decreasing as $K^{1/2}$, where K is the scaling parameter (see section on scaling). However, if the capacitor areas are not scaled, the value of the capacitors will increase as the scaling factor, because the insulator thickness is being reduced, and the total thermal noise contribution will actually decrease as $K^{1/2}$. This noise decrease, is of course at the expense of reduced circuit density, so that a tradeoff between silicon area and dynamic range will be required for a given application. An estimate of reduction of the dynamic range can be made by assuming that the maximum signal decreases with the the supply voltage which scales down as K, which yields a decrease in dynamic range by the square root of the scaling factor, $K^{1/2}$. If it is assumed that between now and 1990 that the scaling will be by a factor of 5 (i.e. supply voltages which are now 10 volts for analog circuits will be reduced to 2 volts), the reduction in dynamic range will be the very moderate level of 7 dB. At this time complex analog filtering circuits (~20 poles) are being fabricated with dynamic ranges on the order of 97 dB. [21]. A reduction in dynamic range to 90 dB remains more than adequate for most signal processing applications.

2. CONCLUSIONS

A variety of analog design techniques have been developed which are compatible with MOS technology and it is expected that these techniques will continue to be important as the technology is scaled. In fact, if progress is not made in the analog interface circuits which is equivalent to the improvements in the digital processors, that the ability for VLSI to make an impact will be greatly diminished.

REFERENCES

1. N. Bernstein, "What to Look For in Input/Output Boards," Electronics, January 19, 1978, pp. 113-119.
2. D.A. Hodges , P.R. Gray and R. W. Brodersen, "Potential of MOS Technologies for Analog Integrated Circuits," IEEE J. Solid State Circuits, Vol. SC-13, June, 1978, pp. 285-294.
3. J.L. McCreary and P.R. Gray, "All-MOS Charge Redistribution Analog-Digital Conversion Techniques-Part 1," IEEE J. Solid State Circuits, Vol. SC-10, pp. 371-379, Dec. 1975.
4. R.W. Brodersen, P.R. Gray, and D.A. Hodges, "MOS Switched-Capacitor Ladder Filters," Vol 67, No. 1, Jan., 1979, pp. 61-75.
5. D. Senderowicz, D.A. Hodges and P.R. Gray, "A High Performance NMOS Operational Amplifier," IEEE J. Solid State Circuits, Deca, 1978.
6. D.D. Buss, D.R. Collins, W.H. Bailey and C.R. Reeves, "Transversal Filtering Using Charged Coupled Devices," IEEE J. Solid-State Circuits, Vol SC-8, Apr., 1973, pp. 138-146.
7. C.R. Hewes, R.W. Brodersen and D.D. Buss, "Applications of CCD and Switched-Capacitor Filter Technology," Proceedings of the IEEE, Oct., 1979, pp.1403-1415.
8. D.L. Fried, "Analog Sampled-data Filters," IEEE J. Solid State Circuits, (Corresp.), SC-7, pp. 302-304, Aug. 1972.
9. B.J. Hosticka, R.W. Brodersen, and P.R. Gray, "MOS Sampled Data Recursive Filters Using Switched-Capacitor Integrators," IEEE J. Solid State Circuits, vol. SC-12, pp.600-608, Dec., 1977.
10. J.T. Caves et. al., "Sampled Analog Filtering Using Switched- Capacitors as Resistor Equivalents," IEEE J. Solid-State Circuits, Vol. 6, SC-12, Dec., 1977, pp. 592-599.
11. G.M. Jacobs, D.J. Allstot, R.W. Brodersen and P.R. Gray, "Design Techniques for Switched-Capacitor Ladder Filters," IEEE Trans. on Circuits and Sys., Vol. CAS-25, No. 12, Dec., 1978, pp. 1014-1021.
12. P.R. Gray and D.A. Hodges, "All-MOS Analog-Digital Converter Techniques," IEEE Trans on Circuits and Systems, CAS-25, No. 7, July, 1978, pp482-489.

13. A.R. Hamade and E. Campbell, "A Single-chip 8-bit A/D Converter," ISSCC Dig. Tech. Papers, Feb. 1976, pp.154-155.
14. J.F. Albarron, "A Charge-Transfer Multiplying D to A Converter," IEEE J. Solid-State Circuits, Vol. SC-11, Dec., 1976, pp.772-779.
15. B. Fotouhi and D.A. Hodges, "High Resolution A/D Conversion in MOS-LSI," IEEE J. Solid-State Circuits, Vol. SC-14, No. 6, Dec., 1979, pp. 920-925.
16. A.G.F. Dingwall, "Monolithic Expandable 6 Bit 20 MHz CMOS/SOS A/D Converter," IEEE J. Solid-State Circuits, Vol. SC-14, No. 6, Dec., 1979, pp. 926-937.
17. R. Poujois, J.M. Ittel and J. Borel, "A Low Drift MOSFET Operational Amplifier ARZ," European Solid-State Circuits Conf. Dig. Tech. Papers, 1976, pp. 14-15.
18. Y.S. Lee, L.G. Heller, and L.M. Terman, "A 1 mV MOS Comparator," European Solid-State Circuits Conf. Dig. Tech. Papers, 1977, pp. 164-166
19. W.C. Black, D.J. Allstot, S.H. Patel and J.B. Wieser," CMOS PCM Channel Filter," Int. Solid-State Circuits Conf. Dig. Tech. Papers, Feb., 1980, pp. 84-85.

VLSI TECHNOLOGY IMPLICATIONS FOR SIGNAL PROCESSING

R. W. Brodersen

Department of Electrical Engineering
and Computer Sciences
University of California, Berkeley, CA 94720

ABSTRACT

The implementation of digital signal processing algorithms has until recently been almost exclusively performed with the use of high speed bipolar technology. There has recently been much interest in the capabilities of MOS technology because its performance has improved to the point that sophisticated algorithms can now be implemented on signals at up to audio bandwidths. However, when projections are made to near the fundamental limits of MOS technology (i.e. VLSI densities) it is found that its processing power will only have improved to the point where it is equal to the capabilities of today's advanced bipolar technologies. Therefore, the real advantage of MOS technology will not be vastly increased processing power, but will be the low cost, size and power consumption that can be obtained if a complete system can be integrated onto a few integrated circuits.

1. INTRODUCTION

Very large scale integrated circuits (VLSI) is the name given to integrated circuits (I.C.'s), which by advances in I.C. technology, have become larger than some arbitrarily chosen size. The point at which circuits become VLSI instead of just LSI is unclear. This confusion is apparent if one observes the wide range of definitions for VLSI which have been proposed, which range from as low as 1000 logic gates on one integrated circuit (chip) [1], up to circuits which contain more than 100,000 gates [2]. The reason for this indecision is that VLSI is not a step function improvement in the capability of I.C. technology, but rather just the next stage in its continuing development. This continuity makes it possible to use the experience of the recent past to project the nature and timetable of future developments.

A critical assumption which must be made is which present day technology to use for the projections. Most (but by no means all) believe that the metal-oxide-semiconductor (MOS) technology has the most promise for the future because of the relatively simple structure of the MOS device. The more complex bipolar technology and devices, are not as able to exploit the advantages in density of shrinking (scaling) the minimum feature size. However, in order to actually provide new capabilities for signal processing, it is speed, not density, which is required. Increased density is important in that it provides the possibility of reducing the size, cost, power and weight of systems which are now implemented with bipolar technology; but even in 10 years, when MOS technology is near its fundamental limits, its processing speed will not be as high as today's advanced bipolar technologies. As will be discussed below, this lack of capability even occurs if it is assumed that architectures and algorithms (such as parallel processing and pipelining) are used, that can convert increased complexity into increased processing power.

The reason high speed is so important is that signal processing is the computation of mathematical algorithms on "real world" signals in real time; i.e. the processing rate is equal to or greater than the data rate of the incoming signal. The real time requirement has a very fundamental effect on the hardware implementation since processing speed even for the lowest bandwidth signals is usually the most difficult performance criteria to meet. Until recently the only means for obtaining adequate speed in cost sensitive applications has been the use of analog circuitry, even though these techniques severely limit the complexity of the signal processing algorithms which can be implemented. It is expected that the capabilities of VLSI technology will finally make it possible for digital techniques to be useful in other than low volume, expensive systems which can make use of high speed bipolar technology. However, it should be mentioned that this has been the prediction of researchers for at least the last 10 years and except in very limited application areas that the analog approach has been found to yield adequate performance at a much lower cost, less power and smaller size. Even now analog techniques have been developed which use a discrete time representation of signals and can be implemented using MOS technology [3]. Thus these analog methods can make use of many of the advances in digital signal processing theory as well as the advances in I.C. technology. In fact to a very important extent progress in digital signal processing is tied to advances in analog techniques, since the "real world" signals must be amplified, filtered and A/D converted in the analog domain before they can be processed digitally.

For signal processing probably the most important system consideration is the bandwidth of the signal to be processed. A graph of the bandwidths of a variety of signals is given in Fig. 1. The bandwidths in this figure cover the enormous range of 10 orders of magnitude in frequency. At the low end are the seismic signals which do not extend much below 1 Hz because of the absorption characteristics of the earth. At the other extreme are the microwave signals which are not used much above 30 GHz because of the difficulties in performing even the simplest forms of signal processing at higher frequencies.

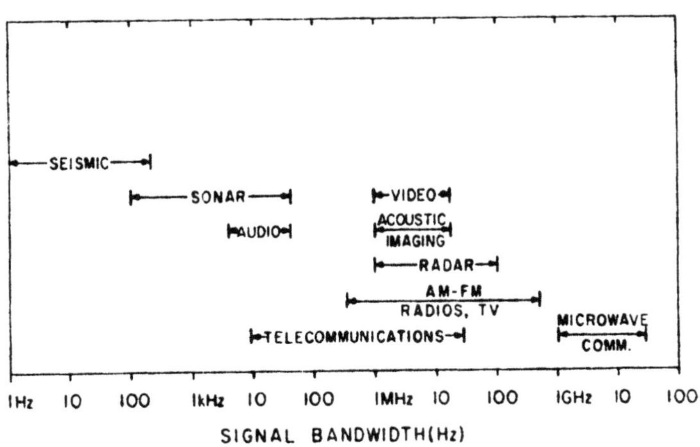

Figure 1: Bandwidths of signals used in signal processing applications.

In order to perform signal processing over this range of frequencies a variety of techniques have been developed, which are almost exclusively analog above 10 MHz and digital below 100 Hz, as shown in Fig. 2. In the overlap region a tradeoff must be made between the accuracy and flexibility of a digital approach and the low cost, power and size of analog techniques.

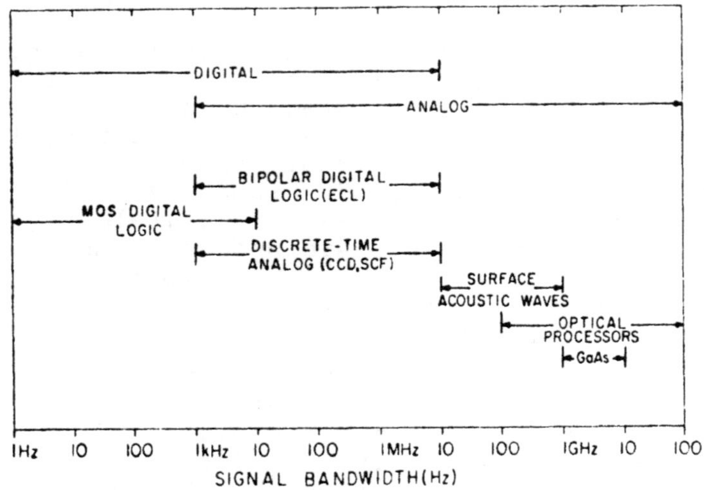

Figure 2: The signal bandwidths which can be processed by present day technologies.

The fastest moving boundary in Fig. 2 is the upper limit of the MOS-LSI digital logic, which as the technology progresses to VLSI (as will be shown in a later section) should be able to process signals by the year 1990 at bandwidths of 1-10 MHz. This will be accomplished by means of the utilization of greatly increased density along with moderately increased device speeds. However, as can be seen in this figure bipolar digital techniques are already able to process at these rates. Therefore VLSI will not make possible increased processing rates over what can now be achieved with processors built with bipolar technology, but rather that the reduced cost, size and power requirements of the MOS-VLSI signal processors will be the primary advantage. This will make it possible for signal processing techniques to make an impact in cost sensitive areas such as consumer products, which has until now been unable to afford the costs of using sophisticated signal processing techniques.

There are a number of circuits which indicate that MOS technology has the capability at this time (1980) to process signals with bandwidths up to 10 KHz as is indicated in Fig. 2 [4-8]. The most impressive of these chips are the signal processors for telecommunication applications which were developed by Bell laboratories [4] and Nippon Electric Co. [5], both of which have the capability to implement over 40 second order filters at a sampling rate of 8 KHz, with a digital word length which yields a dynamic range on the order of 80 dB.

2. SCALING

One of the most important reasons for the increasing dominance of MOS technology is that as the device dimensions are reduced (scaled), improvements are obtained in essentially all performance characteristics of the resultant integrated circuits. Therefore, in order to predict the capabilities of the technology for the future an understanding of scaling theory is required.

The three terminals of a MOS transistor (MOSFET) are the source, drain and gate as shown in Fig. 3. The gate of a MOSFET, which is shown pictorially in Fig. 4, is typically a layer of deposited poly-crystalline silicon (polysilicon) which is heavily diffused with

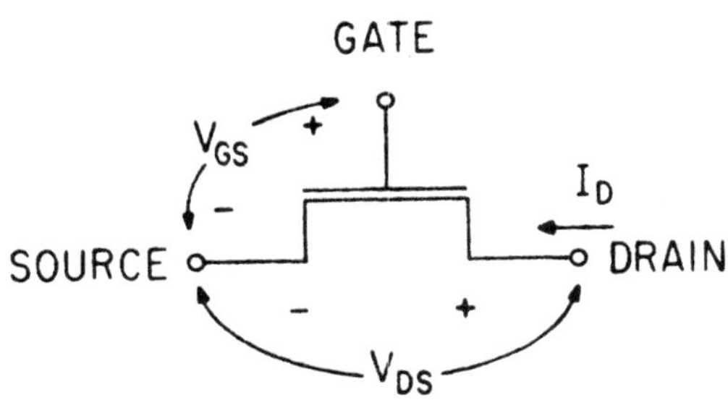

Figure 3: MOSFET showing voltage and current conventions.

impurities so that it is a reasonably good electrical conductor. It acts as one plate of a capacitor with the low conductivity silicon substrate the other plate. The source and drain are heavily doped regions which form p-n junctions (diodes) in the silicon as well as forming high conductivity paths. Under all normal operation the voltage on the source and drain regions with respect to the substrate is of the polarity which reverse biases the source and drain diodes. Under these conditions there is only extremely small current flow into the substrate. In order for there to be any current flow from the source to the drain it is necessary to apply a voltage to the gate. This attracts carriers to the silicon-silicon dioxide interface and thereby forms a conducting channel between the source and drain regions. The gate of a MOS transistor therefore acts as a switch between the source and drain which is open or closed depending on the voltage applied to the gate.

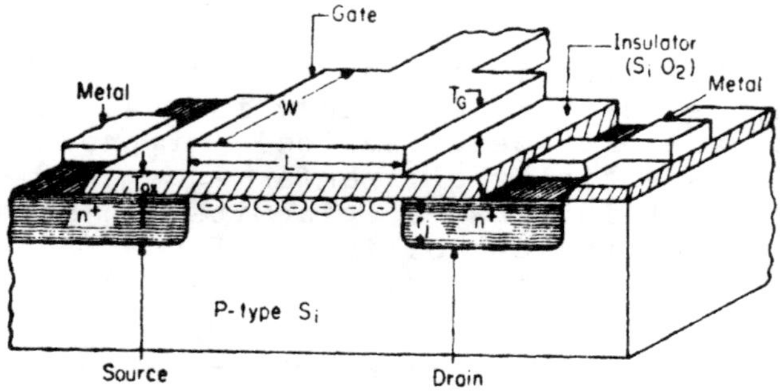

Figure 4: Drawing of a silicon-gate MOSFET.

In Fig. 4 the physical dimensions are labled, which under standard scaling rules are all reduced proportionally.The cross-hatched area is the insulator (SiO_2) which has a thickness, T_{ox};r_j is the junction depth of the source and drain diffusions (the shaded area); and the width and length of the gate are W and L respectively. The metal leads in Fig. 4 (which are typically aluminum) can connect this transistor to others as well as to the package. If all of the physical dimensions are scaled down by a factor of K, then the density (number of circuits per unit area) of circuit elements can increase as K^2. This increase in density has in recent years been the greatest contributor to the continuing trend of increased complexity in integrated circuits.

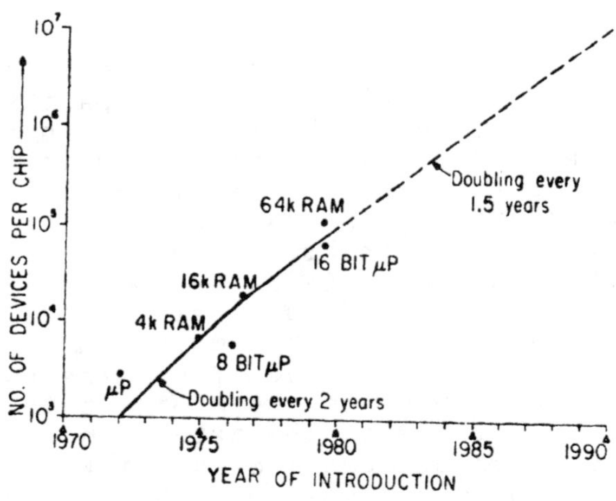

Figure 5: Increase in the number of transistors on an integrated circuit.

For the past 20 years it has been noted that the number of devices on a single integrated circuit has doubled every year [9]. This relationship is established by plotting the number of transistors per chip versus their year of introduction as in Fig. 5. It was recently noted by Gordon Moore of Intel Corp. who first noted this trend, that in recent years their seems to be "slowing" to a doubling in complexity every year and a half, which is indicated by the dashed line in this figure [10]. From the extrapolation it can be seen that the complexity of integrated circuits is expected to increase by 100 times in the next 10 years to chips which contain up to 10 milion devices.

The simplest form of the current voltage equations (See Fig. 3) of a MOSFET in which electrons carry the current in the channel (an n-channel or NMOS device) are given by:

$$I_{ds} = U_n \cdot C_{ox} \cdot (W/L) V'_{gs} \cdot V_{ds}; \quad V'_{gs} > 0, V_{ds} > V'_{gs} \qquad (1a)$$

$$I_{ds} = 0 \quad ; \quad V'_{gs} < 0 \qquad (1b)$$

where ,

$V'_{gs} = V_{gs} - V_t$

U_n = electron mobility

C_{ox} = oxide capacitance per unit area The effective gate voltage V'_{gs} takes into account a series of charge layers beneath the gate which can be modeled as an offset voltage, V_t, in series with the applied gate voltage, V_{gs}.

An important parameter for signal processing is the delay due to the finite resistance of a MOSFET to charge the parasitic capacitances to which it is connected. A rough estimate of this time (typically referred to as the gate delay) can be found by calculating the resistance of the channel of a MOSFET using Eq. 1 to obtain

$$R = \frac{V_{ds}}{I_{ds}} = \frac{L}{W \cdot U_n \cdot C_{ox} \cdot V'_{gs}} \qquad (2)$$

and since the gate capacitance, C_g, that a transistor must drive is given by

$$C_g = C_{ox} \cdot L \cdot W \qquad (3)$$

The gate delay $T_d = R \cdot C_g$ can be found

$$T_d = \frac{L^2}{U_n V'_{gs}} \qquad (4)$$

In the past few years since MOS circuits were made to operate from single 5 volt supplies, the voltage, V'gs, has been kept constant as the channel length has been scaled down by K. This results in a delay time which is inversely proportional to the square of the scaling factor, K:

$$T_d^{scaled} = \frac{(L/K)^2}{U_n V'_{gs}} = \frac{T_d}{K^2} \tag{5}$$

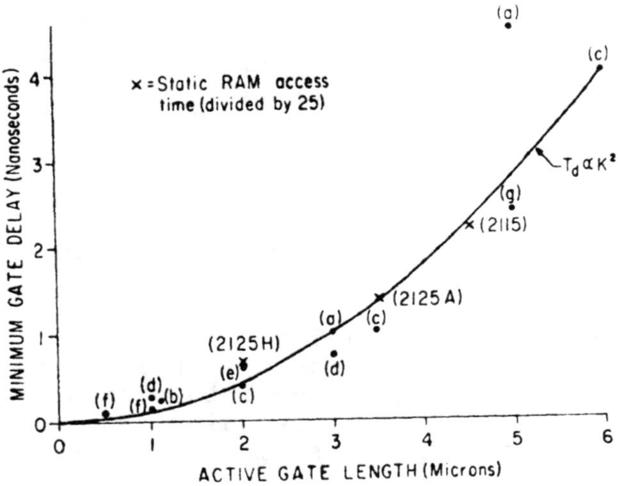

Figure 6: Minimum gate delay and static RAM access times as a function of active channel length, L.

In Fig. 6 is plotted the minimum delay versus channel length, L, as determined by ring counters for a variety of scaled devices (the points labeled a-g correspond to references 11-17 respectively). A ring counter is a loop of an odd number of inverters, N, which oscillates at a frequency proportional to 2N times the delay of one inverter. This is a minimum delay value since the load of each inverter is only one other inverter. Also in this figure, for comparison to the ring counter results, are three points which correspond to access times of commercially available static random access memories. As seen in this figure they also follow the speed increase predicted by Eqn. 5, which is plotted as the solid line in Fig. 6.

In order to indicate the rate at which scaling is proceeding, in Fig. 7 the active channel length, L, is plotted for a variety of commercial processes as a function of the year of their introduction. If the past trend continues, then in 1990 the channel length should be approximately .4 microns which is approaching some fundamental limits of MOS technology [18].

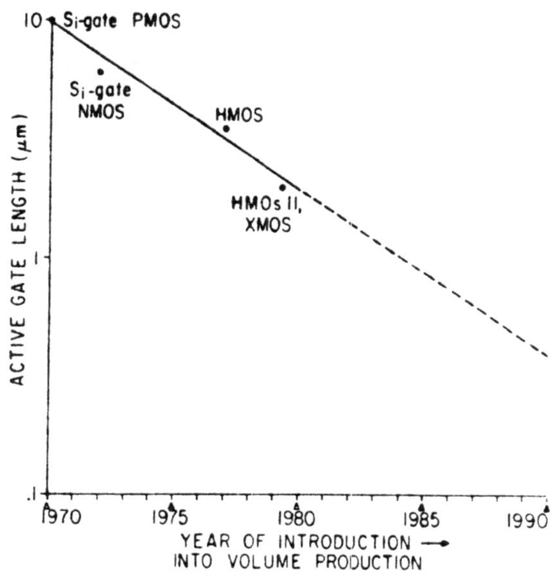

Figure 7: Rate of decrease of the minimum active channel length, L, for commercial processes.

3. IMPLICATIONS OF SCALING

In the last section it was shown that the device technology has been following a predictable trend of scaling to smaller dimensions, and it is expected that this trend will continue. This extension of MOS technology to VLSI complexity has some important implications for those who would like to use it for signal processing applications. In the following sections a few of these implications will be discussed.

3.1. NEW VOLTAGE STANDARD As mentioned previously when MOS technology advanced to the point at which 5 volt only circuits could be used it became almost a necessity for any new component to retain 5 volt compatibility. In Fig. 8 a plot is given of the supply voltage versus channel length that were used for the delay times plotted in Fig. 6. There are several reasons for this inflexibility. First, is that even at the present levels of integration the power supply can be one of the most costly components of the system as well as being the most unreliable. Therefore, the use of the smallest number of the most standard power supplies is desirable from the viewpoint of the system designer. Since in the most complex systems there will be (and will continue to be) a relatively large amount of circuits using unscaled MOS circuits or other technologies which require 5 volts, it is undesirable to use MOS parts that require additional voltage levels.

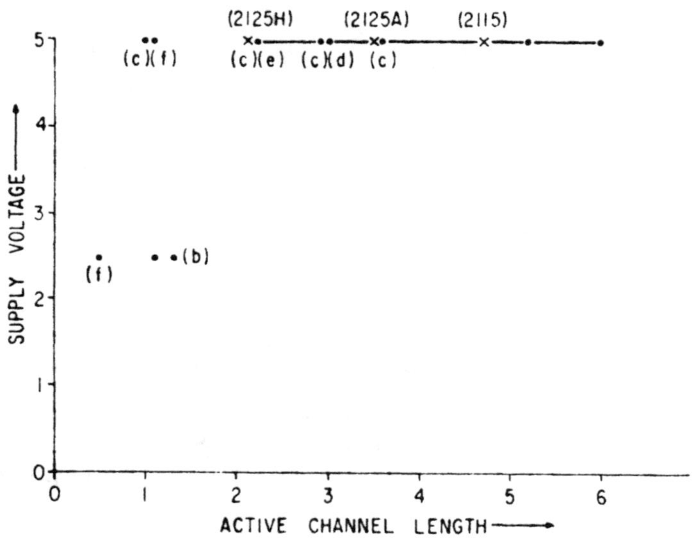

Figure 8: Supply voltage versus the active channel length, L.

A major inconsistency between scaling the physical dimensions, while retaining a constant supply voltage is that the electric fields across the insulator beneath the gate are increased. These increased fields result in electron injection and tunneling into and through the oxide as well as considerably enhancing the phenomenom of oxide wear-out. All of these effects adversly impact the reliability of circuits made using a scaled technology, if the operating voltages are too high. In Fig. 9 a plot of oxide thickness versus channel length for a number of recently developed processes is shown. If it is assumed that the oxide thickness continues to scale with channel length as projected by the line in this figure, then the oxide thickness for a

.4 micron device will be on the order of 80 angstroms. This would result in oxide electric fields of 7×10^6 with a 5 volt supply. In order to retain present day reliability, the oxide electric field should be at least four times less than the breakdown electric field (which is approximately $(10)^7$ V/CM for thin oxides). Thus the new supply voltage standard will need to be 2 volts or less by the end of the decade in order for VLSI circuits to be widely used.

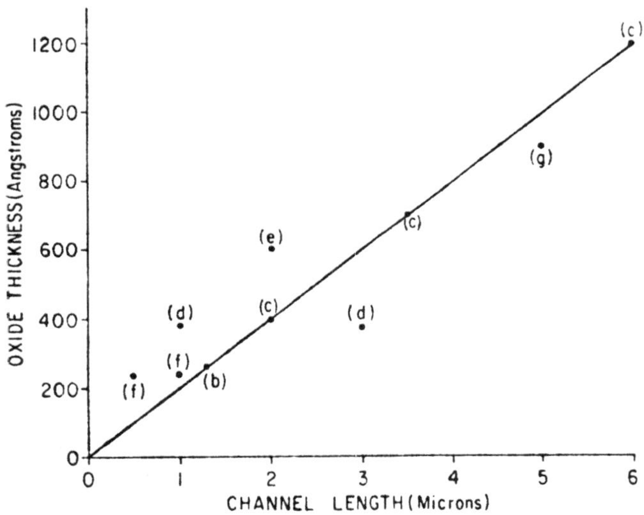

Figure 9: Scaling of the insulator thickness with active channel length, L.

3.2. PROCESSING BANDWIDTH Extrapolation of the trend of scaling of the channel length plotted in Fig. 7 indicates that there should be a decrease by a factor of 5 by the year 1990. Since at .4 micron channel lengths the technology is approaching some fundamental limits, either fundamentally new technologies must appear or the progress that the I.C. industry has experienced will drastically be reduced. In either case it is not productive to attempt to project past 1990. In that year from Eqn.5 we see that the gate delay should decrease by 5^2 or 25 times. However, this assumed a constant supply voltage, which as discussed above will result in reliability problems. If it is assumed that the new supply voltage standard is reduced two and a half times to 2 volts, then from Eqn. 4 it can be seen that the speed will also decrease by that amount. The increase in bandwidth of signals which can be processed due to increases in device speed will therefore be a factor of 10.

The other approach to increasing bandwidth as discussed above is to exploit the increased number of devices available on a VLSI circuit. From Fig. 5 it is seen that 100 times more devices will be available in 1990. Because of the difficulty in efficiently utilizing increased complexity, the processing rate will rise at a slower rate than the increase in the number of devices. One estimate for this effect is that the processing power increases as the .7 power of the number of circuits [19]. This would imply that the increased density of VLSI would result in a factor of $100^{.7}$ or 25 times more processing power. The total improvement in processing bandwidth from now to 1990 should be on the order of 25·10 or 250 times. If the processing bandwidth of MOS-LSI at this time is taken to be 10 KHz, the bandwidth in 1990 should therefore be in the range of 1-10 MHz (allowing for errors in the projections). As mentioned earlier this is just equivalent to the present capability of bipolar processors, so the real advantage of the VLSI circuits will be in the low cost, size and power of the VLSI approach.

3.3. ALGORITHMS AND ARCHITECTURES As discussed in the previous section the actual increase in processing rate for VLSI even at the fundamental limits of MOS technology will not be as fast as current bipolar processors, there can not be much allowance for inefficient architectures.

It is therefore important that architectures be developed that can efficiently use the increased number of devices which will be available. Architectures which can do this make use of concurrent processing in a variety of forms, such as parallel processing and pipelining, in which simultaneous processing occurs in one data path. In order to use these type of architectures the signal processing algorithms should contain as little branching and decision making steps as possible. In general it will probably not be true that the best algorithms will be those which contain the minimum number of multiplies or other arithmetic operations, since they may be trading off arithmetic computation for more complex control. A better criteria would be those algorithms that can be implemented by parallel processing or ones which have a minimum of branching so that pipelining can be used. For example, the extra data manipulation that the fast Fourier transform requires may make that a less desirable algorithm than a straight-forward calculation of the discrete Fourier transform, especially if only a small number of coefficints are required.

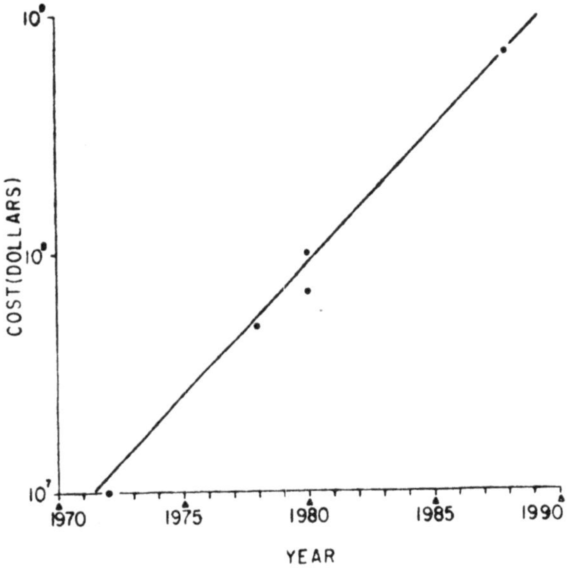

Figure 10: Capital equipment costs of high volume production lines.

3.4. ACCESS TO VLSI While most performance parameters improve as the technology is scaled there is one parameter that rapidly gets worse and it is the capital costs of starting up a new production line. As indicated in Fig. 10 the cost of a state-of-the-art, high volume production line (greater than 100,000 wafer starts per month) has been increasing almost as fast as the increase in complexity of the circuits which are being fabricated. This is having some profound effects on the nature of the semiconductor industry since only the largest manufacturers are able to afford this capital investment. This has a direct effect on the availability of VLSI chips for signal processing applications, since in general these chips are only needed in small volumes (at least small compared to memory and microprocessor chips) and it is not economical for a large manufacturer to produce them. This means that system houses which do not have internal fabrication capability will find it very difficult to obtain VLSI chips designed for specific signal processing applications. Therefore the vertically integrated system /

I.C. manufacturer will be the primary developer of VLSI signal processing circuits. There will of course continue to be general purpose chips that the I.C. manufacturers will continue to make available, however systems developed based on these chips will have performance which falls further and further behind circuits which can directly use architectures that can exploit the increased number of devices that VLSI will make available.

Simultaneous with the pressures for system houses to develop in-house processing capability, there are also pressures which are forcing I.C. manufacturers to develop system expertise. A problem which has been voiced by a top level manager of a components manufacturer is that "besides memory I don't have the slightest idea how to take advantage of VLSI" [9]. In the search for new markets some I.C. manufacturers are recognizing that signal processing applications have enormous potential. Recently there has been much interest in speech processing (recognition and synthesis) by the I.C. industry. Projections for this market segment are that total sales will reach the range of $1.3 to $1.8 billion by the late 1980's [21]. In order to attract personnel with systems backgrounds to the I.C. industry, some of the major manufacturers are organizing system oriented sub-divisions with a charter to develop speech processing products.

The other motivation for I.C. manufacturers to get into the system buisiness is that because of the high cost of capitalization, the VLSI circuits will be very costly. Therefore there is a strong economic advantge to selling systems which can be sold at much higher prices (and profit margins) than individual components.

<u>3.5.</u> <u>INTERFACE CIRCUITS</u> Since in the real world, signals are analog and continuous in time in order to use a digital signal processor there is a need for analog interface circuits which perform filtering, amplification, analog-to-digital and digital-to-analog conversion. In the past the interface circuits have often been an insignificant fraction of a complete system. However, with the projected densities and low cost of VLSI circuits, it is anticipated that for many system applications the digital portion of the system will be implemented on a single chip, which places new importance on the complexity of the interface circuitry which connects that chip to the outside world. Even now, with the density of digital circuitry achievable with state of the art LSI technology, the chip count,

power dissipation and size of simple systems can be dominated by the peripheral interface circuitry. For example, a not untypical data acquisition board for a microprocessor which is commercially available contains 10 chips and costs $100. Clearly further reduction in the digital circuitry in the microprocessor, which can now be implemented in a single chip for less than $10 will have small impact on a system which makes use of such a data acquisition board. Therefore, to truly realize the advantage of even present day technology, the importance of integration of interface circuits is clear. Extensive progress has been made in the development of interface circuits which use MOS-LSI technology and thus make possible the integration of the analog circuits onto the same chip as the digital processing. Also, because the density advantages of MOS carry over into analog circuits, sophisticated analog processing circuits can be implemented on a single I.C.. This makes it possible for the analog interface circuits to exploit the low cost, size and power advantages of integration.

The primary emphasis in the development of VLSI technology has been on the optimization of the performance of digital logic circuits. This unfortunately is inconsistent with improving (or even maintaining) the performance of the analog interface circuits. It would be a considerable limitation to the usefulness of VLSI technology if techniques are not found which make it possible for the interface circuits to improve in performance along with the digital technology. One approach which holds promise, trades off the requirements of analog accuracy with high speed digital operations, such as in the "interpolative" coders which have been developed for telecommunication applications.

3.6. DESIGN METHODS A critical problem related to the increase of device complexity is how a chip with an increase in the number of devices by two orders of magnitude can be designed in a reasonable time at a reasonable cost. Also after it is designed the layout must be checked to insure that it is correct (verified). A standard method used at this time for verification is to trace the circuit by hand. This method is clearly inadequate to handle circuits which contain 1-10 million devices (see Fig. 5). In fact, there are some who feel that the design problem is so difficult that only general purpose circuits are viable, because their general purpose character will allow them to be made in sufficient volume to justify the design costs. The problem with this approach (in addition to the performance problem discussed above) is that it becomes increasingly more difficult to define a chip that will be useful to large numbers of users as the chip complexity is increased.

Fortunately, there are some methods which are intermediate between the past procedure of optimization of almost every device and the inefficiency of a general purpose chip. Three of these techniques are gate arrays (masterslice); standard cell layout; and logic arrays. An important advantage of these design approaches is that one does not need to be very knowledgable of I.C. technology in order to design a chip. This results in some degree in inefficiency in the use of the technology so that for these design methods, the processing bandwidths will be even lower than the 1-10 MHz projected above.

3.6.1. Gate arrays: The gate array or masterslice technique makes use of a chip which has a fixed array of very simple gates which are interconnected in a custom manner using the various levels of interconnect available. Since the location of the logic gates are fixed the only step which remains in the design is the interconnection of the available gates. High levels of computer assistance can be used to assist in this wiring as well as in verification.

Probably the most aggressive use of gate array techniques for signal processing applications is by Fujitsu Ltd. who have developed a family of MOS gate arrays which contain up to 3900 gates (15-20,000 devices) in a technology which has a minimum gate delay of 8 nanoseconds. By comparison with Fig. 5 and 6 is seen that these arrays are well below the capability of state-of-the-art technology, but they compensate for this deficiency by the reduced design time and costs. It also should be pointed out that even the most advanced, general purpose microprocessor can not even come close to implementing the real time signal processing functions (such as a 2400 baud modem) for which these gate arrays are being used.

3.6.2. Standard cell layout: In this technique a library of logic functions is developed which contains logic elements such as NOR's, NAND's and flip-flops. These standard cells are typically rectangular and have standard locations for the input and output conections so that a high degree of computer assistance can be used for placement of the cells and routing of the interconnect lines. Standard cell layouts allow more flexibility than is available in the gate array approach yet retains sufficient structure that a high level of computer assistance can still be used. The use of standard cells is basically the extension to silicon of the techniques that have been in use for a number of years in design using small scale integrated circuit logic in printed circuit boards.

Since the computer does the routing of the interconnections (often with considerable assistance by the designer), the verification problem is considerably reduced since a logic simulator can be used to check the logical operation of the circuit. Also, simulation of the circuit's speed performance can be performed by high level timing simulators since the device level simulation has already been performed on the standard cells. The major disadvantage of this approach is that due to the rigidity of the shape and placement of the building block structures there is a high premium paid for the area required for interconnections which results in inefficient use of silicon area, and long interconnection lines, which will yield increasing performance degradation (longer delay times) as the technology is scaled.

3.6.3. Logic Arrays: It is possible to directly implement random logic with read only memories (ROM) which becomes particularly attractive for VLSI since computer automation can be used in the design and layout. A particularly efficient use of ROM's is in a programmable logic array (PLA) configuration, in which the output of one ROM forms the input to another. There are powerful minimization programs which have been developed that reduce the size of the ROMs in a PLA for a given logic function which can result in PLA's that actually require less area than an optimized random logic layout.

ROM's and PLA's are now extensively used for control functions in large complex circuits. For example, more than 1/3 of the Motorola 68000, 16 bit microprocessor is ROM and PLA. Another recent investigation of PLA's, was in their use for implementing large adders. It was found that the number of bits in the PLA increased as $(6.6 \cdot N)^{2.5}$ where N is the number of bits in the adder [22]. This should be compared to the size of an adder implemented using conventional random logic techniques (ripple carry adder) which requires $35 \cdot N$ transistors. Since each bit of a ROM can consume as much as ten times less area than each transistor of a random logic layout, it is actually more efficient in area to implement up to a 13 bit adder using a PLA.

4. CONCLUSIONS

By using the fact that VLSI is just the continuing development of MOS technology, it is possible to extrapolate the trends of the past ten years to obtain predictions about the capability of MOS technology in the next 10 years. Some of the implications of these predictions were presented. It was shown that even at the limits of scaling that the processing bandwidth of MOS technology will only have improved to the point where it is equivalent to today's advanced bipolar technologies. Because of the increased capital costs of implementing VLSI technology, the access to the technology will become increasingly limited. The result of this trend is that the development of VLSI signal processing systems will be by large vertically integrated system / I.C. manufacturers. Because the VLSI signal processing chips will be able to implement significant system functions, it will be necessary for the system designer to be involved in the chip design. This will be made possible by standardized design methods such as gate arrays, standard cells and PLA's which will allow system designers to design the VLSI chips themselves. This high level of system integration will make it possible for there to be widespread use of sophisticted signal processing algorithms because of the reductions in cost, size, and power of the signal processing systems.

REFERENCES

1. S. Waser, "Survey of VLSI for Digital Signal Processing," Proc. of ICASSP, 1980, Denver Colo., pp. 376-379.
2. J.F. Bucy, "Texas Instruments First Quarter and Stockholders Meeting Report," 1980, p 12.
3. D.A. Hodges ,et. al., "Potential of MOS Technologies for Analog Integrated Circuits," IEEE J. Solid State Circuits, Vol. SC-13, June, 1978, pp. 285-294.
4. J.R. Boddie, et. al., "A Digital Signal Processor for Telecommunications Applications," Int. Solid State Circuits Conf. Tech. Dig., San Francisco, Feb. 1980, pp. 44-45.
5. Y. Kawakami, et. al., "A Single Chip Signal Processor for Voiceband Applications," ibid., pp. 40-41.
6. Y. Chen and D.L. Duttweiler, "A 35,000 Transistor VLSI Echo Canceller," ibid., pp. 42-43.
7. M. Townsend, et. al., "An NMOS Microprocessor for Analog Signal Processing," IEEE J. of Solid State Circuits, Feb. 1980

8. R.W. Blasco, "VMOS Chip Joins Microprocessor to Handle Signals in Real Time," Electronics, Vol. 52, No. 18, Aug. 30, 1979.
9. G. Moore, "VLSI: Some Fundamental Challenges," IEEE Spectrum, April, 1979, pp. 30-37.
10. ibid. p. 33.
11. R.P. Capece, "Special Report: New LSI Processes," Electronics, Sept. 13, 1979, pp. 109-115.
12. H. N. Yu, et. al., "1 micron MOSFET VLSI Technology: Part 1," IEEE Trans Elect. Dev., Vol. ED-26, No. 4, April, 1979, pp. 318-324.
13. R.M. Jecmen, et. al., "HMOS II Static RAMs," Electronics, Sept. 13, 1979, p 124-128.
14. L.M. Dang, "A Simple Current Model for Short Channel IGFET's," IEEE Trans. El. Dev., Vol. ED-26, No. 4, April, 1979, pp. 436-445.
15. P.A. Hart, T.V. Hof and F.M. Klassen, "Device Down Scaling and Expected Circuit Performance," ibid., pp. 421-429.
16. M.T. Elliot, et. al., " Size Effects in E-Beam Fabricated MOS Devices," ibid., pp. 469-475.
17. T. Yamaguchi, M.L. Lust, S. Ragsdale and S. Sato, " A New Submicron Channel/ High Speed MOS-LSI Technology," Ibid, pp. 611-622.
18. B. Hoeneisen and C. A. Mead, "Fundamental Limitations in Microelectronics-I. MOS Technology," Solid State Tech., Vol. 15, 1972, pp. 819-829.
19. R.W. Keyes, "The Evolution of Digital Electronics Towards VLSI," IEEE Trans. on Elect. Dev., Vol. ED-26, No. 4, April, 1979, pp. 271-279.
20. B. LeBoss "Speech I/O is making itself heard," Electronics, May 22, 1980, p. 95-105 .
21. A. Weinberger, "High Speed PLA Adders, IBM Journal of Res. and Dev., Vol. 23, No. 2, Mar., 1979, pp. 163-178.

PARTICIPANTS

ASTRAND B., Royal Institute of Technology, Stockholm, Sweden
AKAVIA G., Technion, Haïfa, Israël
ALLEN Ph., Texas A.M. University, College Station, USA
ALVES MARQUES J., IMAG, Grenoble, France
AASERUD O., Sintef, Trondheim, Norway
AKERS L., University of Nebraska, Lincoln, USA
BRUUN E., Chr. Rovsing A/S, Ballerup, Denmark
BEYNON J., University of Surrey, Guildford, England
BAILLIEU F., ESIEE, Paris, France
BIJJANOVIC, Electro Faculty, Zagreb, Yougoslavia
BRAYTON R., IBM Research, Yorktown Hts, USA
BRUINSMA S., Philips Data Systems, Apeldoorn, The Netherlands
CHARLES G., CNET, Meylan, France
COSTA FREIRE J., Instituto Superior Tecnico, Lisboa, Portugal
CROCKER N., I.C.L., Manchester, England
COT N., Université Paris VI, Paris, France
CERGEL L., Motorola Semiconductor, Genève, Switzerland
COHEN J., Philips, Nijmegen, The Netherlands
CARLSON E., Central Institute for Industrial Research, Oslo, Norway
CARR J., University of Pennsylvania, Philadelphia, USA
CHRISTENSSON S., Chalmers University of Technology, Gothenburg, Sweden
DARUVALA D., Middlesex Polytechnic, Enfield, England
DESCHAMPS J.P., Philips Research Lab., Bruxelles, Belgium
DYSON C., University of Southampton, Southampton, England
ERICSSON T., Linköping University, Linköping, Sweden
ETIENNE R., IMAG, Grenoble, France
EDER A., Institut für Festkörpertechnologie der Frauenhofer-Gesellschaft, München, Germany
GAMBLE H., The Queen's University of Belfast, Belfast, Ireland
GARRETT S., University of South Florida, Tampa, USA
GRAY A., INMOS Ltd, Bristol, England
HAMMAR C., Royal Institute of Technology, Stockholm, Sweden
HEETMAN A., Eindhoven University of Technology, Eindhoven, The Netherlands
HOFFMANN K., Fraunhofer-Institut für Festkörpertechnologie, München, Germany
HICKS P., UMIST, Manchester, England
HOLZMANN D., ITT Intermetall, Freiburg, Germany
HUBBARD B., Jet Propulsion Laboratory, Pasadena, USA
HURST S., University of Bath, Bath, England
HAULIN T., Ellemtel AB, Olvjo, Sweden

IVEY P., Post Office Research, Ipswich, England
JOHANSSON J., S.R.A., Spanga, Sweden
JABLON C., Matra SA, Velizy, France
JACKETS A., Post Office Research, Ipswich, England
JOANNES R., F.N., Herstal, Belgium
KOSCIELNIAK H., University Erlangen-Nürnberg, Erlange, Germany
KASSUR A., Institute of Electron Technology, Warszawa, Poland
KEMHADJIAN H., University of Southampton, Southampton, England
LEA R., Brunel University, Uxbridge, England
LEWIS D., University of Santa Clara, Santa Clara, USA
LOCKEY D., DAI, Bruxelles, Belgium
LUBZENS D., Technion, Haïfa, Israël
MICHEL J.P., L.E.P., Limeil Brévannes, France
MORANT M., University of Durham, Durham, England
MOORMANN H., Siemens AG, München, Germany
MOTTINI F., CII Honeywell Bull, Angers, France
MULDERS L., Den Haag, The Netherlands
MULLER J., Eurosil GmbH, München, Germany
MALOBERTI, University of Pavia, Pavia, Italy
MARTIN A., Philips Research Lab., Eindhoven, The Netherlands
MC CANNY J., Royal Signals & Radar Establishments, Malvern, England
MAGEE T., ITT Standard Telecommunications Labs, Essex, England
MAJELIJNE R., Siemens, Oostkamp, Belgium
MATOS J.,Syracuse University, Syracuse, USA
NEWMANN M., Rutherford and Appleton Labs, Didcot, England
NIELSEN O., Bruel & Kjaer, Naerum, Denmark
OBREBSKA M., IMAG, Grenoble, France
OLDHAM H., G.E.C., Wembley, England
ORTON D., Racal M.S. Ltd, Reading, England
OTTEN R., University of technology Eindhoven, Eindhoven, The Netherlands
PAABO M., Linköping University, Linköping, Sweden
PARTRIDGE S., G.E.C., Wembley, England
PATEL D., University of Surrey, Guildford, England
PAULSON H., A.M.E., Horten, Norway
PROFUMO E., SGS-ATES, Agrate, Italy
RAWLINGS J., Afwal Avionics Laboratory, Patterson, USA
RAY N., Plessey Research Ltd, Towcester, England
RONZI G., SGS-ATES, Agrate, Italy
RUDIN H., ETH Zentrum, Zürich, Switzerland
RUZIC V., RIZ Poluvodici, Zagreb, Yougoslavia
SENN P., CNET, Meylan, France
SEVERI, Lamel Laboratory, Bologna, Italy

SLAVENBURG G., Philips Research Lab., Eindhoven, The Netherlands
MATTHEUS, Bell Tel. Mfg, Antwerpen, Belgium
SCHMIDT P., Instituto Venzolano de Investigaciones Cientificas, Caracas, Venezuela
SHIVA S., University of Alabama in Huntsville, Huntsville, USA
SOLBERG B., Norwegian Defence Research Est., Kjeller, Norway
SORBELLO F., Universita di Palermo, Palermo, Italy
STENBERG L., D.T.H., Lyngby, Denmark
THEUS U, Institut für Theoret. & Technik, Aachen, Germany
TOSUN O., Bogazici University, Istanbul, Turkey
TRELEAVEN P., The University of Newcastle-upon-Tyne, Newcastle-upon-Tyne, England
WAGNER K., Philips Gloeilampenfabrieken, Nijmegen, The Netherlands
WATANABE S., Sony Corporation, Kanagawaken, Japan
ZAGLER W., Institut für Industrielle Elektronik, Wien, Austria
ZINSZNER R., Centre Electronique Horloger, Neuchâtel, Switzerland
CORNELISSEN J., Katholiek Universiteit Leuven, Leuven, Belgium
REYNOLUS L, Katholiek Universiteit Leuven, Leuven, Belgium
FORREZ L., Katholiek Universiteit Leuven, Leuven, Belgium
DEBAKKER J., Katholiek Universiteit Leuven, Leuven, Belgium
CALUWAERTS L., Katholiek Universiteit Leuven, Leuven, Belgium
VERDONCK E., Katholiek Universiteit Leuven, Leuven, Belgium
CLAESSEN R., Katholiek Universiteit Leuven, Leuven, Belgium

LECTURERS

ANCEAU F., IMAG, Grenoble, France
BRODERSEN B., University of California, Berkeley, USA
DE MAN H., Katholiek Universiteit Leuven, Leuven, Belgium
DEMOULIN E., Université Catholique de Louvain, Louvain-la-Neuve Belgium
FELDBRUGGE F., Philips Gloeilampenfabrieken, Eindhoven, The Netherlands
GRASSL G., Siemens, München, Germany
HARTENSTEIN R., Universität Kaiserslautern, Kaiserslautern, Germany
HORBST E., Siemens, München, Germany
JESPERS P., Université Catholique de Louvain, Louvain-la-Neuve, Belgium

MOLNAR E., Washington University, St-Louis, USA
MUDGE C., Digital Equipment Corporation, Maynard, USA
SEQUIN C., University of California, Berkeley, USA
VAN DER GAAG H., Philips Gloeilampenfabrieken, Eindhoven, The Netherlands
VAN de WIELE F., Université Catholique de Louvain, Louvain-la-Neuve, Belgium

SCIENTIFIC ORGANIZING COMMITTEE

Professor P. JESPERS, ASI Director, Université Catholique de Louvain, Louvain-la-Neuve, Belgium
Professor C. SEQUIN, University of California, Berkeley, USA
Professor F. VAN de WIELE, Université Catholique de Louvain, Louvain-la-Neuve, Belgium.

LIST OF PANEL DISCUSSIONS

A. MARQUES
Mosaic : a design methodologies for VLSI custom circuits
R.H.J.M. OTTEN
The genealogical approach to VLSI layout design
M. LEA
VLSI-DLM's : a structurel approach to chip architecture
VLSI chip architectures for associative processors
S.L. HURST
Custom LSI design : the universal-logic-module approach
S. SHIVA
Use of HDL's in an automatic LSI design system
D.C. PATEL
Dynamic logic simulation of LSI circuits
M. PAABO
DIGSIM for VLSI verification
P. TRELEAVEN
A single-chip building block for parallel computers
A.J. MARTIN
A distributed architecture for parallel recursive computation
G. AKAVIA
Communication problems in large systems

J.W. CARR
Possible learning theory chips
P. JESPERS
A character recognition IC.

www.ingramcontent.com/pod-product-compliance
Lightning Source LLC
LaVergne TN
LVHW012322171026
850499LV00003B/72
* 9 7 8 9 0 2 8 6 2 7 8 1 9 *